Fungi and Lignocellulosic Biomass

Fungi and Lignocellulosic Biomass

By Christian P. Kubicek

with figures by

Irina S. Druzhinina and Lea Atanasova

WILEY-BLACKWELL

A John Wiley & Sons, Inc., Publication

Wiley-Blackwell is an imprint of John Wiley & Sons, formed by the merger of Wiley's global Scientific, Technical and Medical business with Blackwell Publishing.

Editorial offices: 2121 State Avenue, Ames, Iowa 50014-8300, USA
The Atrium, Southern Gate, Chichester, West Sussex, PO19 8SQ, UK
9600 Garsington Road, Oxford, OX4 2DQ, UK

For details of our global editorial offices, for customer services and for information about how to apply for permission to reuse the copyright material in this book please see our website at www.wiley.com/wiley-blackwell.

Library of Congress Cataloging-in-Publication Data

Kubicek, C. P. (Christian P.)
Fungi and lignocellulosic biomass / Christian P. Kubicek ; with figures by Irina S. Druzhinina and Lea Atanasova.
p. cm.
Includes bibliographical references and index.
ISBN 978-0-470-96009-7 (hardcover : alk. paper) 1. Lignocellulose–Biodegradation.
2. Fungi–Biotechnology. 3. Biomass energy. I. Title.
TP248.65.L54K82 2012
662′.88–dc23

2012010996

A catalogue record for this book is available from the British Library.

Wiley also publishes its books in a variety of electronic formats. Some content that appears in print may not be available in electronic books.

Set in 10/11.5 pt Times New Roman by Aptara® Inc., New Delhi, India
Printed and bound in Singapore by Markono Print Media Pte Ltd

1 2013

Contents

Preface ix

1 The Plant Biomass 1
1.1 The Structure of Plant Cell Wall 1
1.2 Chemical and Physicochemical Properties of the Major Plant Cell Wall Constituents 3
1.2.1 Cellulose 3
1.2.2 Pectin 6
1.2.3 Hemicelluloses 8
1.2.4 Lignin 11
1.3 Abundant Sources of Carbohydrate Polymers and Their Monomer Composition 13
1.3.1 Agricultural Wastes 13
1.3.2 Forest Product Residues 14
1.3.3 Energy Crops 15
1.3.4 Weedy Lignocellulosic Substrates 15
1.4 Biosynthesis of Plant Cell Wall Polymers 16
1.4.1 Cellulose 16
1.4.2 Hemicellulose Biosynthesis 19
1.4.3 Pectin Biosynthesis 20
1.4.4 Lignin Biosynthesis 23
1.5 Strategies for Manipulating Wall Composition 26
1.5.1 Manipulation of Plant Cell Wall Polymer Composition 26
1.5.2 Manipulation of Plant Lignin Content 27

2 The Actors: Plant Biomass Degradation by Fungi 29
2.1 Ecological Perspectives 29
2.2 The Major Three Mechanisms of Lignocellulose Degradation by Fungi 30
2.2.1 White Rot 31
2.2.2 Brown Rot Fungi 35
2.2.3 Soft Rot Fungi 39
2.3 Plant Cell Wall Degradation by Plant Pathogenic Fungi 40
2.4 Anaerobic Fungi 41

3 The Tools—Part 1: Enzymology of Cellulose Degradation 45
3.1 General Properties and Classification of Enzymes That Hydrolyze Polysaccharides 45

3.2 Fungal Cellulolytic Enzymes 49
3.2.1 Cellulose-Binding Domains 52
3.2.2 Cellobiohydrolases (EC 3.2.1.91) 55
3.2.3 Endo-β-1,4-Glucanases (EC 3.2.1.4) 58
3.2.4 β-1,4-Glucosidases 64
3.3 Nonenzymatic Proteins Involved in Cellulose Hydrolysis 65
3.3.1 GH61 Proteins 65
3.3.2 Swollenin 67

4 The Tools—Part 2: Enzymology of Hemicellulose Degradation 69
4.1 Xyloglucan Hydrolysis 69
4.2 Degradation of the Xylan Backbone 72
4.2.1 GH10 Xylanases 73
4.2.2 GH11 Xylanase 75
4.2.3 GH30 Glucuronoxylan Xylanohydrolases 76
4.2.4 GH3 β-Xylosidases 76
4.2.5 GH43 β-Xylosidases 77
4.2.6 GH54 β-Xylosidases 78
4.3 Degradation of the Galactomannan Backbone 78
4.4 Degradation of Pectin 80
4.4.1 Hydrolytic Pectin Degradation 80
4.4.2 Pectin Degradation by β-Elimination 82
4.5 Accessory Glycoside Hydrolases for Hemicelluloses Degradation 84
4.5.1 Enzymes that Act on Arabinose-Containing Substituents 85
4.5.2 Enzymes that Act on Galactose-Containing Substituents 86
4.5.3 α-Xylosidases 89
4.5.4 α-Fucosidases 90
4.5.5 α-Glucuronidases and Glucuronan Lyases 90
4.5.6 Accessory Hydrolases for Pectin Degradation 91
4.6 Other Accessory Enzymes 92
4.6.1 Feruloyl and p-Coumaroyl Esterases 92
4.6.2 Acetyl- and Methylesterases 93
4.6.3 Pectin Esterases 95
4.6.4 Glucuronoyl Esterases 95

5 The Tools—Part 3: Enzymology of Lignin Degradation 99
5.1 Lignin Peroxidase 101
5.2 Manganese Peroxidase 105
5.3 Versatile Peroxidase 108
5.4 Dye-Oxidizing Peroxidase 109
5.5 Laccases 110
5.6 Enzymes Generating Hydrogen Peroxide 115
5.6.1 Glyoxal Oxidase 115
5.6.2 Other FAD-Dependent Oxidases 116
5.7 Cellobiose Dehydrogenase 116
5.8 Enzymes Essential for Oxalic Acid Formation 117
5.9 Glycopeptides 118

6 Catabolic Pathways of Soluble Degradation Products from Plant Biomass 119
6.1 Uptake of Mono- and Oligosaccharides 119
6.2 Metabolism of D-Glucose and D-Mannose 121
6.3 Catabolism of D-Galactose 122
6.4 Catabolism of Pentoses 125
6.5 Catabolism of Hexuronic Acids 127

7 Regulation of Formation of Plant Biomass-Degrading Enzymes in Fungi 129
7.1 The Cellulase Inducer Enigma 129
7.2 Inducers for Hemicellulases 133
7.3 Transcriptional Regulation of Cellulase and Hemicellulase Gene Expression 134
7.3.1 Activators of Cellulase and Hemicellulase Gene Expression 135
7.3.2 Specific Repressors of Cellulase and Hemicellulase Gene Expression 138
7.3.3 Wide Domain Regulators of Cellulase and Hemicellulase Gene Transcription 138
7.3.4 Regulation of Cellulase and Hemicellulase Transcription at the Level of Genome Accessability 144
7.3.5 Signal Transduction to Cellulase and Hemicellulase Gene Expression 145
7.4 Regulation of Ligninase Gene Expression 146

8 The Fungal Secretory Pathways and Their Relation to Lignocellulose Degradation 149
8.1 The Fungal Secretory Pathway 149
8.1.1 It Starts at the Endoplasmic Reticulum 149
8.1.2 Quality Control in the ER 152
8.1.3 Golgi Apparatus 154
8.1.4 The Plasma Membrane 156
8.1.5 Vacuoles 158
8.2 Protein Glycosylation 159
8.2.1 N-Glycosylation 159
8.2.2 O-Glycan Synthesis 161
8.3 Strategies for Improvement of the Fungal Secretory Pathway 161
8.3.1 Folding and UPR 163
8.3.2 Engineering of Protein Glycosylation 163

9 Production of Cellulases and Hemicellulases by Fungi 165
9.1 Fungal Producer Strains 165
9.1.1 Trichoderma Reesei 165
9.1.2 Other Fungal Producer Strains 167
9.1.3 Thermophilic Fungi 169
9.2 Strain Improvement 170
9.2.1 Strain Breeding by Classical Mutagenesis 170
9.2.2 Strain Improvement by Targeted Gene Manipulation 171

9.2.3 Strain Improvement by Complementation with Heterologous Enzymes 176
9.2.4 Strain Improvement by Protein Engineering 179
9.3 Cellulase Production 180
9.3.1 Cellulase Fermentation 180
9.3.2 Cellulase Downstream Processing 182

10 Production of Fermentable Sugars from Lignocelluloses 185
10.1 Pretreatment Technologies 186
10.1.1 Physical Pretreatment 186
10.1.2 Chemical Pretreatment 186
10.1.3 Physicothermal Pretreatment 189
10.1.4 Solvent Treatment 191
10.1.5 Biological Pretreatment 193
10.1.6 Summarizing Considerations on Pretreatment 195
10.2 Hydrolysis 195
10.2.1 Enzymatic Hydrolysis 195
10.2.2 Chemical Hydrolysis 200

11 Lignocellulose Biorefinery 201
11.1 Ethanol 202
11.1.1 Yeast Strain Improvement 203
11.1.2 Alternatives to *S. cerevisiae* as an Ethanol Producer 210
11.1.3 Process Improvement 212
11.2 *n*-Butanol 212
11.3 Advanced Biofuel Alcohols 213
11.3.1 Isobutanol 213
11.3.2 Syngas 214
11.3.3 Fuels from Isoprenoid and Fatty Acid Pathways 215
11.4 Lactic Acid 215
11.5 Succinic Acid 217
11.6 Xylitol 222
11.7 1,3-Propanediol 222
11.8 Polyhydroxyalkanoate 223
11.9 Other Products 223
11.10 Refinement by Chemical Processes 225
11.10.1 Furfural 225
11.10.2 Levulinic Acid 225
11.10.3 Uses of Lignin 225
11.10.4 Extraction of Chemicals from Lignocelluloses Biomass 226

Acknowledgments 229
References 231
Index 285

Preface

The advent of the third millennium has been and is still characterized by an increasing concern about the dependency of the human society on oil reserves and on the consequences for our planet as a whole from the rising carbon dioxide levels. A large part of this fossil carbon is used for generation of energy, for which alternatives such as nuclear, solar-electric, solar-thermal, hydroelectric, or geothermal have been proposed or developed and may well function at the individual and smaller community level. However, at the time of this writing, replacement of fossil fuels for long-distance road transport or aviation is not in sight. In addition, about 10% of fossil carbon is currently used by the petrochemical industry for the production of components required for the manufacture of a wide array of goods that form an essential part of our everyday life.

The only reasonable alternative to these major problems of our society is the use of carbon sources that are permanently available in large amounts, and which can be used in a carbon dioxide neutral way. Quantitatively, plant biomass is by far the only carbon source that can fulfill these requirements: it arises by carbon dioxide fixation during photosynthesis, and its dry weight consists mainly of three polymers (cellulose, hemicelluloses, and lignin) whose monomer constituents (hexose and pentose sugars and phenylpropan compounds) can be converted to useful starting materials for industry by fermentation or biotransformation (the so-called biorefinery concept). Owing to the above-cited problems with energy, most of the research on the use of plant biomass has gone into liquid biofuels, particularly ethanol.

A key step in this plant biomass for "biofuels/biorefineries" concept is the production of the above-named monomeric components in a sufficiently high concentration using technologies that do not produce hazardous by-products. Enzymatic hydrolysis is the only means that can theoretically fulfill this purpose and has been investigated to this end since the early 1960s. These studies have revealed that particularly fungi can form cellulolytic, hemicellulolytic, and ligninolytic enzymes, and some of them (notably *Trichoderma reesei*, the current paradigm of cellulase and hemicellulase research) have been used with success for the production of enzymes used in the hydrolysis of plant cell wall material.

While the basic path of the road from biomass to biofuels/biorefineries thus seems to be straightforward, and has led to the presentation of several demonstration plants in the United States, Canada, and Europe, there are still several large stones to remove from this road. One of the biggest of them is to render the price for enzymatic hydrolysis and subsequent production of compounds like ethanol compatible or better lower than the prices for liquid fossil fuels, which requires improvement of several steps such as activity and composition of the enzymes used, the hydrolysis process, yield of the desired product, and appropriate uses for side products and not-used components from the hydrolysate (e.g., xylose or lignin). Obviously, solutions to overcome these bottlenecks must come from an interdisciplinary treatment of these processes, involving contributions from botany, microbiology, biochemistry, biotechnology, and biochemical engineering.

The focus of this book is on the fungal enzymes that are required and applied for lignocellulose hydrolysis, but does not limit this treatment only to a description of the enzyme inventory (Chapters 3–5). Instead, it attempts to relate this main focus to the other areas that influence the process as a whole, such as composition and availability of the plant biomass (Chapter 1) and the different modes of how fungi make this material available for their own lifestyle (Chapter 2), as well as a treatment of the biochemical pathways for the metabolism of the arising monomers (Chapter 6), the regulation of enzyme formation (Chapter 7), and the cell biology of their secretion (Chapter 8). Finally, the last three chapters deal with the selection of appropriate producer strains and their fermentation (Chapter 9), the pretreatment and hydrolysis of the plant biomass (Chapter 10), and with the processes for production of biofuels and biorefineries (Chapter 11).

This book has been written in an attempt to serve both as a professional reference for all people who work in this area and as an introduction into the field for all those who are generally interested in the topic, from academic institutions and research teams to teachers, as well as graduate and postgraduate students. Toward this purpose, I have refrained from going into too much detail with many aspects, but have instead provided an extensive list of both original and review references that can be used to obtain a yet deeper information on individual topics. Still, I have to apologize to all those colleagues whose work is not cited, and to those I refer only by citation of a review. This is not due to neglect but only due to the necessity to keep the references within a reasonable size.

Fungi and Lignocellulosic Biomass

Chapter 1
The Plant Biomass

1.1 The Structure of Plant Cell Wall

When we talk of plant biomass in terms of its use for biofuel and other biorefineries, we mostly mean the plant cell wall that makes up for more than 50% of the plants dry weight. This most outside located structure of the plant cell is also its most distinguishing feature, and because of its rigidity an essential component for their sedentary lifestyle. This rigidity also provides the strength to withstand mechanical stress and forms and maintains the plants shape. Despite this rigidity, nevertheless, the cell wall is a dynamic and metabolically active entity that plays crucial roles in growth, differentiation, and cell-to-cell communication and acts as a pressure vessel that prevents overexpansion when water enters the cell (Raven et al., 1999).

Plant cell walls typically consist of three layers: the "primary cell wall" (a rather thin but continuously extending layer that is produced by growing cells), the "secondary cell wall" (a thick layer that is formed inside the primary cell wall after termination of cell growth), and the "middle lamella" (the outermost layer that forms an interface between secondary walls of adjacent plant cells and glues them together) (Figure 1.1).

The primary cell wall consists of the polysaccharides cellulose, hemicellulose, and pectin (Rose et al., 2004). The cellulose thereby aggregates to microfibrils that are covalently linked to hemicellulosic chains and form a cellulose—hemicellulose network that is embedded in the pectin matrix. The secondary wall is formed in some plants between the plant cell and primary wall when either a maximum size or a critical point in development has been reached and makes the plant cells rigid. It is made up from cellulose, hemicelluloses (mostly xylan), and lignin. The latter is a complex polymer of aromatic aldehydes that fills the spaces between cellulose, hemicellulose, and pectin components of the cell wall. Because of its hydrophobic nature, it drives out water and so strengthens the wall. In wood, three layers of the secondary cell wall, referred to as the S_1, S_2, and S_3 lamellae, are found that result from different arrangements of the cellulose microfibrils (Mauseth, 1988; Figure 1.1). The first outermost layer—the S_1 lamella—has both left- and right-handed microfibril helices; in contrast, the S_2 (middle) and

Fungi and Lignocellulosic Biomass, First Edition. Christian P. Kubicek.

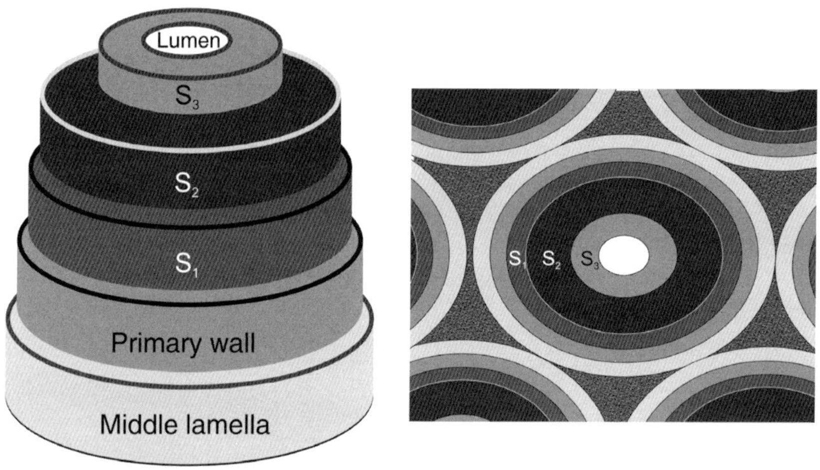

Figure 1.1. Organization of the different layers of the plant cell wall.

S_3 (innermost) lamellae only comprise a single helix of microfibrils, although with opposite handedness to each other. During formation of the secondary cell wall, lignification takes place in the S_1 and S_2 but not S3 lamellae and also in the primary wall and middle lamella (Levy and Staehlin, 1992; Reiter, 2002; Popper, 2008). This arrangement allows the cellulose microfibrils to become embedded and fixed within the lignin, similar to steel rods that become embedded in concrete (Figure 1.2).

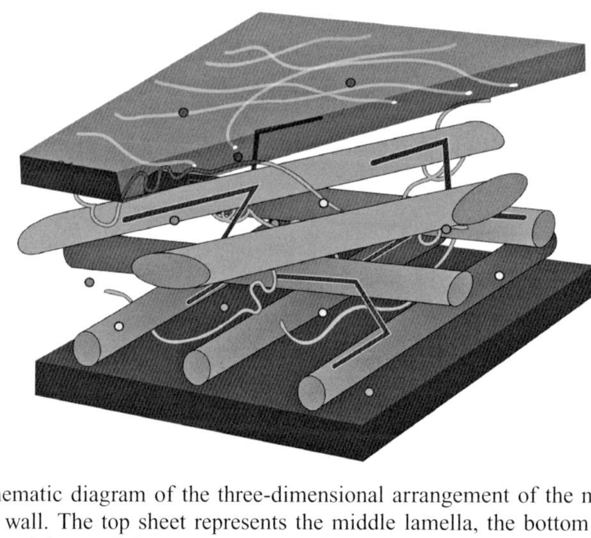

Figure 1.2. Schematic diagram of the three-dimensional arrangement of the main polymers in the primary plant cell wall. The top sheet represents the middle lamella, the bottom sheet represents the plasma membrane; and the area in between represents the primary cell wall. Bright gray threads symbolize pectin, dark gray rectangular lines indicate hemicelluloses, small globes indicate soluble proteins, and the gray tubes indicate the cellulose microfibrils.

In addition, structural proteins (1–5%) are found in most plant cell walls; they are usually classified as hydroxyproline-rich glycoproteins (HRGPs), arabinogalactan proteins (AGPs), glycine-rich proteins (GRPs), and proline-rich proteins (PRPs) (Albenne et al., 2009). The function of these proteins is not well understood. However, it is likely that the glycan moieties in these proteins can form hydrogen bonds and salt bridges to the cell wall polysaccharides, and thereby contribute to the mechanical strength to the wall. The relative composition of carbohydrates, secondary compounds, and proteins varies between plants and between the cell type and age (Levy and Staehelin, 1992; Reiter, 2002; Popper, 2008).

The secondary cell wall may also contain additional layers of lignin in xylem cell walls, and suberin in cork cell walls, that confer rigidity and contribute to the exclusion of water.

1.2 Chemical and Physicochemical Properties of the Major Plant Cell Wall Constituents

1.2.1 Cellulose

As noted earlier, cellulose is one of the principal components of both primary and secondary plant cell walls and reaches its highest abundance (40%) in the secondary cell walls. Cellulose consists of unbranched, unsubstituted 1,4-β-D-glucan chains that can reach degrees of polymerization of 2,000–6,000 and 2,000–10,000 residues in primary and secondary walls, respectively. The CH_2OH side group is arranged in a *trans-gauche* position (a term that described the separation of two vicinal groups by a 60° torsion angle) relative to the O5–C5 and C4–C5 bonds. Because of the absence of coiling or branching, the molecule adopts an extended, rod-like conformation, aided by the equatorial conformation of the glucose residues. The multiple hydroxyl groups on the glucose from one chain can form hydrogen bonds with oxygen molecules on the same or on a neighboring chain (Figure 1.3), and so hold the chains firmly together side by side and form microfibrils with high tensile strength. This strength is one of the major sources of rigidity to the plant cell wall (Klemm et al., 2004; O'Sullivan, 1997).

Carl Naegeli suggested in 1858 that cellulose has a crystalline structure (reviewed by Wilkie, 1961), and this was experimentally verified 80 years later by Meyer and Misch (1937). It consists of two parallel glucan chains that are bound into sheets by hydrogen bonding and

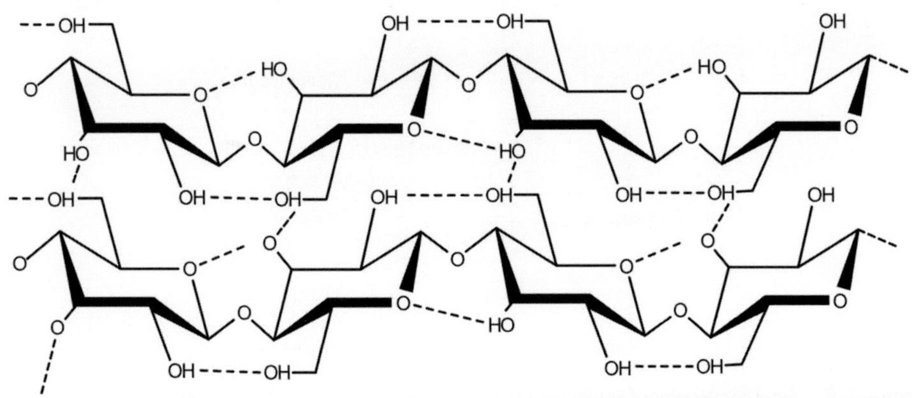

Figure 1.3. The chemical structure of cellulose. Dotted lines represent hydrogen bonds.

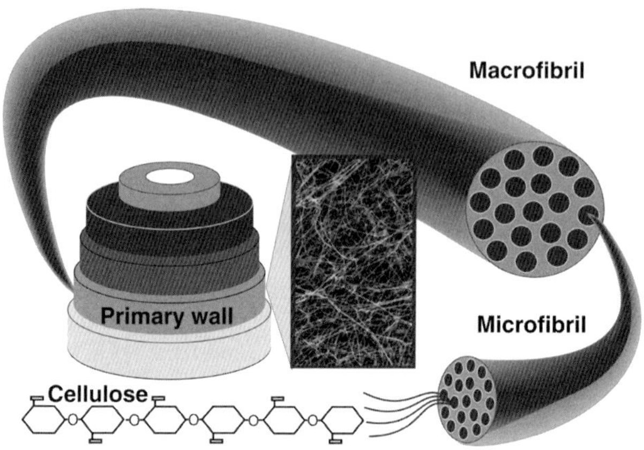

Figure 1.4. Generation of cellulose micro- and macrofibrils.

Van der Waals forces, and which are then stacked to needle-shaped fibers. The morphological hierarchy is defined by elementary fibrils, microfibrils, and microfibrillar bands. The length of these structural units is between 1.5 and 3.5 nm (elementary fibrils), 10 and 30 nm (microfibrils), and on the order of 100 nm (microfibrillar bands) (Figure 1.4). The length of the microfibrils is on the order of several hundred nanometers, which typically contain 30–40 parallel chains that form a diameter of 3.5–4 nm. The microfibrils form characteristic helices that differ as a function of the composition of the cell wall layer (see Section 1.1 above) and according to the plant type as well. As an example, cotton fibers have a lower orientation of the cellulose microfibrils (helix angle $\approx 18°$) as bast fibers (helix angle $\approx 4°$–$5°$), which results in decreased elasticity, a higher elongation at breakage, and less fiber strength. This adaptation of the mechanical properties of wood to environmental conditions through corresponding helix angles is fascinating and has yet to be rivaled in technical composite materials (O'Sullivan, 1997).

Despite the uniform chemical structure, however, cellulose occurs in the form of seven polymorphs (I_{α}, I_{β}, II, III_{I}, III_{II}, IV_{I}, and IV_{II}) that can be interconverted, as shown in Figure 1.5 (O'Sullivan, 1997). Among them, cellulose I is the form that is found in nature and therefore called "native cellulose." The polymorphic forms II, III_{I}, and III_{II} arise from artificial treatments, which are, however, relevant to the process of biomass pretreatment

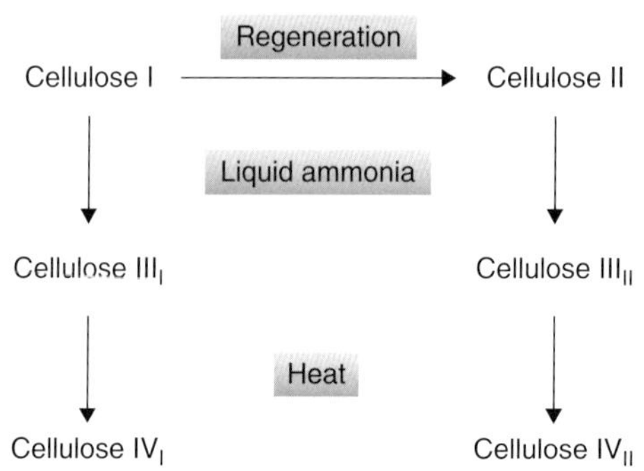

Figure 1.5. Interconversion of polymorphs of cellulose. L and g mean liquid and gaseous aggregate state, respectively.

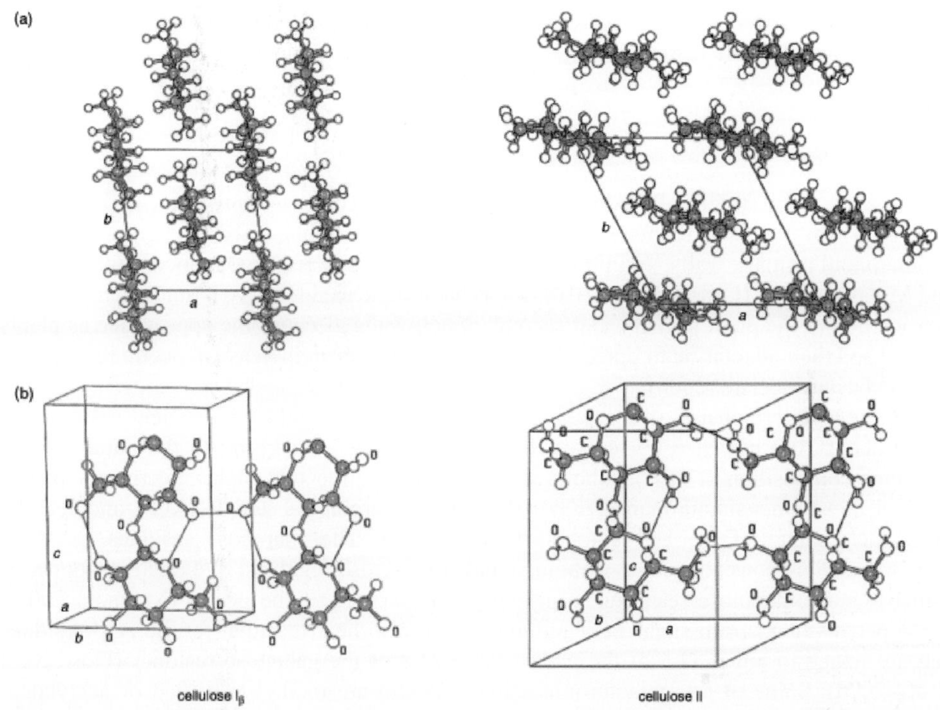

Figure 1.6. Arrangements of cellulose chains in cellulose Iβ and cellulose II. (Reprinted with permission from Klemm D, Heublein B, Fink HP, Bohn A. Cellulose: Fascinating biopolymer and sustainable raw material. Angew Chem Int, 2005; 44(22): 3358–3393. Copyright 2005, with permission from John Wiley & Sons, Inc.)

(see Chapter 10) and shall therefore be explained here as well. Polymorphs IV_I and IV_{II} arise by heating celluloses III_I and III_{II}, respectively, to 206°C in glycerol (Hess and Kissig, 1941; Gardiner and Sarko, 1985) and will thus not be treated here.

Cellulose I consists of two distinct crystalline forms (I_α and I_β), which differ from each other in their intramolecular bonding patterns (Wada et al., 1993). The crystal structure of cellulose I can be described by a monoclinic unit cell (space group $P2_1$), which contains two cellulose chains in a parallel orientation with a twofold screw axis. The two forms I_α and I_β can be found alongside each other, their ratio depending on the origin of the cellulose. I_α has triclinic and I_β has monoclinic unit cells, in which two new intramolecular, chain-stiffening hydrogen bonds inside neighboring molecular layers have also been described (O'Sullivan, 1997). The two crystalline forms occur in different proportions depending on the source, I_α being the dominant form in bacteria and lower organisms, whereas the more stable I_β form dominates in higher plants (Yamamoto and Horii, 1994).

Apart from the thermodynamically less stable cellulose I, cellulose II (Figure 1.6) is the most stable structure of technical relevance. It can be formed from cellulose I by treatment with aqueous sodium hydroxide or by dissolution of the cellulose and subsequent precipitation/regeneration. Its monoclinic crystal structure is characterized by the specific unit cell geometry with a modified H-bonding system. It is not yet understood how the parallel chain arrangement of cellulose I undergoes transition into the antiparallel orientation of cellulose II without an intermediate dispersion of cellulose molecules (O'Sullivan, 1997).

1.2.2 Pectin

Pectin is the most structurally complex family of polysaccharides in nature, making up to 35% of primary walls in dicots and nongraminaceous monocots, 2–10% of grass and other commelinoid primary walls, and up to 5% of walls in woody tissue (Mohnen, 2008; Caffall and Mohnen, 2009; Harholt et al., 2010). It has been suggested to serve a fundamental role in the function of the plant primary and secondary cell walls, because the appearance as plants on land and their adaptation to upright growth correlate with an increase in pectin in their cell walls (Matsunaga et al., 2004).

Pectin generally consists of a backbone of α-1,4-linked D-galacturonic acid residues, of which about 20% can also be replaced by other residues. According to the nature of the monomer composition of this backbone and also of that of the side chains, pectins are distinguished into homogalacturonans (HGs), rhamnogalacturonans I and II (RG-I and RG-II), xylogalacturonan (XGA), and apiogalacturonan (AP). The rhamnose residues can bear long side chains consisting of L-arabinose and D-galactose residues, and small amounts of D-fucose and D-mannose resulting in a "hairy" appearance of the pectin (Mohnen, 2008). Some pectins (e.g., from sugar beet and apple) can also bear terminal ferulic acid residues that are linked to either *O*-5 of the arabinose or *O*-2 of the galactose residues (Oosterveld et al., 2000). Some of the galacturonic acids in pectin are methyl esterified or acetylated. The nonmethylated D-galacturonic acid sequences interact with Ca^{2+} and thereby cross-link different pectic acid chains, which contributes to the firmness of the plant tissue (Cafall and Mohnen, 2009).

HG is the most abundant pectic polysaccharide. It is a linear homopolymer of about 100 α-1,4-linked galacturonic acid residues that makes up for about 65% of all pectin in plant cell walls (Figure 1.7a). It can bear methyl ester groups at the C-6 carboxyl group, and—albeit less—acetyl groups at *O*-2 or *O*-3.

RG-I represents 20–35% of pectin. It contains a backbone of alternating α-D-galacturonic and α-L-rhamnose residues (Figure 1.7b), and a varying number of different sugars and oligosaccharides as side chains. Twenty to eighty percent of the rhamnosyl residues in the RG-I backbone can contain side chains of α-1,5-linked L-arabinan with C2- and C3-linked arabinan side chains, β-1,4-linked D-galactans with a degree of polymerization of up to 47, β-1,4-linked D-galactans with C3-linked L-arabinose or arabinan side chains, and β-1,3-linked D-galactan with β-6-linked galactan or arabinogalactan side chains (Schols et al., 1990). The side chains may further contain α-L-fucose, β-D-glucuronic acid, and 4-*O*-methyl-β-D-glururonic acid residues.

RG-II is the structurally most complex pectin and makes up for up to 10% of it (Figure 1.7c). Its structure consists of an HG backbone of at least eight α-1,4-linked α-D-galacturonic acid residues, which further contain side branches that can involve up to 12 different types of sugars in more than 20 different linkages. RG-II usually exists in plant walls as RG-II dimers cross-linked by a 1:2 borate diol ester between the apiosyl residues in side chain A of two RG-II monomers. RG-II dimerization cross-links HG domains resulting in a macromolecular pectin network. Mutations that result in even small modifications of the structure of RG-II lead to severe growth defects, suggesting that RG-II in the wall is crucial for normal plant growth and development (Mohnen, 2008).

Two other substituted galacturonans, XGA and AP, are only minor pectin components. XGA, which has been found mainly in reproductive plant tissues, is an HG substituted at *O*-3 with a β-linked xylose, which can sometimes bear additional β-linked xylose residues at *O*-4. AP is an HG substituted at *O*-2 or *O*-3 with D-apiofuraose and has been found in aquatic monocots (Cafall and Mohnen, 2009).

The way by which the pectic polysaccharides are linked to each other is still in debate. Most of the available data support the assumption that HG, RG-I, and RG-II are linked via their backbones. However, it has also been suggested that pectins may be covalently linked to, or tightly associated with, other types of wall polysaccharides such as xyloglucans and xylans. This hypothesis is supported by the finding of xylose residues in some pectins and that a mutation in genes of pectic biosynthesis influences also the xylan content of the cell walls (Orfila et al., 2005). This suggests that pectins may serve to hold at least some hemicelluloses in the wall.

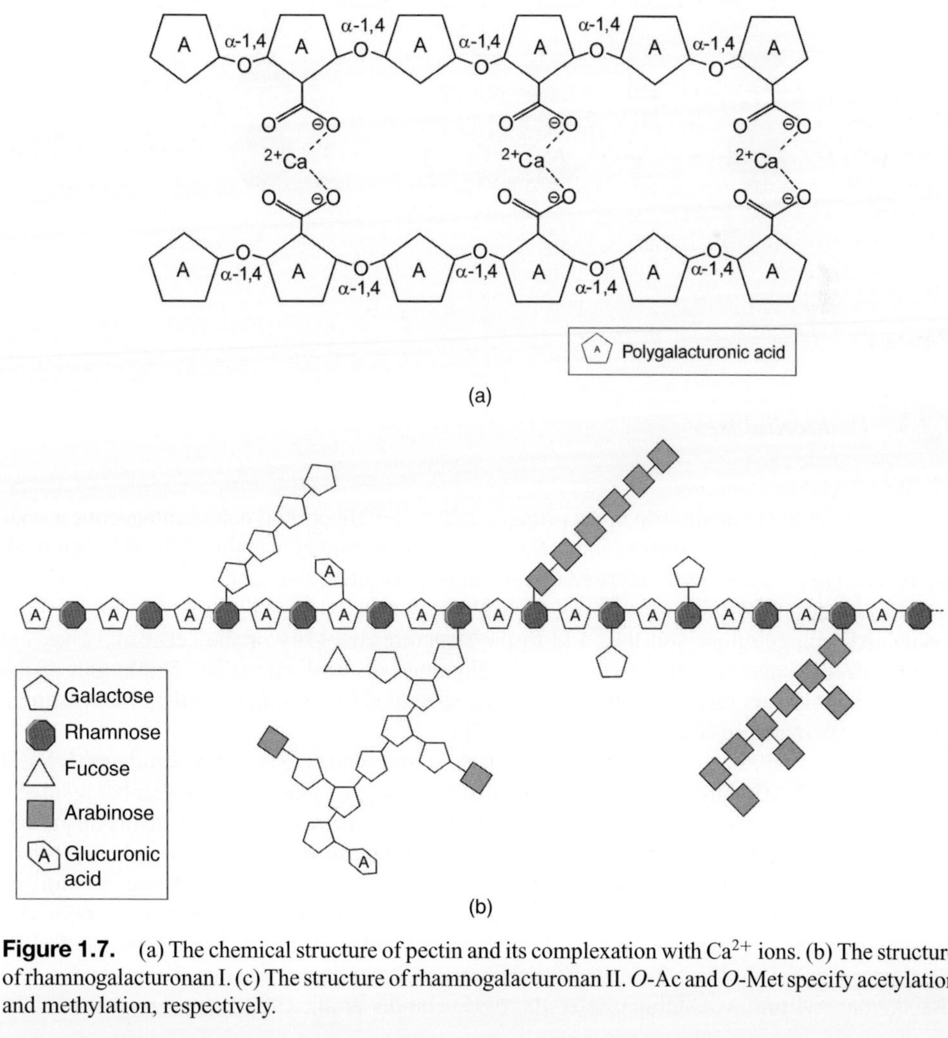

Figure 1.7. (a) The chemical structure of pectin and its complexation with Ca^{2+} ions. (b) The structure of rhamnogalacturonan I. (c) The structure of rhamnogalacturonan II. *O*-Ac and *O*-Met specify acetylation and methylation, respectively.

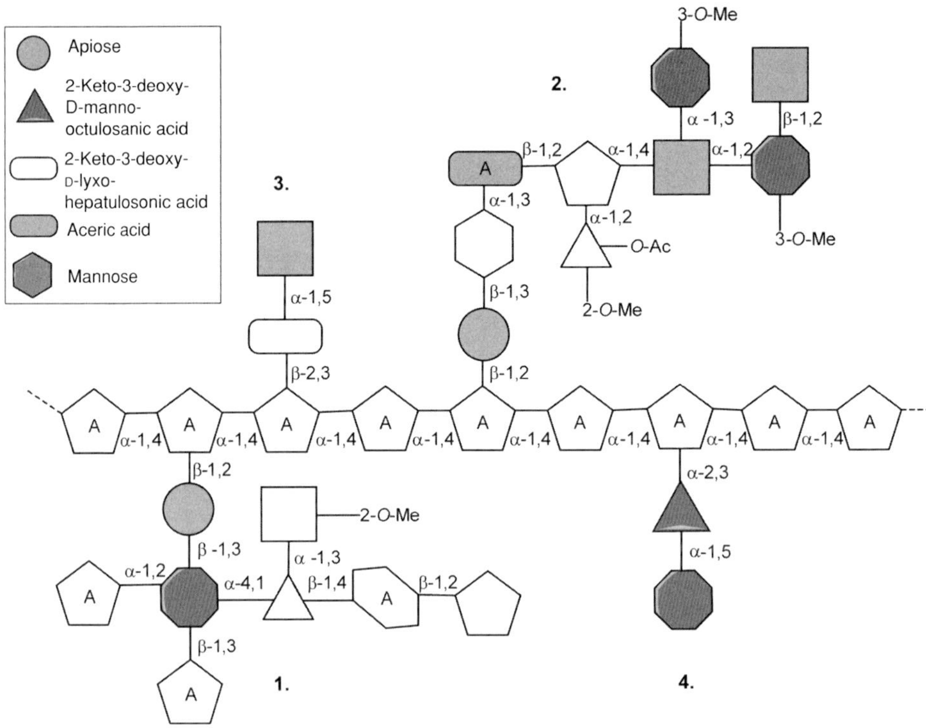

(c)

Figure 1.7. (*Continued*)

1.2.3 Hemicelluloses

Xyloglucan

The predominant hemicellulose in the primary cell wall of dicots and nongraminaceous monocots is xyloglucan, which may account for up to 20% of the dry weight of the primary wall. Xyloglucan has a backbone composed of 1,4-linked β-D-glucose residues, of which up to 75% are substituted at O6 with mono-, di-, or triglycosyl side chains. Xyloglucans are strongly associated with cellulose and thus add to the structural integrity of the cell wall. They are also believed to play an important role in regulating cell wall extension. The length of the xyloglucan polymers enables them to cross-link several cellulose microfibrils, thus creating a rigid network structure (Hayashi and Kaida, 2011).

Xyloglucans have been identified to occur in two types, that is, type XXXG and type XXGG (Figure 1.8): XXXG consists of repeating units of three β-1,4-linked D-glucopyranose residues, substituted with D-xylopyranose via an α-1,6-linkage, which are separated by an unsubstituted glucose residue. In xyloglucan type XXGG, two xylose-substituted glucose residues are separated by two unsubstituted glucose residues. The structural features of these, as well as some other types of xyloglucans, have been discussed in detail by Vincken et al. (1997). The xylose residues in xyloglucan can further be substituted with α-1,2-L-fucopyranose-β-1,2-D-galactopyranose and α-1,2-L-galactopyranose-β-1,2-D-galactopyranose disaccharides and *O*-linked acetyl groups (Maruyama et al., 1996; Hantus et al., 1997; Vincken et al., 1997).

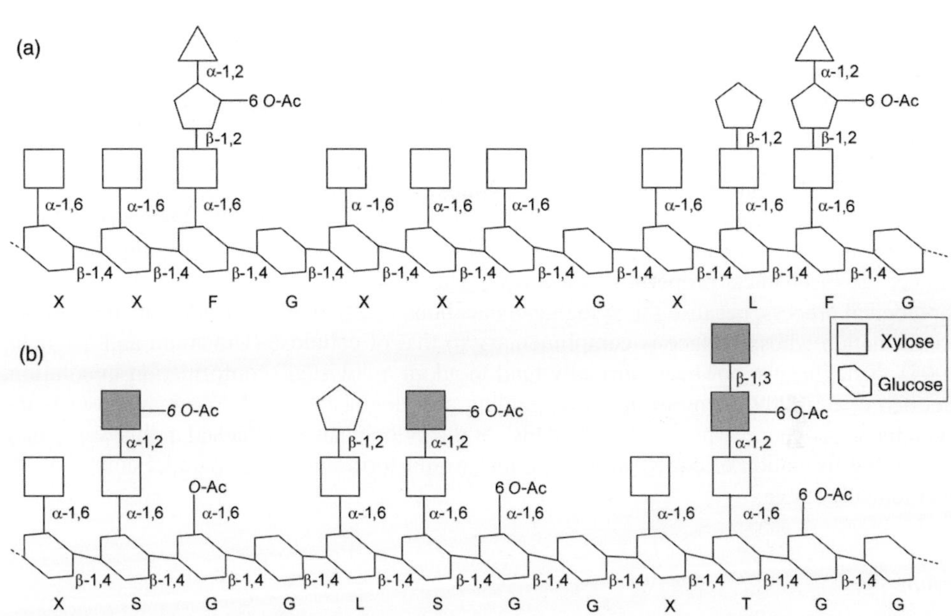

Figure 1.8. The chemical structure of xyloglucan. (a) XXXG-type; (b) XXGG-type. Symbols not shown are used as in Figure 1.7.

In cell walls of dicotyledons, the xyloglucans are partially replaced by glucuronoarabinoxylan, which has a linear β-1,4-linked D-xylopyranosyl backbone with both neutral and acidic side chains attached at intervals along its length. The acidic side chains are terminated with glucuronosyl or 4-*O*-methyl glucuronosyl residues, whereas the neutral side chains are composed of arabinosyl and/or xylosyl residues (Darvill et al., 1980). The structures of xyloglucans from several plants in the subclass Asteridae were characterized by nuclear magnetic resonance (NMR) spectroscopy and mass spectrometry to determine how their structures vary in different taxonomic orders (Hoffman et al., 2005). The structure of xyloglucan has been shown to differ also in a tissue-specific manner, for example, fucosyl residues are typically absent from seed xyloglucans but present on the xyloglucans in the vegetative portions of the same plant.

A single letter nomenclature is used to simplify the naming of xyloglucan side chain structures (Table 1.1). For example, a capital **G** represents an unbranched glucopyranose residue. A capital **F** represents a glucopyranose residue that is substituted with a fucose-containing

Table 1.1. Xyloglucan one letter code.

G	-4)-β-D-Glc$_p$-(1-
X	α-D-Xyl$_p$-(1–6)-β-D-Glc$_p$-(1-
L	β-D-Gal$_p$-(1–2)-α-D-Xyl$_p$-(1–6)-β-D-Glc$_p$-(1-
F	α-L-Fuc$_p$-(1–2)-β-D-Gal$_p$-(1–2)-α-D-Xyl$_p$-(1–6)-β-D-Glc$_p$-(1-
S	α-L-Ara$_f$-(1–2)-α-D-Xyl$_p$-(1–6)-β-D-Glc$_p$-(1-
T	α-L-Ara$_f$-(1–3)-α-L-Ara$_f$-(1–2)-α-D-Xyl$_p$-(1–6)-β-D-Glc$_p$-(1-
J	α-L-Gal$_p$-(1–2)-β-D-Gal$_p$-(1–2)-α-D-Xyl$_p$-(1–6)-β-D-Glc$_p$-(1-

From http://www.ccrc.uga.edu.

trisaccharide. The Complex Carbohydrate Research Center at the university of Georgia has developed a searchable ^{1}H NMR database (http://cell.ccrc.uga.edu/world/xgnmr/index.html) to facilitate the rapid identification of enzymatically generated xyloglucan subunit structures.

In the plant cell wall, the xyloglucans are arranged in such a way that they coat the surface of the cellulose microfibrils (some regions binding directly to the cellulose surface, other regions are not in direct contact with the cellulose but form cross-linking tethers, and still some regions of xyloglucan are entrapped within the cellulose microfibrils; Mellerowicz et al., 2008). This results in a limited aggregation of the cellulose chains and especially the tethers impact the mechanical properties of the wall. The binding of cellulose is likely a complex topological process, because the xyloglucan backbone must, to this end, adopt a "flat ribbon" conformation whose surface is complimentary to that of cellulose (Umemura and Yuguchi, 2005). Xyloglucans, however, normally tend to adopt a "twisted" conformation in solution. Bootten et al. (2009) demonstrated that binding of xyloglucan to cellulose may untwists the xyloglucan backbone, which—if both "ends" of the xyloglucan are attached in this way—may lead to the formation of coiled structures, that would form duplex antiparallel coils that are energetically stable.

Xylan

Xylan is the major hemicellulose polymer in cereals and hardwood. It always contains a β-1,4-linked D-xylose backbone, to which differently structured side chains can be attached, thus resulting in a high variety of xylan structures. Although most xylans are branched structures, linear polysaccharides have also been isolated. The xylans of cereals often contain large quantities of L-arabinose and are consequently termed arabinoxylans. In contrast, hardwood xylans contain large amount of D-glucuronic acid linked to the backbone and are named glucuronoxylans (Scheller and Ulvskov, 2010; Figure 1.9).

L-Arabinose (either single residues or short chains) is connected to the xylan backbone via either α-1,2- or α-1,3-linkage. These side chains can also contain D-xylose in a β-1,2-linkage to L-arabinose and D-galactose, which can be either β-1,5-linked to L-arabinose or β-1,4-linked to D-xylose. Acetyl residues can be attached to *O*-2 or *O*-3 of the D-xylan backbone, but the

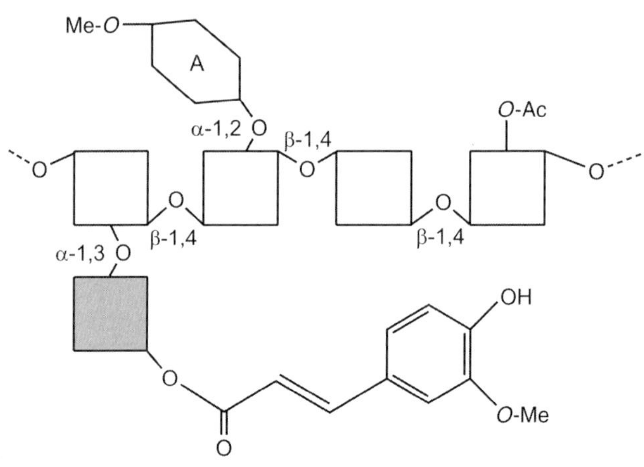

Figure 1.9. The chemical structure of xylan. Symbols not shown are used as in Figure 1.7.

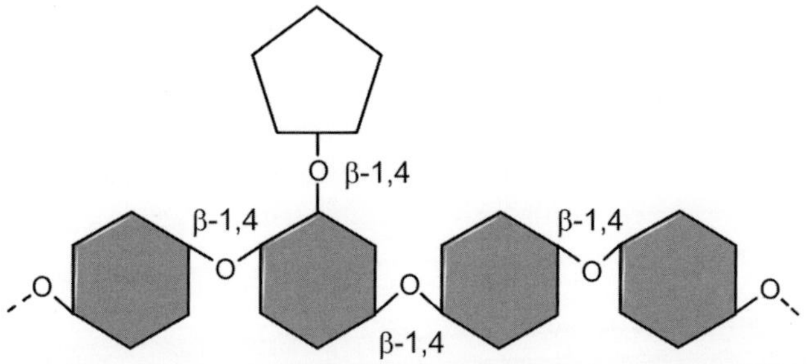

Figure 1.10. The chemical structure of galactomannan. Symbols not shown are used as in Figure 1.7.

degree of acetylation differs strongly depending on the origin. Glucuronic acid and its 4-*O*-methyl ether are attached to the xylan backbone via an α-1,2-linkage, whereas feruloyl and *p*-coumaroyl residues can be attached at the *O*-5 of terminal L-arabinose residues (Ebringerova and Heinze, 2000).

Galactomannans

Galactomannans and galactoglucomannans comprise a second group of hemicellulolytic structures, which form the major hemicellulose fraction of the gymnosperms cell walls (12–15%). They consist of a backbone of β-1,4-linked D-mannose residues, which can bear α-1,6-linked D-galactose residues (Figure 1.10) in ratios between 1:1 and 5:1 depending on the source (Dey, 1978). Therein the galactosyl side chain hydrogen interacts to the mannan backbone intramolecularly and so stabilizes the structure. Acetyl groups can be present but are irregularly distributed in glucomannan. Also some of the mannosyl units of galactoglucomannan are partially substituted by *O*-acetyl groups (Moreira and Filho, 2008).

1.2.4 Lignin

Lignin is found in all vascular plants, and is—after cellulose—the most abundant carbon source on earth. Lignin is characterized by a complex structure derived from oxidative coupling of three primary hydroxycinnamyl alcohols, that is, *p*-coumaryl, coniferyl, and sinapyl alcohol, which render it extremely recalcitrant to degradation. The corresponding phenylpropanoid units in the lignin polymer are usually denoted as *p*-hydrophenyl (H), guaiacyl (G), and syringyl (S) units, respectively, based on the methoxy substitution on the aromatic rings (Figure 1.11). Although gymnosperm lignin (=softwood) contains mostly G units and very low levels of H units (G/S/H = 96:*t*:4), angiosperm lignin (=hardwood) is composed of similar levels of G and S units with traces of H units (G/S/H = 50:50:*t*). Monocotyledons (e.g., grasses) contain all the three units in a ratio of G/S/H = 70:25:5 (Davin and Lewis, 2005; Calvo-Flores and Dobado, 2010).

Lignification is achieved by cross-linking reactions of the monomer with the growing polymer or by polymer–polymer coupling via radicals generated by oxidase enzymes (see Chapter 5). Endwise reactions coupling a lignin monomer to the growing polymer result in the formation of β-linked structures. Further, coupling between two preformed lignin oligomers or polymers results in 5–5 and 5-*O*-4 linked structures. Finally, end groups arise from coupling

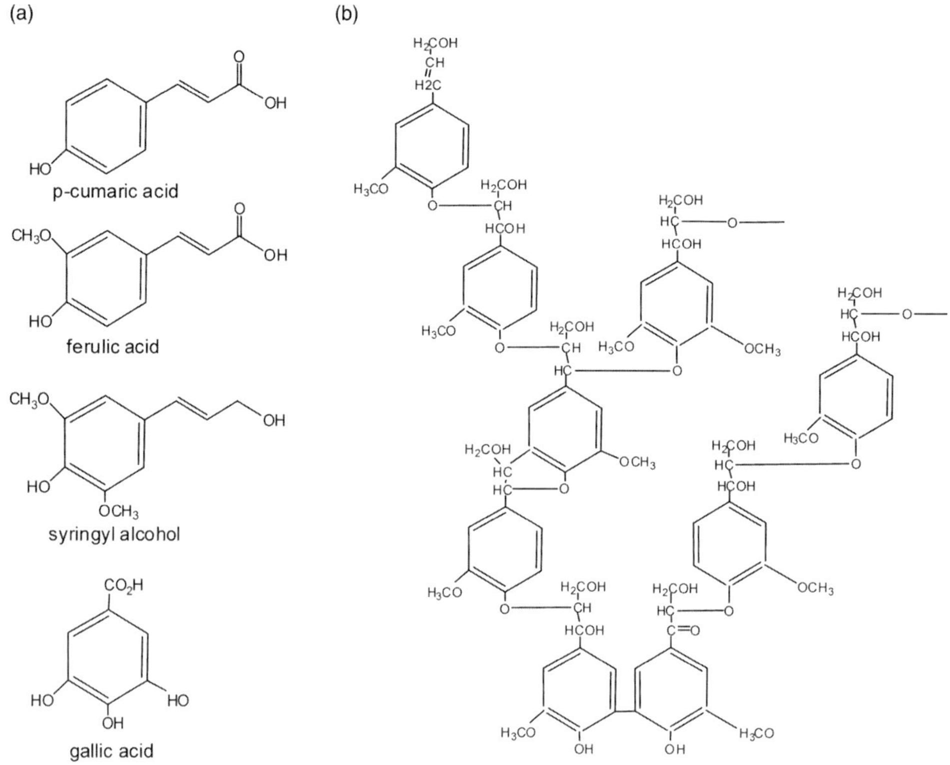

Figure 1.11. The chemical structure of lignin. (a) The monomers, from which lignin has been formed, are given on the left side of the figure. (b) Only a small part, yet indicating all occurring bonds between different aromatic rings, is shown.

reactions that are not at the side chain β-position of the monomer. The relative abundance of the different linkages largely depends on the relative contribution of the monomers to the polymerization process during lignin biosynthesis; for example, the β-*O*-4 (arylglycerol-β-aryl ether) coupling of a monolignol with the growing lignin oligomer/polymer creates the most abundant structural unit, involving generally about 50% and 80% of the phenylpropanoid units in softwood and in hardwood lignin, respectively (Alder, 1977). The β-5, 5-5, and 4-*O*-5 structures account for roughly 10%, 25%, and 4%, respectively, in softwood lignin (Brunow, 2001). Acylated structural units, such as 4-propoxy-sinapyl-γ-acetate, are found at high levels in some lignins, and in grass lignins, the hydroxycinnamic acids can be esterified at the γ-position of the propyl side chains. The amount of β-1 structures in softwood lignin is about 2%. The 5-5 unit is frequently etherified with additional monolignol via intramolecular reaction of quinone methide intermediates. The resulting dibenzodioxocin, an eight-member cyclic ether unit, as well as the 5-5 and 4-*O*-5 units may serve as branching points in softwood lignin. Similarly, some β-1-linkages seem to be part of spirodienone substructures (Boerjan et al., 2003; Wong, 2009). The lignins from the herbaceous plants sisal (*Agave sisalana*), kenaf (*Hibiscus cannabinus*), abaca (*Musa textilis*), and curaua (*Ananas erectifolius*) are extensively acylated at the gamma-carbon of the lignin side chain (up to 80% acylation) with acetate and/or *p*-coumarate groups and preferentially over S units. The structures of all these highly acylated lignins are characterized by a very high S:G ratio, a large predominance of β-*O*-4′ linkages (up to 94% of all linkages), and a strikingly low proportion of traditional beta-beta′ linkages, which indeed are completely absent in the lignins from abaca and curaua (del Río et al., 2008).

Lignin interacts with the cellulose fibrils, creating a rigid structure strengthening the plant cell wall. They also form several types of covalent cross-links to hemicelluloses (for review, see Fry, 1986): one is formed by diferulic acid bridges between lignin and arabinoxylans, pectin polymers, or xylan and lignin. Another type is an ester linkage between lignin and the glucuronic acid residues in xylan, which has been observed, for example, in beech wood. A third type (Rizk et al., 2000) involves a protein- and pH-dependent binding of pectin and glucuronoarabinoxylan to xyloglucan, and it is dependent on the presence of fucose on the xyloglucan (for details see de Vries and Visser, 2001).

1.3 Abundant Sources of Carbohydrate Polymers and Their Monomer Composition

With the aim of producing ethanol and other biorefinery products from biomass without interfering with the food and feed chains, plant biomass that cannot be used for this purpose must be used ("second generation biofuels"). Materials that are currently considered as potentially useful include agricultural residues (currently considered the most likely feedstock to be adopted), forest harvest residues, and finally the dedicated breeding of energy crops. Their content in cellulose, hemicelluloses, and lignin is given in Table 1.2.

1.3.1 Agricultural Wastes

Cellulosic wastes, including waste products from agriculture (straw, stalks, leaves) and forestry, wastes generated from processing (nut shells, sugarcane bagasse, sawdust) and organic parts of municipal waste, could all be potential sources. However, it is also important to consider the crucial role that decomposing biomass plays in maintaining soil fertility and texture; excessive withdrawals for bioenergy use could have negative effects.

Kim and Dale (2004) presented a calculation about how much bioethanol could be produced from agricultural residues worldwide: to avoid conflicts between human food use and industrial use of crops, they considered only the wasted crop, which is defined as crop lost in distribution,

Table 1.2. Cell wall compositions of different plant lignocellulose sources.

		Cellulose (%)	Hemicellulose (%)	Lignin (%)
Hard wood	Birch	40	23	21
	Willow	37	23	21
	Aspen	51	29	16
Soft wood	Spruce	43	26	29
	Pine	46.4	8.8	29.4
Agro residues	Wheat straw	38.2	21.2	23.4
	Corn stover	37.5	22.4	17.6
Crop	Switch grass	31	20.4	17.6
	Miscanthus	40	18	25

Data taken from Chandel and Singh (2011).

as feedstock. Overall, they arrived at about 73.9 Tg of dry wasted crops in the world that could potentially produce 49.1 GL of bioethanol. Asia is the largest potential producer of bioethanol from crop residues and wasted crops, mainly rice straw, wheat straw, and corn stover, and could produce up to 291 GL/year of bioethanol. The next highest potential region is Europe (69.2 GL of bioethanol), in which most bioethanol comes from wheat straw. Corn stover is the main feedstock in North America, from which about 38.4 GL/year of bioethanol can potentially be produced. Options for dealing with the straw debris so far are (a) leaving them on the field, (b) plowing it back into the soil, (c) burning, or (d) removing from the land depending on the decision made by landowner. In recent years, however, burning has become discredited due to increased concern over the health effects of smoke from burning fields (Kerstetter and Lyons, 2001). In the case of wheat straw, full removal may lead to soil erosion, and the fraction of wheat straw that must be left on the field depends on the weather, crop rotation, existing soil fertility, slop of the land, and tillage practices. According to Kim and Dale (2004), 60% ground cover could be maintained in order to ensure prevention of soil erosion that roughly requires 1.7 Mg wheat residue per hectare. Globally considered rice straw would be the leader by producing 205 GL/year of bioethanol, and the next highest potential feedstock is wheat straw, which can produce 104 GL/year of bioethanol (Kim and Dale, 2004).

Another, albeit, less abundant biomass source could be leaf litter: Simmons et al. (2008) emphasized that leaf biomass can represent 5–15% of the total aboveground biomass in a year. While this figure is small, however, this process occurs annually and represents 25–60% equivalent of total biomass at harvest. Leaves also present the additional advantage compared with stemwood that they are easier to process.

1.3.2 Forest Product Residues

Current forest product manufacturing techniques produce large amounts of residues–chips and particles—from milling processes. They are currently used for the manufacture of particle board and paper products. In addition, wood residues accumulate that arise during forest products processing. Mabee and Saddler (2010) estimated that they represent approximately 50% of the biomass energy consumption in the United States. According to their calculations, 331 million dry tons could be dedicated to biofuel production on a sustainable basis.

The average forest residues generation rate for temperate northern countries is estimated to be about 12%. In order to be truly renewable, the removal of forest biomass must be carried out in a fashion that limits impacts on local ecosystems, in accordance with the principles of sustainable forest management. Estimates for Canada have calculated that the dry residues that may be sustainably removed from forest harvest operations range between 9.8 (Mabee et al., 2006) and 46 (Wood and Layzell, 2003) million dry tons per annum. In addition, residues from forest products manufacture have been conservatively estimated to range between 2 and 5.4 million dry tons per annum (Mabee et al., 2006; Wood and Layzell, 2003). However, use of this material is in strong competition with new ventures such as pelletization, and the perspectives for using a large proportion of this for bioethanol production are small.

An additional, nonrenewable source of forest biomass is the wood killed by plant pathogenic insects and microbes ("disturbance wood," Schmitz and Gibson, 2003). Because these sources of forest biomass are nonrenewable, they are not included in the estimates of renewable biomass supply, but could constitute an additional significant amount of biomass for biofuel production (Mabee and Saddler, 2010).

1.3.3 Energy Crops

As lignocellulosic biomass becomes more widely established, forest and agricultural residues will no more be available in the desired quantities, and other feedstocks be required. Three energy crops, widely considered as feasible options in this direction, are *Miscanthus giganteus*, *Panicum virgatum* elephant grass, and hybrid poplar (Chandel and Singh, 2011).

M. giganteus is a grass native to Asia and can grow up to 12 ft (3.7 m) tall without significant addition of water or fertilizers. It is a sterile hybrid between *M. sinensis* and *M. sacchariflorus*, and because of its rapid growth (more than 3.5 m in one growth season), low mineral content, and high biomass yield (up to 25 tons per hectare), it has been proposed as a renewable biomass in Europe since the early 1980s.

P. virgatum, commonly known as switchgrass, is a perennial warm season bunchgrass native to North America, where it occurs naturally from 55°N latitude southward into the United States and Mexico. It is one of the dominant species of the central North American tallgrass prairie, and it is also found in remnant prairies, in native grass pastures, and naturalized along roadsides. It is similar to *M. giganteus* with respect to cold and drought tolerance and water use efficiency (Chandel and Singh, 2011). Switchgrass is a diverse species, with striking differences between different plants, thus providing a range of valuable traits for breeding programs. It has two distinct forms: the lowland cultivars, which tend to produce more biomass (they may grow up to ≥2.7 m), and the upland cultivars that are generally shorter (≤2.4 m tall), more cold tolerant and of northern origin. In 1992, the US Department of Energy started a research program focused on developing switchgrass as a sustainable bioenergy feedstock (Sanderson et al., 1996). Switchgrass was chosen as a biomass-based renewable energy source crop because it had high forage yield and seed production at different regional cultivar testing fields in several states in the United States. Switchgrass cv. Alamo has been ranked as the highest yielding cultivar in most yield trials conducted and is relatively amenable to genetic transformation and subsequent regeneration into mature plants (McLaughlin and Kszos, 2005; Burris et al., 2009). Additionally, it can be grown and cultivated on marginal lands, thereby alleviating concerns about competing with land for cultivation of food crops. The best-case land-use scenario for switchgrass is its cultivation on agriculturally depleted soils that no longer support agriculturally important row crops (for review see Keshwani and Cheng, 2009; and Moon et al., 2010).

Hybrid poplar, a member of the willow family, is a cross between North America's cottonwoods, aspens, and Europe's poplars. Poplar is considered as a model example of a short-rotation woody crop and can produce 9–15.7 dry tons per hectare (4–7 dry tons per acre) annually over a 6–10-year rotation (Alig et al., 2000). Other trees that could be grown specifically for biofuel production are eucalyptus, loblolly pine, willow, and silver maple. Eucalyptus, native to Australia but grown throughout the world, has been studied extensively in California and Florida, and appears to be amenable to high-density cultivation in plantation farms (Rockwood et al., 2008).

1.3.4 Weedy Lignocellulosic Substrates

Weedy lignocelluloses have also been discussed as promising future biomass feedstocks (Huber and Dale, 2009). These include wild sugarcane (*Saccharum spontaneum*), a perennial weedy grass with worldwide distribution but which is especially abundant in Asian countries like India, or water hyacinth (*Eichhornia crassipes*), which is a free-floating perennial aquatic plant native to tropical South America. For a full account on these and many more plants, see

Huber and Dale (2009). A drawback of the targeted cultivation of these plants, however, is that they all are weeds that may be different to control and thus threaten other crop production.

1.4 Biosynthesis of Plant Cell Wall Polymers

1.4.1 Cellulose

Cellulose biosynthesis has been subject of many excellent and recent reviews (Somerville, 2006; Crowell et al., 2010; Guerriero et al., 2010; Harris and de Bolt, 2010; Wightman and Turner, 2010, Endler and Persson, 2010). It occurs by a membrane-bound cellulose synthase (CESA) protein complex consisting of a CESA hexamer called "rosette" that can be visualized in the inner leaflet of freeze-fractured plasma membranes (Mueller and Brown, 1980). A CESA gene has been first cloned from cotton, and subsequent analysis of the *Arabidopsis thaliana* genome sequence revealed the presence of 29 genes (Richmond and Somerville, 2000). The CESAs are in fact an ancient gene family and conserved from algae to higher plants (Crowell et al., 2010).

The rosette hexamer contains multiple CESA catalytic subunits, of which three (CESA1, CASE3, and CESA6) are involved in the formation of the primary cell wall, whereas three other isoforms (CESA4, CESA7, and CESA8) synthesize cellulose for the secondary walls. The occurrence of putative zinc-finger domains at the N-terminus of the CESA proteins and the analysis of the redox states of the cysteine residues present in them have suggests that they are responsible for the dimerization of the CESA subunits (Doblin et al., 2002). The requirement for three isoenzymes to assemble a functional CESA complex has been a matter of discussion and several models have been proposed: one of them suggests that they might have distinct binding sites for specific intra- and interparticle interaction for assembling the hexameric rosette (Doblin et al., 2002; Timmers et al., 2009). Alternatively, it has been discussed that the three subunits may catalyze distinct chain initiation and elongation reactions or add adjacent glucose residues in opposite orientations (Simmons et al., 2006; Guerriero et al., 2010).

The rosette complex synthesizes cellulose at the plasma membrane, to which it is delivered via actin filaments and cortical microtubules (reviewed in detail by Crowell et al., 2010). However, as explained by Guerriero et al. (2010), the possibility that cellulose biosynthesis already begins in the Golgi apparatus in higher plants has not been strictly ruled out as yet. Indeed, noncrystalline chains of β-(1$\rightarrow$4)-glucan may be synthesized in the Golgi by catalytic subunits of the CESA that have not yet been assembled in the form of rosettes. Alternatively, the oligosaccharides that are formed in the Golgi could be too short to form microfibrillar strings, or interact in the Golgi with carbohydrates or an aglycone that would prevent microfibril formation (Guerriero et al., 2010). Following delivery, the CESA catalytic subunits appear to be inserted into the plasma membrane by means of eight transmembrane helices. As a consequence of this even number of helices, both the N-terminus and the C-terminus of the catalytic subunits face the same side of the plasma membrane. It is generally (although not univocally) assumed that cellulose synthesis occurs on the cytoplasmic side of the plasma membrane (Guerriero et al., 2010).

Cellulose synthesis proceeds by the attachment of glucosyl residues to the nonreducing terminus of the acceptor glucan chain. The substrate for this attachment is uridine-diphosphate (UDP)-glucose, which can be provided by the cytosolic enzyme UDP-glucose pyrophosphorylase (Figure 1.12). However, it is also possible that the provision of UDP-glucose comes from sucrose synthase, which converts sucrose and UDP into fructose and UDP-glucose

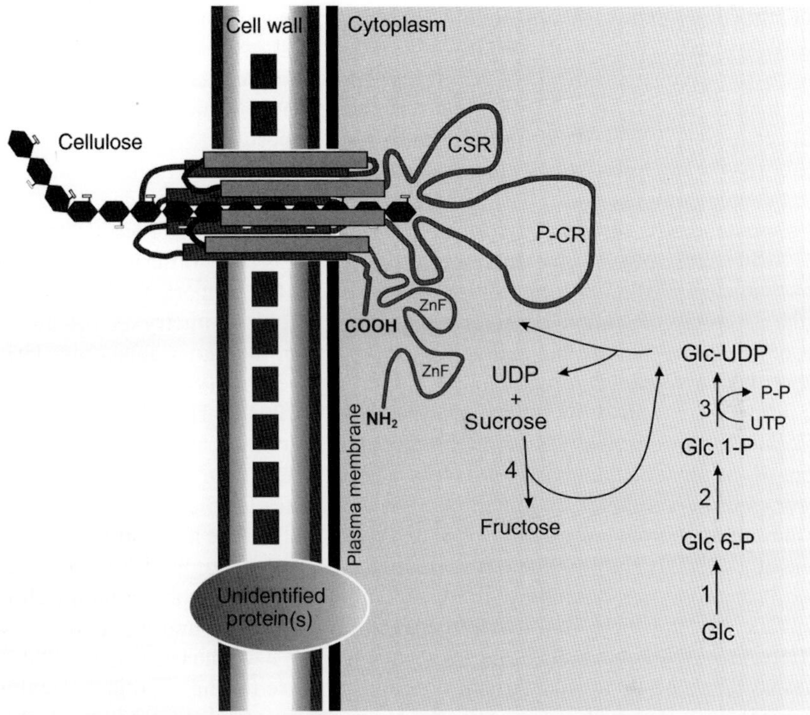

Figure 1.12. Biosynthesis of cellulose by the catalytic dimer hypothesis for cellulose synthase. Precursor biosynthesis in the cytosol from glucose and fructose involves five enzymes: 1, hexokinase/glucokinase; 2, phosphoglucomutase; 3, UDP-glucose pyrophosphorylase; 4, sucrose synthase. Two CESA proteins form a complex that synthesizes a single β-1,4-D-glucan chain. This dimerization also results in a channel composed of 16 membrane-spanning domains. Two zinc fingers (plant-specific conserved region, P-CR; class-specific region, CSR) are potential interaction sites. The catalytic site resides within the loop structures on the cytoplasmic side close to the membrane. (Modified and complemented from Carpita, 2011.)

(Amor et al., 2005). This theory is supported by the fact that a membrane-bound sucrose synthase can be found at sites of very active cellulose synthesis, but must still be viewed with caution because UDP-glucose also serves as a substrate for the various glycosyltransferases involved in the biosynthesis of noncellulosic cell wall carbohydrates (Simmons, 2006; Guerriero et al., 2010).

Initiation of cellulose biosynthesis is assumed to require a primer (although evidence for this *in vivo* is still lacking), which most likely is sitosterol-β-glucoside (Peng et al., 2002). The glycosyltransferase that initiates cellulose chain synthesis then uses UDP-glucose as a donor and sitosterol-β-glucoside as an acceptor. Further polymerization is believed to proceed by the direct addition of the monomers to the carbohydrate chains. However, the theory of Albersheim et al. (1997) that the sugars are first transferred to an S or T residue of the enzyme cannot be excluded at the moment. Guerriero et al. (2010) also presented a third alternative, that is, the polymerization of cellulose chains in higher plants could be initiated on lipid acceptors. Cello-oligosaccharides would then be transferred to the nonreducing end of the elongating glucan chains by CESA or a transglycosylase rather than a glycosyltransferase. This hypothesis bears the attractivity that it is compatible with the proposed role of sitosterol-β-glucoside in cellulose biosynthesis (Peng et al., 2002).

Whatever the exact mechanism is, the elongating cellulose chains are then translocated across the membrane into the cell wall, either via a pore that is formed by the eight transmembrane helices of the CESAs or through adjacently located porins.

Assembly of the microfibrils from the arosen cellulose chains then occurs at some distance from the enzyme complex itself. This process is in fact still poorly understood: it may be occurring spontaneously through the formation of interchain hydrogen bonds when a number of chains sufficient to form an elementary microfibril such as those observed in primary walls (ca 3 nm diameter) are in close proximity. Alternatively, the formation of microfibrils may require the assembly of complete rosettes with their sixfold symmetry, or the involvement of accessory proteins that are not functional or present in the Golgi apparatus. The use of *Arabidopsis* CESA mutants strongly supports the hypothesis that the rosette structures are required for the formation of microfibrils (Guerriero et al., 2010). The CESA complexes have been observed to move in the plasma membrane, and a biophysical model suggests that this movement of the enzyme complex is driven by the polymerization and crystallization events (Diotallevi and Mulder, 2007). In this model, the aggregates of microfibrils form simultaneously as they are spun from groups of rosettes that are spatially close in a confined environment. Bessueille et al. (2009) showed that CESA is located in lateral patches of the plasma membrane that bear properties resembling lipid rafts, cholesterol, and sphingolipid-rich parts of the membrane that lower the free energy between the membrane and the adjacent aqueous phase (Rietveld and Simons, 1998). Guerriero et al. (2010) have thus speculated "that groups of rosettes co-localized in raft-like structures are responsible for the coordinated synthesis of multiple microfibrils that coalesce shortly after having been extruded from individual enzyme complexes to form fibrils or ribbons of cellulose."

The actual stoichiometry of the CESA catalytic subunits in the rosettes has not been experimentally demonstrated as yet, but it is assumed that the rosette would be able to synthesize 36 β-(1→4) glucan chains that would co-crystallize to form a microfibril (Delmer, 1999), which corresponds with the size of microfibrils isolated from most primary walls (~3.5–4 nm) (Delmer, 1999). This model is based on the rationale that six globules of the rosette each have six CESA catalytic subunits. However, the number of active catalytic subunits per rosette has never been experimentally demonstrated. Thus, the above rationale that individual cellulose microfibrils consist of 36 chains is exclusively based on the observed sixfold symmetry of the rosette structures. It is as well possible that the different rosettes contain a lower number of catalytically active subunits and consequently form thinner microfibrils. In fact, the minimum number of chains to form a crystalline structure is a total of 16 chains (i.e., 4×4 package). This number of chains is compatible with a lateral size of 2–2.5 nm for an individual microfibril and consistent with the width observed for microfibrils from primary walls (Guerriero et al., 2010).

In this regard, one must also consider the fact that the synthesis of a β-1-4-D-glucan chain implies a big steric problem: to make a (1→4)-β-D linkage means that each glucosyl residue is turned 180° with respect to each neighbor. Thus, the *O*-4 position of nonreducing terminal sugar of the acceptor chain is displaced several angstroms upon addition of each successive unit, and the site of catalysis must thus also move several angstroms within the protein. In addition, the acceptor chain must swivel 180°, or the catalytic or acid–base amino acids must toggle between two forms to account for the displacement (Carpita, 2011). To overcome this problem, models have proposed that two CESAs associate to form two opposing catalytic sites (Capita, 2011). The requirement of two CESAs to synthesize one chain would in fact reduce the stoichiometry to 18 chains, which is close to what is observed experimentally (*vide supra*). Regardless of the precise number of chains that form individual microfibrils, their packing into a crystalline structure is believed to occur spontaneously, driven by hydrogen and van der Waals bonding (Crowell et al., 2010).

There is strong evidence from mutants affected in cellulose biosynthesis that the rosette complex also contains other proteins than the CESAs. Examples are a membrane-bound sucrose synthase, a membrane-bound endo-β-(1→4)-glucanase (Korrigan in *Arabidopsis*), annexins, actin, tubulin, a putative lipid transfer protein, and others (for a detailed review, see Mølhøj et al., 2002). However, the precise roles of these proteins have not yet been determined and we cannot exclude that the observed effects of the respective mutations were indirect. In fact, purification of the rosette complex from *Arabidopsis* to physical homogeneity disproved the association of other proteins than AtCesA4, 7, and 8 (Atanassov et al., 2009). It is possible that the above named proteins would only be transiently associated with the complex, that is, be present only at a specific stage of the cellulose biosynthesis where they are needed (Guerriero et al., 2010).

1.4.2 Hemicellulose Biosynthesis

Structural similarities between the β-1,4-linked glucan chains of cellulose and the backbones of the various β-linked hemicellulosic polysaccharides led to the prediction that "CESA like" (CSL) genes might encode glycan synthases that are involved in the biosynthesis of these polysaccharides (Lerouxel et al., 2006). In contrast to cellulose synthase, the hemicelluloses synthases are located in the Golgi. Similar to the CESAs, however, the catalytic domains of the Golgi polysaccharide synthetases are also believed to face the cytosol. This then raises the problem of how the glycan chains are extruded. There is a general belief that this could occur in a similar way as cellulose synthesis (see earlier) through the formation of pores in the membrane. However, it appears that some members of the CslA, B, and C groups do not possess enough TM helices to allow the formation of such a pore. It has therefore been discussed that the catalytic sites of these enzymes are located in the Golgi lumen, which leaves the mechanism of further transport of the glycans open (Sandhu et al., 2009).

The *CSLs* are a family of genes that have sequence similarity to the *CESA* genes and appear to be present in all plant genomes (Figure 1.13). Certain *CSL* subfamilies are common to all plants, whereas other subfamilies are present only in specific groups of plants (Keegstra and Walton, 2006). Availability of a plant CESA sequence allowed annotation of a large number of genes with relatively low sequence similarity as CESA-*like* (*Csl*). Sequences in the *Csl* class are further divided into subgroups A–H and J. Groups B and G are specific to nongrass species, whereas groups F, H, and J are found only in grasses (Sandhu et al., 2009). In comparison with the biosynthesis of cellulose, however, that of hemicelluloses is far less well understood. Candidates for xylan synthase, feruloyl transferase, and arabinosyl transferase were tentatively identified by bioinformatic approaches, using a comparison of gene expression patterns of rice and *Arabidopsis* with a number of tissue expression libraries (Mitchell et al., 2007), and by the use of reversed genetics. All the β-glycan chain-forming enzymes identified thus far by these approaches belong to the Csl family (Sandhu et al., 2009).

Two groups of *Csl* genes, *CslF* and *CslH*, which are found only in grasses, have been shown to catalyze (1→3),(1→4)-β-D-glucan biosynthesis (Carpita, 2011). Heterologous expression of a rice *CslF* in *Arabidopsis*, a species that does not make (1→3),(1→4)-β-D-glucan, results in small amounts of the β-D-glucan in the cell walls (Burton et al., 2006). However, considerably greater amounts of the (1→3),(1→4)-β-D-glucan result when a *CslH* is coexpressed with *CslF*,

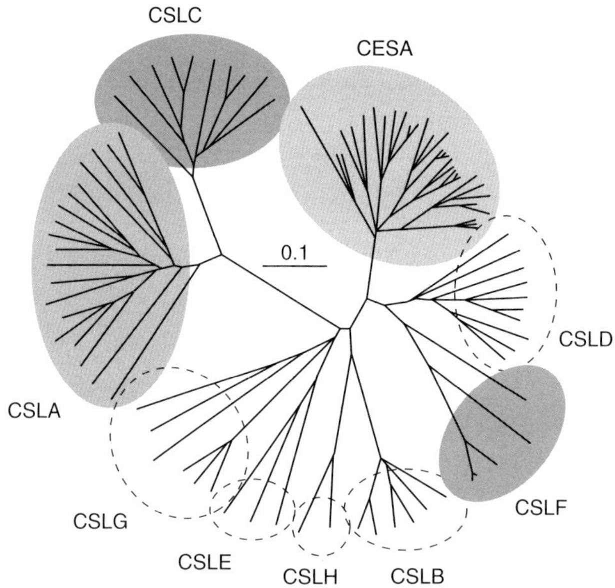

Figure 1.13. Phylogenetic identification of function of plant β-glucan synthases. CESA and CSLC are cellulose synthases; CSLA, β-mannan synthases; and CSLF, a β-1,3/1-4 glucan synthase. (Reprinted from Lerouxel O, Cavalier DM, Liepman AH, Keegstra K. Biosynthesis of plant cell wall polysaccharides—a complex process. Curr Opin Plant Biol, 2006; 9(6): 621–630. Copyright 2006, with permission from Elsevier.)

suggesting a synergistic role for both CslH and CslF in the synthesis of the polysaccharide and that a catalytic heterodimer enhances the activity (Carpita, 2011).

In addition to the backbones discussed earlier, most hemicelluloses also contain side chains, which are added by glycosyltransferases that are usually type II integral membrane proteins that have a catalytic domain facing the Golgi lumen (Lerouxel et al., 2006). The enzymes that add side chains to XyG and glucomannans are the best characterized, so we briefly consider them here.

Among the glycosyltransferases that add sugars onto the backbone galactosyltransferase, fucosyltransferase and glucuronoarabinoxylan have been identified (Zeng et al., 2010; Carpita, 2011).

1.4.3 Pectin Biosynthesis

As with hemicellulose synthases, pectin is also synthesized in the Golgi and transported to the wall in membrane vesicles. Pectin synthesis occurs simultaneously in numerous Golgi stacks in the cell in a process that appears to include a compartmentalization of specific biosynthetic enzymes to drive the construction of increasingly complex pectin polysaccharides through the *cis*, medial, and *trans*-Golgi cisternae (Mohnen, 2008).

Several reviews on pectin biosynthesis have recently been published. For a more detailed summary of the enzymology of pectin synthesis, readers are directed to Mohnen (2008) and Cafall and Mohnen (2009).

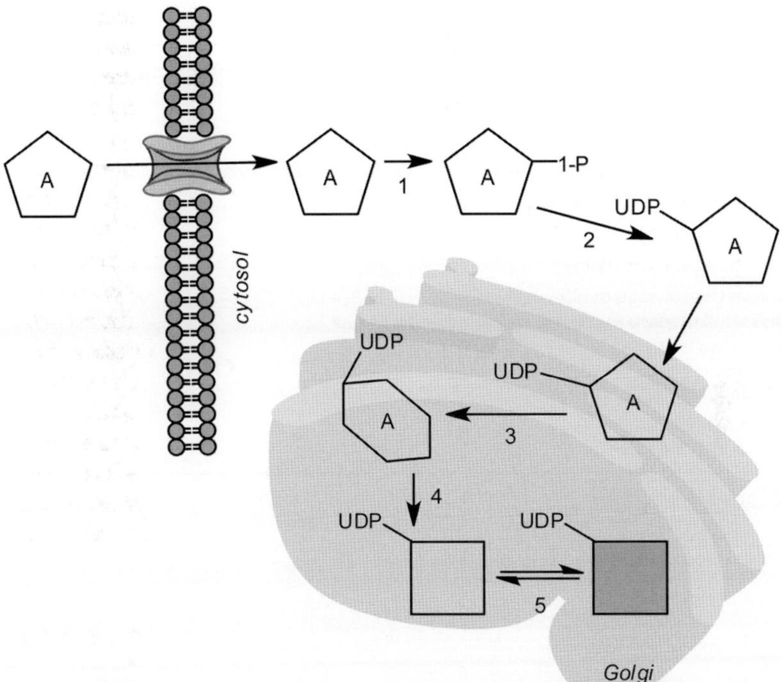

Figure 1.14. A model for galacturonic acid (GalA) recycling during plant cell wall assembly and restructuring from storage tissues: GalA (indicated by a the pentaeders with an A in the center), released from pectins, is transferred to the cytosol, where it is subsequently phosphorylated by galacturonate kinase and then converted to UDP-GalA in the cytosol by the UDP-sugar PPase "Sloppy." UDP-GalA can then be transferred into the Golgi, and can be either incorporated into glycans (not shown) or enter the nucleotide-sugar interconversion pathway, that is, being epimerized to UDP-GlcA (hexaeders with A in center) by UDP-GlcA 4-epimerase. UDP-GlcA is then further converted to UDP-xylose (*empty squares*) by UDP-GlcA decarboxylase. UDP-xylose 4-epimerase interconverts UDP-xylose and UDP-arabinose (*full squares*). (Modified after Yang et al., 2009.)

All evidence to date suggests that pectin is synthesized in the Golgi lumen by membrane-bound, or associated, Golgi-localized glycosyltransferases (GTs) that transfer glycosyl residues from nucleotide-sugars onto oligosaccharide or polysaccharide acceptors. It is not clear, however, how synthesis of any of the pectic polysaccharides is initiated or whether lipid or protein donors are required. Modification of the pectic glycosyl residues occurs by methyltransferases, acetyltransferases, and feruloyl transferases, using *S*-adenosylmethionine (SAM), acetyl-CoA, and feruloyl-CoA as substrates, respectively. Approximately 67 glycosyltransferase, methyltransferase, and acetyltransferase activities are predicted to be required for pectin synthesis (for review, see Caffall and Mohnen, 2009). Yet, conclusive identification of these genes has so far been accomplished only for a few of them.

Diverse biosynthetic pathways lead to the synthesis of the specific nucleotide-sugars required for plant pectin biosynthesis. Nucleotide-sugars may be formed via salvage pathways from sugars recycled from the wall polysaccharides or from sugars supplied to cultured cells (Bar-Peled and O'Neill, 2011; Figure 1.14). Such nucleotide-sugars, or primary sugar

Table 1.3. Predicted and proven *Arabidopsis thaliana* nucleotide-sugar interconverting enzymes required for pectin biosynthesis.

Enzyme	NDP-Sugar[a]	Enzymatic Reaction[b]
CMP-Kdo synthetase (CKS)	CMP-Kdo	E.C.2.7.7.38
GDP-mannose dehydrogenase (GMD)	GDP-4-keto-6-deoxy-Man	E.C.4.2.1.47
GDP-mannose epimerase (GME)	GDP-L-Gal	E.C.5.1.3.18
GDP-mannose epimerase/reductase (GER)	GDP-Fuc	NK
Kdo synthase (KDS)	CMP-Kdo	E.C.2.5.1.55
Myo-inositol oxygenase (MIOX)	D-GlcA	E.C.1.13.99.1
UDP-apiose/UDP-xylose synthase (AXS)	UDP-Api/UDP-Xyl	NK
UDP-glucose dehydrogenase (UGD)	UDP-Gal	E.C.1.1.1.22
UDP-glucose epimerase (UGE)	UDP-Glc	E.C.5.1.3.2
UDP-glucuronic acid epimerase (GAE)	UDP-GalA	E.C.5.1.3.6
UDP-rhamnose synthase (RHM)	UDP-L-Rha	E.C.4.2.1.76
UDP-xylose epimerase (UXE)	UDP-Ara	E.C.5.1.3.5
UDP-xylose synthase (UXS)	UDP-Xyl	E.C.4.1.1.35

NK, not known.
Data adapted from Caffall and Mohnen (2009), enzyme abbreviations are given in parentheses.
[a]The nucleotide-sugar synthesized.
[b]The enzyme commission number of each enzyme activity is listed based on the chemical reaction carried out.

phosphates derived directly from photosynthesis metabolism, are converted into a diverse array of sugar donor molecules by the nucleotide-sugar interconverting enzymes (NIEs) (Table 1.3). Nucleotide-sugars are supplied to wall biosynthetic glycosyltransferases by NIEs that regulate wall polysaccharide biosynthesis and are themselves frequently regulated by elements of the NIE pathway (Seifert et al., 2004). Evidence of NIE regulation of nucleotide-sugar availability is observed in the biological shift from primary wall to secondary wall synthesis: the abundance of nucleotide-sugars and their precursors is coordinately shifted to reflect an upregulation in hemicellulose and cellulose nucleotide-sugar substrates and a downregulation in pectic polysaccharide nucleotide-sugar substrates (Caffall and Mohnen, 2009).

GDP-D-mannose-4,6-dehydrogenase gene (GMD1) is a key component in the synthesis of GDP-fucose that is necessary for the correct synthesis of cell wall RG-II and of fucosylated xyloglucans. UDP-D-4-glucose epimerase (UGE) catalyzes the epimerization of UDP-D-Glc to UDP-D-Gal, which has an effect on the specific incorporation of Gal onto XG side chains, type-II arabinogalactan (β-1,6-galactan) and to a lesser extent RG-I (Seifert et al., 2002). UDP-D-Xyl is synthesized by the decarboxylation of UDP-D-GlcA by UDP-D-Xyl synthase (UXS), and UDP-D-Xyl is in turn utilized for the synthesis of UDP-D-Ara (Harper and Bar-Peled, 2009). Rhamnose biosynthesis (RHM) is responsible for the synthesis of UDP-L-Rha from UDP-D-Glc by three distinct activities: UDP-D-glucose-4,6-dehydratase, UDP-4-keto-6-deoxy-D-glucose-3,5-epimerase, and UDP-4-keto-L-rhamnose 4-keto-reductase (Oka et al., 2007).

The HG backbone is a polymer of α1,4-linked GalA residues and thus likely requires multiple HG:α1,4-GalATs to synthesize the entire complement of HG required throughout plant development. These enzymes (E.C.2.4.1.43) catalyze the transfer of D-GalA from UDP-D-GalA onto a growing stretch of HG via a catalytic domain facing the lumen of the Golgi. Unmethylated UDP-D-GalA was the preferred nucleotide-sugar substrate for elongation of endogenous acceptors in particulate membrane fractions of mung bean (*Vigna radiata*). A subfraction of HG is composed of XGA; HG decorated with xylose residues at the *O*-3 of backbone GalA residues. A xylosyltransferase in *Arabidopsis* was identified that is reported to be an XGA xylosyltransferase, which would be the first specific glycosyltransferase for XGA synthesis (Caffall and Mohnen, 2009).

The modification of pectic polysaccharides by addition of methyl groups at the C-6 carboxyl group or acetyl groups at the *O*-2 or *O*-3 of GalA residues occurs during synthesis in the Golgi apparatus. Pectins are first secreted in a highly methyl esterified form, but after the deposition of pectins in the apoplast, the methyl esters are removed by pectic methylesterases. Methylation occurs by multiple pectin methyltransferases that catalyze the transfer of a methyl group from the donor *S*-adenosyl-methionine (SAM) to the pectic polysaccharide (Caffall and Mohnen, 2009).

Pectin *O*-acetyltransferase (PAT) activity catalyzes the transfer of an acetyl group from acetyl-CoA to a pectic polysaccharide acceptor. Acetyl groups decorate the GalA residues of pectic polysaccharides at the *O*-2 or *O*-3 positions. The functional consequences of acetylation are not clear but may play a role in preventing pectin breakdown by microbial hydrolases.

The biosynthesis of RG-I requires multiple glycosyltransferase activities to synthesize a backbone of [1→2)-α-L-Rha*p*-(1,4)-α-D-Gal*p*A-(1→4] disaccharide repeats that are branched at the C-4 of approximately half of the rhamnose residues by 5-linked and 3,5-linked arabinan, 4-linked and 4,6-linked galactan, as well as type-I and type-II AG. Potentially 34 specific activities may be required to synthesize RG-I backbone (for details, see Cafall and Mohnen, 2009).

The galactan side chains of RG-I are composed primarily of β-1,4-D-galactan with some branches of β-1,6- and β-1,3-Gal. Galactosyltransferases (GalTs) catalyze transfer of D-Gal*p* residues from UDP-D-Gal*p* to an acceptor molecule. Pectin GalTs are hypothesized to catalyze the initiation of the galactan side chains directly onto the backbone rhamnose residues of RG-I, elongate the galactan chains, initiate branch points onto the galactan chains, and elongate the galactan side chains. Synthesis of these structures may require up to ten or more different pectin-specific GalTs (Mohnen et al., 2008). The synthesis of α-1,5-linked arabinan by α-1,5-arabionosyltransferases (AraTs) occurs by the transfer of arabinose residues from UDP-L-arabinose to acceptor molecules in an α-(1,5)-configuration. Up to 18 AraTs have been estimated to be required for the synthesis of the complex branched arabinans in pectin (Mohnen et al., 2008).

The backbone of RG-II is α-1,4-D-GalA-linked HG. GAUT1 has recently been discovered, which catalyzes the addition of GalA residues onto HG oligomers (Sterling et al., 2006), and could, conceivably, synthesize the backbone of RG-II. There are 15 GAUT genes in the GAUT1-related gene family, and it is also possible that one of these may function to specifically synthesize the backbone of RG-II. Further work in this area will be needed to explore this possibility, because the acceptor specificities for the GAUTs have not been determined (Caffall and Mohnen, 2009).

1.4.4 Lignin Biosynthesis

Lignins are synthesized via oxidative coupling between monomers of *p*-hydroxycinnamyl alcohols, namely, the monolignols *p*-coumaryl (*p*-CA), coniferyl (CA), and sinapyl alcohols (SA) (Boerjan et al., 2003; Vanholme et al., 2008). In addition to these building blocks, lignin also incorporates many more compounds, such as products from incomplete monolignol biosynthesis (5-hydroxyconiferyl alcohol, hydroxycinnamaldehydes, and hydroxycinnamic

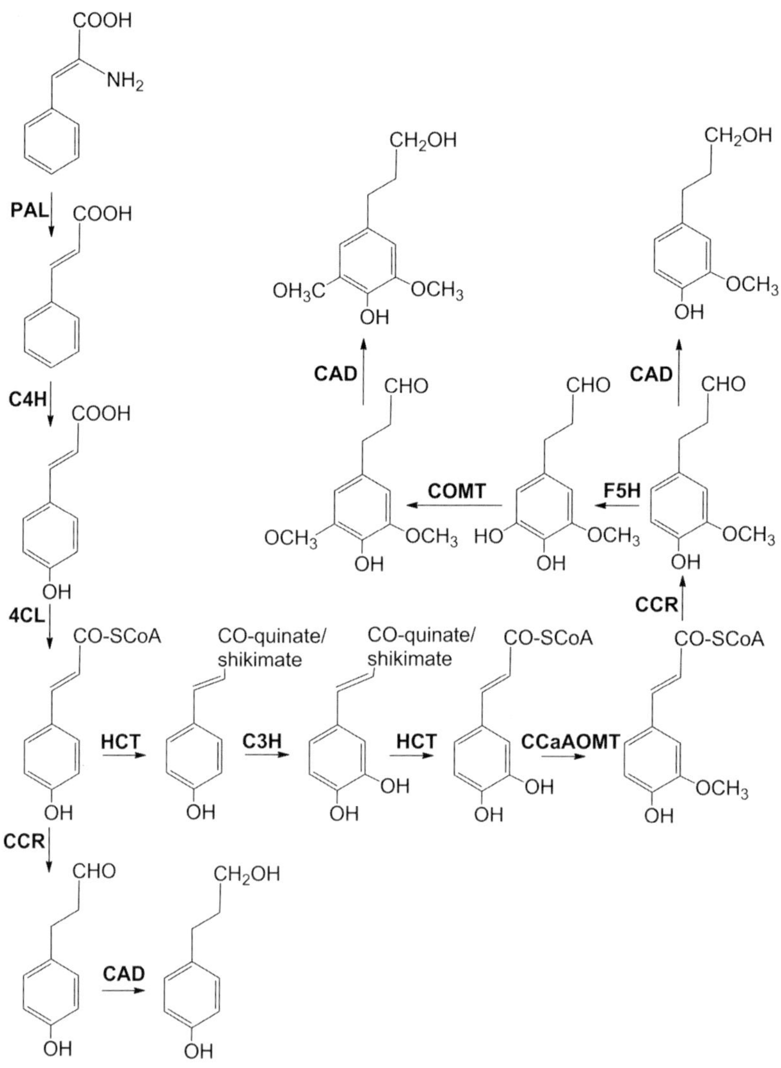

Figure 1.15. Biosynthesis of lignin via the phenylpropanoid pathway. PAL, phenylalanine ammonia-lyase; C4H, cinnamate-4-hydroxylase; 4CL, 4-coumarate:CoA ligase; HCT, hydroxycinnamoyl:CoA transferase; C3H, 5-*O*-(4-coumaroyl)shikimate 3′-hydroxylase; CCoAOMT, caffeoyl-CoA *O*-methyl transferase; CCR, cinnamoyl-CoA reductase; CAD, cinnamyl alcohol dehydrogenase; F5H, ferulate/coniferaldehyde 5-hydroxylase; COMT, caffeate/5-hydroxyconiferaldehyde *O*-methyltransferase.

acids), or derivatives of the classical monolignols (e.g., sinapyl *p*-hydroxybenzoate, coniferyl and sinapyl *p*-coumarate, and coniferyl and sinapyl acetate) (Vanholme et al., 2008).

A general reaction scheme for lignin biosynthesis is shown in Figure 1.15: the precursors are synthesized through the phenylpropanoid pathway, starting from phenylalanine ammonia-lyase (PAL) and leading to the three monolignols through a series of hydroxylations, methylations, and reductions (Simmons et al., 2010). The enzymes catalyzing the lignin biosynthetic pathway

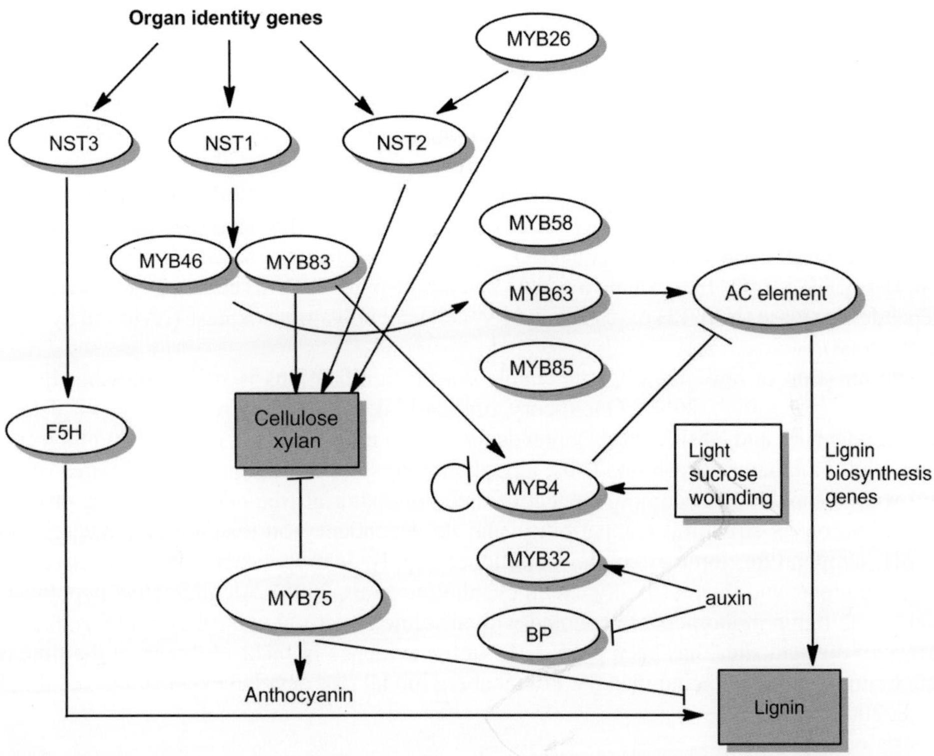

Figure 1.16. Current model of the transcriptional network of secondary cell wall biosynthesis in *Arabidopsis*. Components in oval circles specify regulatory proteins; *arrows* indicate activation or positive influence; vertical bars at the end of a line indicate inhibition or repression. For more details, see the review of Zhao and Dixon (2011).

are presumably all known (see Figure 1.15) and the regulation of their expression has been shown to be governed by several transcription factors of the NAC and MYB gene families (Zhou et al., 2009; Guillaumie et al., 2010; Figure 1.16). There is evidence suggesting that the transcript levels for PAL, C4H, 4CL, HCT, CCoAOMT, and F5H/CAld5H are functionally redundant (Simmons et al., 2010).

Following the biosynthesis of the lignin monomers, they are translocated to the cell wall, where they diffuse until they are oxidized for polymerization within the secondary plant cell wall. This transport is not fully understood, but available evidence suggests that it occurs either as free monolignols or as phenolic glucosides (coniferin, syringin) (Kärkönen and Koutaniemi, 2010). Genes for synthesizing these glycosides have been cloned, but the significance of the "glycoside hypothesis" of monolignol transport is questioned because coniferin and syringin do not accumulate to high levels in angiosperm xylem, and coniferyl and sinapyl alcohols are able to freely diffuse through the plasma membrane (Vanholme et al., 2008).

Once in the cell wall, the lignin precursors are polymerized via radical reaction. The formation of radicals is catalyzed by oxidative enzymes, either H_2O_2-dependent peroxidases or O_2-dependent oxidases/laccases. These enzymes are secreted into the apoplast where they are either soluble or covalently or ironically bound to the cell wall (Kärkönen and Koutaniemi,

2010). The broad substrate specificities and the large gene families of these two classes of oxidases have made it difficult to identify isoforms that are specifically involved in developmental lignification. Coexpression analyses of genes expressed during stages of active lignification or during tension wood formation and gene-family-wide expression have been performed in the attempt to associate individual gene family members with specific processes (Vanholme et al., 2008). Since peroxidases seem to be crucial for lignin biosynthesis, H_2O_2 should be available for peroxidases in the cell wall during lignin formation. Candidate hydrogen peroxide forming apoplastic enzymes are (i) a "cell wall oxidase" that catalyzes the oxidation of NADH to NAD^+, which in turn reduces O_2 to O_2^-, consequently is dismutated to produce O_2 and H_2O_2; (ii) germin-like oxalate oxidases and amine oxidases; (iii) cell membrane NADPH-dependent oxidase (NADPH oxidase); and (iv) cell wall polyamine oxidase (reviewed by Quan et al., 2008).

The question of how monolignols couple during lignifications is still controversially debated (Vanholme et al., 2008). One theory, originally developed by Freudenberg (for details, see Freudenberg and Nash, 1968), contends that lignin monomers are oxidized and then coupled in a combinatorial fashion. Thus, any phenol present in the lignin forming area of the cell wall is capable of becoming subject of the combinatorial free radical-coupling process, depending on its structural compatibility, and its dependency on reaction parameters, such as pH, temperature, ionic strength, monolignol supply, hydrogen peroxide and peroxidase concentrations, and the matrix in general (Vanholme et al., 2008). An alternative hypothesis, that is, the lignin monomers are coupled with absolute structural control by proteins bearing arrays of dirigent sites, has been proposed, but the evidence in favor of this is, at the time of this writing, less convincing than the Freudenberg model (for detailed explanation, see Ralph et al., 2008).

1.5 Strategies for Manipulating Wall Composition

Genetic engineering of crops in order to increase structural carbohydrate content and reduce lignin levels is a promising path that may result in reduced pretreatment severity, facilitate the hydrolysis process, and help recover the maximum amount of sugars. In addition to this, cellulose and hemicellulose degradation enzymes are also being expressed in the cell wall, which decreases the overall cellulase enzyme load during saccharification of biomass (Chandel and Singh, 2011).

1.5.1 Manipulation of Plant Cell Wall Polymer Composition

A major drawback in this area is that, although significant progress has already been obtained in the elucidation of the biochemical mechanisms that ultimately lead to cell wall polymers, the regulation of wall polymer biosynthesis remained largely elusive (Pauly and Keegstra, 2010). As I have explained earlier, UDP-glucose and other hexose-phosphates are the key substrate for biosynthesis of the different nucleotide-sugars required for synthesis of the various wall polymers, and manipulating of the enzymes involved in these steps has therefore been used as one of the earliest targets for altering the cell wall composition of plants. For example, overexpression of the sucrose synthase (for the importance of sucrose in cellulose biosynthesis, see Section 1.4) in poplar leads to a higher proportion and absolute amount (2–6%) of cellulose in fiber cells, whereas growth and biomass remained unaffected (Coleman et al., 2009). The most obvious strategy, that is, overexpression of the genes encoding the nucleotide-sugar conversion enzymes, has so far not yielded plants with altered wall compositions, possibly because of the redundancy of genes (Pauly and Keegstra, 2010).

Another strategy for modifying the composition of cell walls would be to up- or down-regulate the levels of glycosyltransferases and glycan synthases involved in polysaccharide biosynthesis. Attempts in this direction have so far been made only in model plants like *A. thaliana* and tobacco, which showed that it is principally possible to make dramatic alterations in the hemicellulose composition of a plant without serious consequences for its ability to grow and reproduce (Pauly and Keegstra, 2010).

1.5.2 Manipulation of Plant Lignin Content

As will be described in Chapter 10, lignin is one of the most important negative factors in the conversion of plant material to bioethanol. Analysis of lignin mutants showed that altering the expression of individual genes of its biosynthetic pathway has far-reaching consequences on plant metabolism: reduced lignin contents are typically associated with dramatic changes in the soluble phenolic pools, and different species accumulate various storage and detoxification

Table 1.4. Possible targets for genetic engineering of plants to generate high yielding and less recalcitrant biomass.

Selected Crop Traits	Approach Targeted	Effects Observed
Photosynthesis	Overexpression of phosphoenolpyruvate carboxylase, fructose-1,6-bisphosphatase and sedoheptulose-1,7-bisphosphatase	Increased CO_2 fixation leads to increased dry weight of biomass, enables water resistance
Cell wall composition	Specific cytochrome P450 enzymes, caffeic acid *O*-methyltransferase	Increased cellulose amount, less lignin
Starch composition	Starch enzymes, pullulanase	Alteration in starch structure and increased amount of starch
Stress tolerant	Signal transduction, transcription factors, effector genes	Development of stress-tolerant varieties
Cellulose degrading enzymes	Cellulase expression, β-glucanase expression	Cellulase and β-glucanase production in the cell wall
Grain yield	Enhanced ADP-glucose pyrophosphorylase activity; deregulation of endosperm ADP-glucose pyrophosphorylase activity; stimulation of photosynthesis and carbon metabolism	Improved seed weight and biomass yielding crops
Male sterility and plastid transformation	Engineering cytoplasmic male sterility via chloroplast genome by expression of β-ketothiolase	Impact on the development of routine biofuel crops

Adapted from Chandel and Singh (2011).

products. In addition, alterations in lignin metabolism also influence the biosynthesis of other cell wall polymers. Knowledge about these broader effects is essential to fully comprehend how gene function and cell wall properties are linked, how these cell wall properties are elaborated, and how they relate to the quality of raw material destined for agro-industrial uses (for reviews, see Simmons et al., 2010; Pauly and Keegstra, 2010; Chandel and Singh, 2011).

Manipulation of the genes involved in lignin biosynthesis obvious is the first candidate to try: downregulation of PAL, HCT, C3H, CCoAOMT, CCR, and CAD (for abbreviations, see legend to Figure 1.15); enzymes have been shown to impact the monomeric composition and amount of lignin present. Downregulation of C3H in *Medicago sativa* dramatically shifted lignin composition and structure, and produced a material that was easily digested. Downregulation of 4CL1 in *Populus tremuloides* (poplar) produced a 45% decrease in lignin and which consequently increased the relative cellulose content as well as some of the hemicellulose components (Simmons et al., 2010).

Another approach is the interference with lignin polymerization: the phenolic hydroxyl, *para* to the side chain, is critically important for oxidative coupling and polymerization. Lignin precursor phenolic groups in *ortho* configuration become methylated by the *O*-methyltransferases CCoAOMT and COMT, which results in formation of guaiacyl and syringyl monomers. Marita et al. (2003) have used downregulation of COMT, which is responsible for the methylation of 5-hydroxconiferyldehyde to decrease the flux to sinapyl alcohol. This resulted in the accumulation of 5-hydroxyconiferaldehyde in the cell and its reduction to 5-hydroxyconiferyl alcohol, which becomes incorporated into lignin leading to new types of structures in the lignin polymer. Other means to change lignin structure involved the incorporation of coniferyl ferulate into lignins, which could open the way to obtain easily cleavable ester linkages into its backbone, thus facilitating its depolymerization and solubilization (Grabber et al., 2008).

Torney et al. (2007) reviewed the genetic engineering approaches to improve bioethanol production from maize, which are focused on increasing stress tolerance, the rate of photosynthesis, grain yield, and production of biomass conversion enzymes *in planta* (Table 1.4). These approaches could also be incorporated for the improvement of weedy crops in terms of increased biomass weight, cell wall composition, and biomass conversion assisted by enzyme expression *in planta* (Chandel and Singh, 2011).

Chapter 2

The Actors: Plant Biomass Degradation by Fungi

2.1 Ecological Perspectives

The recycling of photosynthetically fixed carbon by the action of microbial plant cell wall degraders is a pivotal biological process that is integral to one of the major geochemical cycles in the biosphere by releasing 10^{11} tons of monosaccharides annually (Brown, 1979). This is understandable when it is considered that the half-life of cellulose—in the absence of (micro)organisms and at neutral pH—is estimated to be several million years (Wilson, 2011). The biological degradation of cellulose and hemicelluloses is therefore—besides destruction by fire—globally of prime responsibility for most of the carbon turnover.

Since most of the cellulose is present in plant cell walls, whose structure (as I have described in Chapter 1) is destined to confer recalcitrance, only a small fraction of all microorganisms are specialized to be capable of doing this job. While a large number of microorganisms with cellulolytic and hemicellulolytic activities are now known, their relative importance as lignocellulose degraders in their ecosystem is opaque. There are two obstacles in studying which microorganisms participate in this activity and to which extent: first, our knowledge on the occurrence of microbes in any ecosystem is largely based on cultivation-dependent approaches, and owing to the fact that only a very small fraction of organisms are monitored by these techniques, it is likely that we still overlook important players in the field. Coniferous forest ecosystems are one of the rare examples where DNA-dependent analyses have been performed to quantify the contribution of fungi and bacteria in carbon cycling. This is due to the fact that coniferous forests display a prominent role in the global carbon cycle (Myneni et al., 2001). Thereby, it was found that fungi quantitatively dominate in the decomposition of above ground biomass, while the importance of bacteria increases with the depth of the soil. Yet fungi still comprise a large proportion of this soil microbial biomass and have a dominant role in decomposition and nutrient cycling in soil (Buée et al., 2009). Baldrian et al. (2011) assessed the fungal diversity in six different forest soils using tag-encoded 454 pyrosequencing of the nuclear ribosomal internal transcribed spacer-1. They mostly detected fungi

Fungi and Lignocellulosic Biomass, First Edition. Christian P. Kubicek.

belonging to the Dikarya subkingdom (Ascomycota and Basidiomycota), Agaricomycetes being the dominant fungal class, while *Ceratobasidium* sp., *Cryptococcus podzolicus, Lactarius* sp., and *Scleroderma* sp. were also abundant.

Second, it is likely (and in fact has been demonstrated in some cases) that microorganisms cooperate in the degradation of lignocelluloses, either in synergistic or in subsequently acting ways (i.e., that one microorganisms feeds on the degradation products of another). It is known that the spatial and temporal development of fungal communities in wood follows diverse and complex pathways, and is influenced by stress aggravation (worsening of abiotic environmental conditions), stress alleviation (improvement in abiotic conditions), disturbance, and combat (interspecific competition for space rather than directly for nutrients) (Heilmann-Clausen and Boddy, 2005). Thus, the community change that can be observed are not necessarily determined by their role in lignocelluloses degradation but is probably mainly effected by the fungi themselves.

2.2 The Major Three Mechanisms of Lignocellulose Degradation by Fungi

The above considerations make it less surprising that most, if not all, multicellular fungi are able to degrade lignocelluloses, at least to some extent. However, they use different mechanisms that are distinguished by the way how these organisms make the celluloses and hemicelluloses accessible for degradation, by the enzymes involved, and by their behavior during attack and degradation of wood. These different mechanisms are generally classified as white rot, brown rot, and soft rot.

White rot is the only mechanism of these three that can completely degrade the wood structure because it is also capable of depolymerizing lignin by cleaving the bond between the Cα and Cβ atoms in it. White rot is typically associated with the decay of hardwood and the resulting material either has a uniformly white appearance or displays selectively decayed areas or pockets of white rot.

Brown rot, on the other hand, is the most prevalent attack on coniferous wood, and since mankind has made an intensive use of materials made from coniferous wood, it also has a significant negative impact on the wooden products in the Northern hemisphere (Schmidt, 2006). Brown rot fungi remove cellulose and hemicellulose from the wood, but cannot attack lignin, or degrade only a very small part of it, which therefore accumulates in the degraded material as a brittle matrix and is responsible for the brown appearance of the decayed wood. Some brown rot fungi can provide their own source of moisture and nutrients, and cause decay in otherwise relatively dry timber (called "true dry rot"). Examples are damage caused by *Serpula lacrymans* in the Northern Europe or by *Meruliporia incrassata* in North America. Such fungi have water-conducting strands and are capable of carrying water and nutrients from soil into the wooden parts of buildings. In the course of the fungal attack, the decaying wood becomes dark and crumbly, finally brittle and eventually crushes into powder. Consequently, this is a very serious infestation of a building, which is hard to combat, and can lead to a collapse of the building.

Both brown and white rot are almost entirely a specific trait of basidiomycetes. This is believed to be due to the evolution of basidiomycetes in the Upper Devonian period, which paralleled the evolution of vascular plants (Eriksson et al., 1990). In contrast, the third way of fungal wood degradation—soft rot—is distributed among both asco- and basidiomycetes.

Fungi that attack wood by this mechanism acts preferentially on wood with low lignin content and usually require a higher moisture content than white or brown rot fungi and also need direct contact with soil.

A summary of these three strategies to use wood biomass as a substrate and the resulting phenotypical differences are summarized in Table 2.1 and will be explained in more detail below.

2.2.1 White Rot

On a balance, white rot is the most frequently found wood-rotting process that acts on both hard- and softwoods. As mentioned above, it is a feature of basidiomycetes and has been studied with several of them in detail (Table 2.1). In addition, some ascomycetes (members of the *Diatrypaceae* and *Xylariaceae*) have been reported to produce white rot of hardwoods (Eriksson et al., 1990). Although white rot is classified by the morphology arising from the attack (such as white pocket, white-mottled, and so on. (Hubert, 1931), two mechanistically different patterns are distinguished today: (i) simultaneous (nonselective) delignification and (ii) selective delignification (sequential decay) (Martinez et al., 2005). However, these two patterns are not necessarily mutually exclusive, but rather represent two modes with a broad range of overlap: many white rot fungi cause both types of rot and the relative degree of their utilization can even vary between different species of the same taxon (Martinez et al., 2005).

Simultaneous delignification acts mainly on hardwood and degrades cellulose, lignin, and hemicellulose at the same time. It has been found in the *Trametes versicolor, Irpex lacteus, Phanerochaete chrysosporium, Heterobasidion annosum,* and *Phlebia radiata*, and in some ascomycetes such as *Xylaria hypoxylon* (Martinez et al., 2005). Simultaneous lignifying fungi start their attack from the cell lumen and then progress toward the middle lamella. The enzymes that catalyze this degradation remain bound to the fungal hyphae or are maintained closely to the cells because the ligninolytic enzymes need intracellularly produced metabolites for provision of hydrogen peroxide (see below and Chapter 5). The decay is therefore confined to small areas around the fungal hyphae and simultaneous delignification thus leaves substantial amounts of wood undecayed.

In contrast, in selective delignification, lignin and hemicelluloses are degraded before cellulose is attacked. This mode of decay occurs both on hardwood and softwood, and the degradation starts from the middle lamella and extends to the secondary plant cell wall. It is specific to certain basidiomycetous species such as *Ganoderma australe, Phlebia tremellosa, Ceriporiopsis subvermispora,* several *Pleurotus* spp., and *Phellinus pini* (Martinez et al., 2005).

Lignin is chemically difficult to degrade because the mechanism responsible for its biosynthesis (see Chapter 1) involves free radical coupling reactions of phenolic cinnamyl alcohols and thus results in a polymer interconnected through diverse carbon–carbon and ether bonds that are not hydrolyzable. This feature of the structure of lignin has important consequences for its biological degradation: although the bonds in lignin can be cleaved by strong nucleophiles, organisms cannot make use of such a mechanism because a strong nucleophile buried in the active site of an enzyme would be unable to make close contact with the highly variable surface of lignin. Therefore, biological delignification must make use of less specific oxidative mechanisms (Hammel and Cullen, 2008). The extracellular oxidative enzymes thought to be involved in lignin depolymerization, therefore, include an array of oxidases and peroxidases that are responsible for generating highly reactive and nonspecific free radicals that can attack lignin.

Table 2.1. Comparison of white rot, brown rot, and soft rot fungi.

	White Rot		Brown Rot	Soft Rot
Appearance		Bleached, spongy; loss of strength	Brown, brittle, powdery; loss of strength	Soft in wet, brown and crumbly in dry environments
Wood type and condition	SR	Hardwood	Mostly softwood	Mostly hardwood
	SD	Hard- and softwood	Forest ecosystems, wood in service	Forest ecosystems, wet wood
Cell wall polymer concerned		Cellulose and hemicellulose lignin	Cellulose and hemicellulose lignin slightly modified	Cellulose and hemicellulose, lignin slightly modified
Anatomy	SR	Starting from the lumen	Degradation in distance from hyphae	Degradation in proximity of hyphae
	SD	From the middle lamella and secondary wall		Starting from the lumen
Fungi	SR	*Trametes versicolor*	*Gloeophyllum trabeum*	*Chaetomium globosum*
		Phanerochaete chrysosporium	*Laetiporus sulfureus*	*Thielavia terrestris*
		Heterobasidium annosum	*Piptoporus betulinos*	*Paecilomyces spp.*
	SD	*Ganoderma australe*	*Postia placenta*	*Trichoderma reesei*
		Phlebia tremelosa, P. pini	*Serpula lacrymans*	
		Ceriporiopsis subvermispora		
		Pleurotus spp.		

Data in table adapted from Martinez et al. (2005).
SR, selective rot; SD, simultaneous delignification.

The nonspecific nature and extraordinary oxidation potential of the peroxidases has also attracted considerable interest in its application in various biotechnological processes such as fiber bleaching and the remediation of organopollutant-contaminated soils and effluents (Asgher et al., 2008). The majority of enzymes that perform this attack use hydrogen peroxide as the oxidizing substrate (Figure 2.1), and—as mentioned above—are thus depending on accessory enzymes that provide this short half-lifed agent. A detailed description of the lignin depolymerizing enzymes is given in Chapter 5 and omitted here.

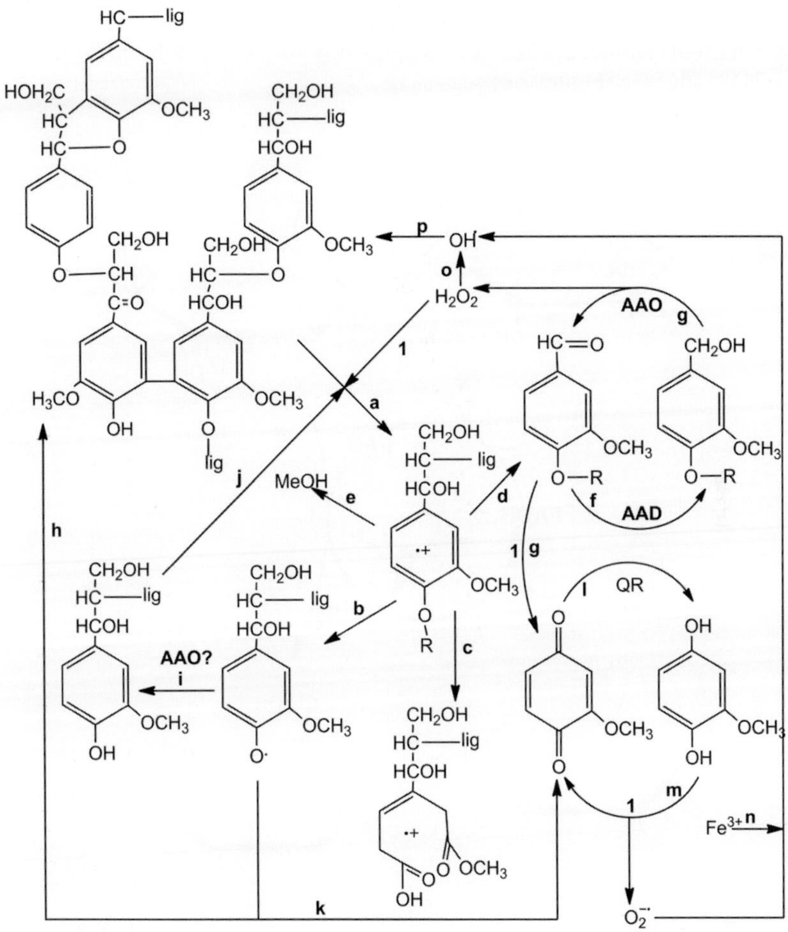

Figure 2.1. The complex interaction of enzymatic and nonenzymatic reactions in lignin degradation by fungi. Initial steps: (a) generation of aromatic radicals by lignin peroxidase, manganese peroxidase and laccase; (b) C4–ether breakdown; (c) aromatic ring cleavage; (d) C–C-cleavage; and (e) demethoxylation. (f, g) Hydrogen peroxide formation by aryl alcohol oxidase (AAO) from the aromatic aldehydes formed in the reactions before or synthesized by the fungus itself, using a redox cycle that also involves aryl alcohol dehydrogenases (AAD); (h) repolymerization of lignin from its phenoxy radical degradation products; (i) reduction to phenolic products, which prevents (j) reoxidation of phenolic compounds by laccases or peroxidases; (k) C–C cleavage of phenoxy radicals, which yields *p*-quinones; (l, m) oxygen activation in redox cycling reactions involving quinine reductases, laccases, and peroxidases; (n) thereby, ferric ions are reduced; (o) and hydroxyl radicals formed. (p) The latter can initiate further attack on lignin. (Modified from Martinez et al., 2005.)

Cellulose is degraded by hydrolysis of the β-1,4-glycosidic bonds by extracellular enzymes that are traditionally classified as endoglucanases (EC 3.2.1.4), which hydrolyze cellulose to glucooligosaccharides, and cellobiohydrolases (EC 3.2.1.91), which release cellobiose from crystalline cellulose. Some researchers also claimed the existence of exoglucanases that release glucose from cellulose and glucooligosaccharides, but most likely these enzymes contained β-glucosidase contaminants that flawed the product pattern. In any case, none of the cellulase enzymes that have been classified in the CAZyme database (see Chapter 3) has been shown to produce glucose as a product. Therefore, the products of the synergistic enzymatic action of endo-β-1,4-glucanases and cellobiohydrolases are cellobioses (and to a lesser extent cellooligosaccharides), which are subsequently further hydrolyzed by β-glucosidases (EC 3.2.1.21) to glucose (Figure 2.2).

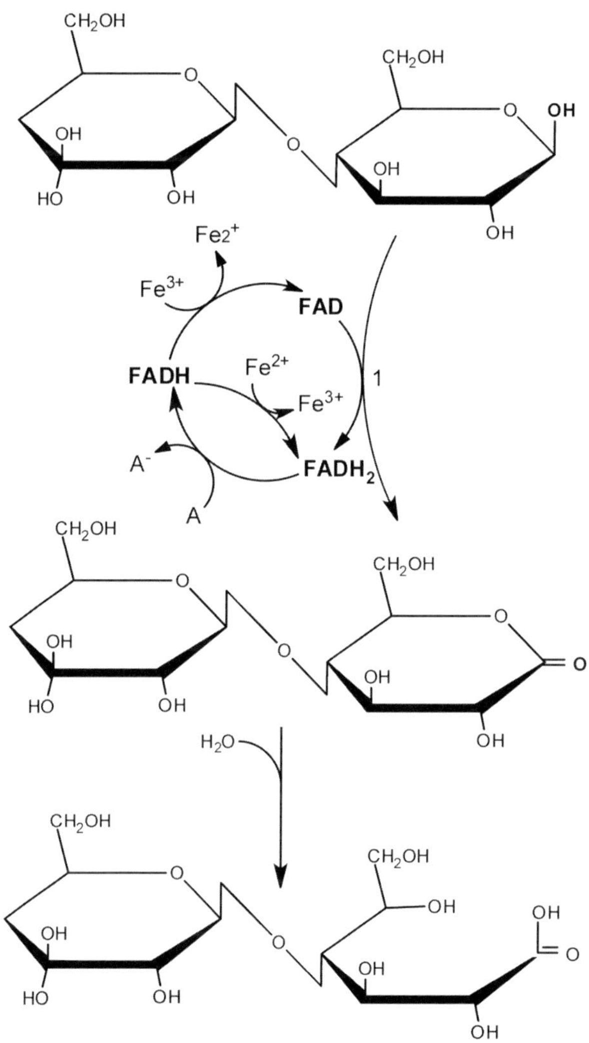

Figure 2.2. Reaction scheme for cellobiose dehydrogenase.

In 1950, Elwyn Reese and coworkers introduced the C1–*Cx* concept for cellulose hydrolysis by cellulases (Reese et al., 1950). This hypothesis proposed that a C1-component must render cellulose accessible to the attack of cellobiohydrolases and endoglucanases by splitting the hydrogen bonds between the chains, yet without exerting any hydrolytic function itself. This concept remained unproven for a long time, however, until Eriksson et al. (1969) proposed that an exoglucanase would play the C1 part, which was accepted by most researchers in the field. However, during the last decade, a number of proteins have however been identified that contribute to the overall rate of cellulose hydrolysis by cellulases without having any hydrolytic activity themselves. The properties and mode of action of the cellulolytic enzymes is described in detail in Chapter 3.

There is also an enzyme, cellobiose dehydrogenase, that can perform oxidative attack on cellulose (see Chapter 5; Figure 2.2). Cellulose degradation by *Phanerochaete chrysosporium* has been shown to be enhanced by external addition of cellobiose dehydrogenase (see Baldrian and Valásková, 2008, for references). Phillips et al. (2011) showed that cellobiose dehydrogenase enhanced cellulose degradation in the ascomycete *Neurospora crassa* by coupling the oxidation of cellobiose to the reductive activation of copper-dependent polysaccharide monooxygenases (PMOs) that catalyzed the insertion of oxygen into C–H bonds adjacent to the glycosidic linkage. These PMOs displayed different regiospecificities and resulted in polysaccharides oxidized either at the reducing or nonreducing end of a glucan chain. Since cellobiose dehydrogenases and proteins with high similarity to the above described PMOs are found throughout both ascomycete and basidiomycete fungi, it is possible that the oxidative cellulose degradation described by Phillips et al. (2011) may be widespread throughout the fungal kingdom.

Cellobiose dehydrogenase may also be involved in cellulose decomposition in another way: as it reduces Fe^{3+} to Fe^{2+} or Cu^{2+} to Cu^{+} during oxidation of cellobiose, the so reduced metal ions could react with H_2O_2 (which can be formed by the enzyme itself) to generate hydroxyl radicals that may modify and depolymerize plant cell wall polymers (Baldrian and Valásková, 2008).

Although hemicelluloses are relatively complex polysaccharides, their enzymatic degradation is well understood, and moreover widely the same in white rot and soft rot fungi. The hydrolysis of hemicelluloses occurs by the concerted action of endoenzymes cleaving internally the main chain, exoenzymes liberating monomeric sugars, and auxiliary enzymes that remove the side chains of the polymers or oligosaccharides, thus leading to the release of various mono- and disaccharides depending on the hemicellulose type. The enzymes involved are described in detail in Chapter 4.

2.2.2 Brown Rot Fungi

Brown rot fungi act mainly on coniferous softwoods and can only degrade cellulose and hemicellulose but are not able to depolymerize lignin (although some are able to modify it by demethylation; Eriksson et al., 1990). This usually involves a first breakage in the cellulose chain, followed by the removal of first the side chain and then main chain hemicelluloses (arabinan and galactan, and xylan and mannan, respectively). Finally, cellulose is completely hydrolyzed and lignin demethylated (Eastwood et al., 2011). The primary attack of the cellulose causes a loss of mechanical strength of the wood, followed by complete degradation of cellulose and hemicellulose. As a result of this process, brown-rotten wood consists almost entirely of (modified) lignin, which is reddish-brown in appearance and eventually cracks into brick-like pieces that give this mechanism its name. Well-studied brown rot basidiomycetes are *Postia placenta*, *Fomes fomentarius*, *Serpula lacrymans*, and *Gloeophyllum trabeum* (Table 2.1). They grow mainly in the cell lumen of the woody cells. Brown rot fungi have an important role in coniferous ecosystems because they contribute significantly to formation of humus, which can be used by other microorganisms (Lundell et al., 2010). A study on the evolution of the *Gloeophyllales*, most of which have been demonstrated to produce a brown rot mode of wood decay and are preferentially found on coniferous substrates, suggested that they arose in the Cretaceous, after the origin of *Pinaceae* conifers (Garcia-Sandoval et al., 2011). This finding implies an ancient plant–fungus specialization.

The molecular mechanisms involved in brown rot are still only partially understood, but available evidence suggests that they are due to both an oxidative and also a hydrolytic attack. The involvement of an oxidative, nonenzymatic attack is at least during the incipient decay necessary because brown rot fungi attack the cellulose in wood directly without first removal of lignin or hemicelluloses, and extracellular enzymes are too large to penetrate the plant cell walls to their place of action and can thus not be involved (Halliwell, 1965; Koenigs, 1974). The recent genome sequences of the brown rot fungus *P. placenta* (Martinez et al., 2009) and the dry rot fungus *Serpula lacrymans* (Eastwood et al., 2011) revealed interesting details: genes encoding peroxidases like those present in white-rot fungi and needed for lignin breakdown (see Chapter 5) are completely absent. Some enzyme families related to lignocelluloses degradation exhibited parallel patterns of losses and expansion in both brown rots and ectomycorrhizas. Both brown rot species lacked the major "GH7 endoglucanase and cellobiohydrolase" genes, and also the genes for "endoglucanase GH61" were strongly reduced (see Chapter 3 for a full description of these genes and proteins). On the other hand, genes encoding "GH5" endoglucanases and pectinases were expanded in both brown rot species and thus may aid in intercellular enzyme diffusion (Eastwood et al., 2011). Brown rot fungi must therefore use diffusible low molecular weight chemicals that degrade cellulose in a nonenzymatic way.

This diffusible chemical may be a hydroxyl radical: the detection of hydrogen peroxide (H_2O_2) production by several brown rot fungi (catalyzed by oxidase enzymes such as glyoxal oxidases and copper radical oxidases) led to the proposal of a degradation pathway based on the Fenton reaction (Figure 2.3), which produces a hydroxyl ion radical. The hydroxyl radical has a half-life of nanoseconds (Eastwood et al., 2011) and is the most powerful oxidizing agent of living cells. In addition, certain states of hypervalent iron in which the radical remains associated with iron were also considered by some authors as potential oxidizing agents (for review, see Baldrian and Valásková, 2008). Consistent with such a role of iron, Martinez et al. (2009) observed the upregulation of genes encoding iron reductases, quinone reductases, and several structurally divergent oxidases that could be involved in extracellular generation of Fe(II) and H_2O_2 during growth of *P. placenta* on cellulose. As reviewed by Baldrian and Valásková (2008) in detail, the hydroxyl radicals thus formed can abstract hydrogen atoms from the sugar subunits of cellulose or other polysaccharides (Ek et al., 1989), thereby producing transient carbon-centered radicals that react rapidly with O_2 to give peroxyl radical species. If the ROO• carries a hydroxyl group on the same carbon, it will eliminate •OOH (Halliwell and Gutteridge, 1999). If there is no α-hydroxyl group present, the molecule undergoes a variety of oxidoreductions, some of which can result in the cleavage of the cellulose chain (Kirk et al., 1991).

However, because Fe^{2+} is usually absent in oxygenated environment, mechanisms must exist for maintaining the necessary reduction of Fe^{3+}. The secretion of hydroquinones are considered as one potential contributor because they have been shown to reduce Fe^{3+}, thus generating

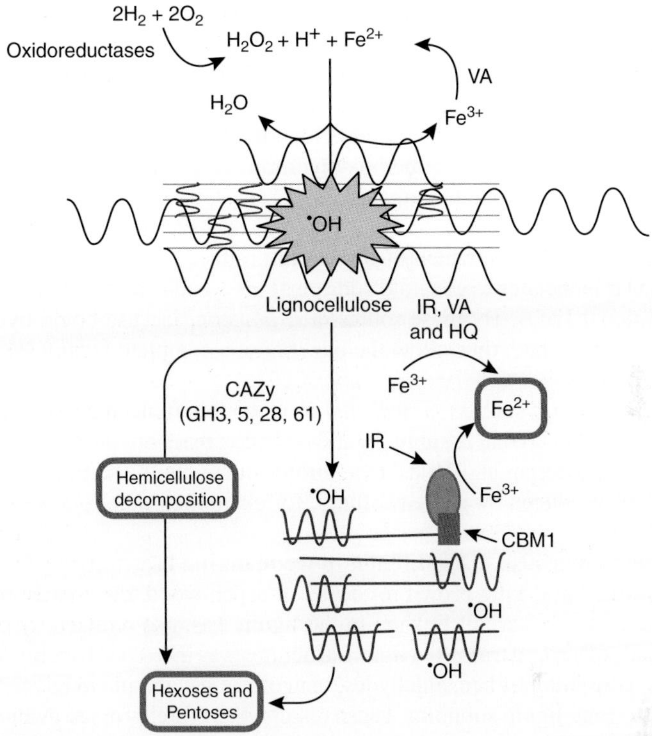

Figure 2.3. Schematic overview of the proposed mechanism of wood decay by brown rot fungi. Scavenging mycelium colonizes a new food source, inducing VA production and expression of oxidoreductase enzymes, which drive hydroxyl radical attack on the lignocellulose composite. CAZy gain access to the weakened composite structure and break down accessible carbohydrates. Cellulose-binding iron reductase targets OH-generating Fenton's reaction on cellulose chains, releasing chain ends for hydrolysis and assimilation. IR, iron reductase; HQ, hydroxyquinones; CBM, cellulose-binding module. (From Eastwood et al., 2011, reprinted with permission from AAAS.)

perhydroxyl radicals and Fe^{2+}. Enzymes potentially capable of forming hydroquinones include intracellular benzoquinone reductases and extracellular sugar dehydrogenases such as cellobiose dehydrogenase, both of which have been shown to use quinones as alternate electron acceptors. The quinones can be found in wood extracts or be produced by the fungus itself. In fact *Gloeophyllum* spp. and *P. placenta* were shown to secrete two hydroquinones: 2,5-dimethoxyhydroquinone (2,5-DMHQ) and 4,5-dimethoxycatechol (4,5-DMC) (for review, see Baldrian and Valásková, 2008).

A second low-molecular weight compound that could play a role in nonenzymatic degradation of cellulose is extracellular oxalic acid, which is produced by brown rot fungi such as *P. placenta* in sufficient concentrations to acidify colonized wood to pH 2, a pH that by itself can slowly hydrolyze cellulose (Green et al., 1991). In fact, oxalate in a certain range of concentrations greatly enhanced cellulose degradation by a Fenton-type oxidation *in vitro* (Tanaka et al., 1994) and a less oxalate-producing mutant of the brown rot fungus *P. placenta* exhibited a reduced ability to cause wood mass loss than its parent strain (Micales and Highley, 1991).

The occurrence of oxalate in the immediate environment of brown rot fungi poses a problem to the Fenton chemistry theory, because oxalate is a strong chelator of Fe^{3+}, and the reduction potential of the resulting Fe^{3+} trioxalate complex is too negative to react readily with methoxyhydroquinones (Wei et al., 2010). In the attempt to explain how hydroxyl radicals could nevertheless be formed, the latter authors identified two laccases genes in the genome of *P. placenta* that were able to oxidize methoxyhydroquinones to semiquinone radicals, which have much lower reduction potentials than their parent hydroquinones and are therefore expected to be better reductants of Fe^{3+}. They also reduce O_2 to generate perhydroxyl radicals ($HOO^{\cdot}$) and superoxide ($O_2^{\cdot -}$), which dismutate to produce H_2O_2. Furthermore, $HOO/O_2^{\cdot -}$ can reduce some Fe^{3+} chelates to generate additional Fe^{2+} and can oxidize some Fe^{2+} chelates to generate additional H_2O_2. By these routes, a *P. placenta* laccase could bypass the disadvantageous effects of oxalate, thus allow the generation a complete Fenton system even in its presence (Figure 2.4).

Further, brown rot fungi such as *G. trabeum* produce low molecular weight glycopeptides (ranging from 1.5 to 12 kDa) that are able to catalyze redox reactions and produce free hydroxyl radicals (Figure 2.5, Baldrian and Valásková, 2008). Interestingly, similar peptides have also been detected to be produced by white rot fungi, for example, *Phanerochaete chrysosporium* (Tanaka et al., 1996).

Hydroxyl radicals may also disrupt lignin in wood during brown rot by *P. placenta*: Yelle et al. (2011) showed that during brown rot decay of aspen wood, the content of the principal arylglycerol-β-aryl ether interunit linkage in the lignin was also reduced by more than half, while the frequency of several truncated lignin structures was increased roughly fourfold. These new end groups, consisting of benzaldehydes, benzoic acids, and phenylglycerols, accounted for 6–7% of all original lignin subunits. These results, therefore, provide evidence that brown rot by *P. placenta* also results in significant ligninolysis, which might enable infiltration of the wood by polysaccharide hydrolases even though the partially degraded lignin remains *in situ*.

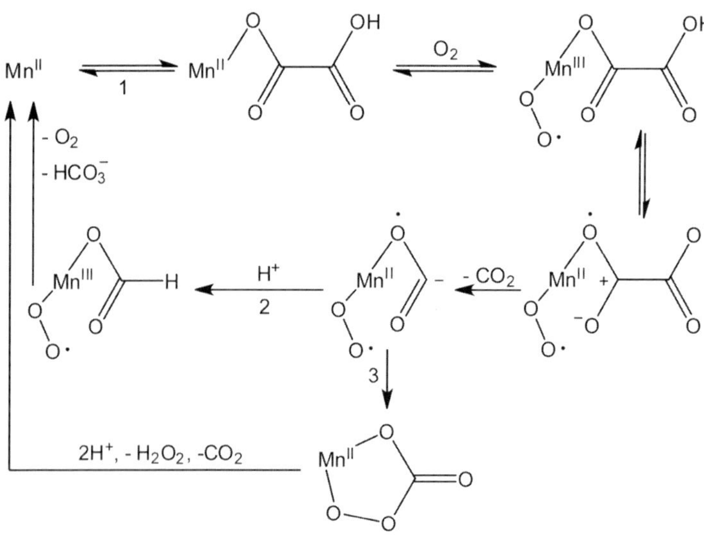

Figure 2.4. Divergent free-radical mechanisms for manganese-dependent oxalate oxidation and decarboxylation. (1) Introduction of free oxalic acid and (2) oxalate decarboxylase. (Modified after Escutia et al., 2005.)

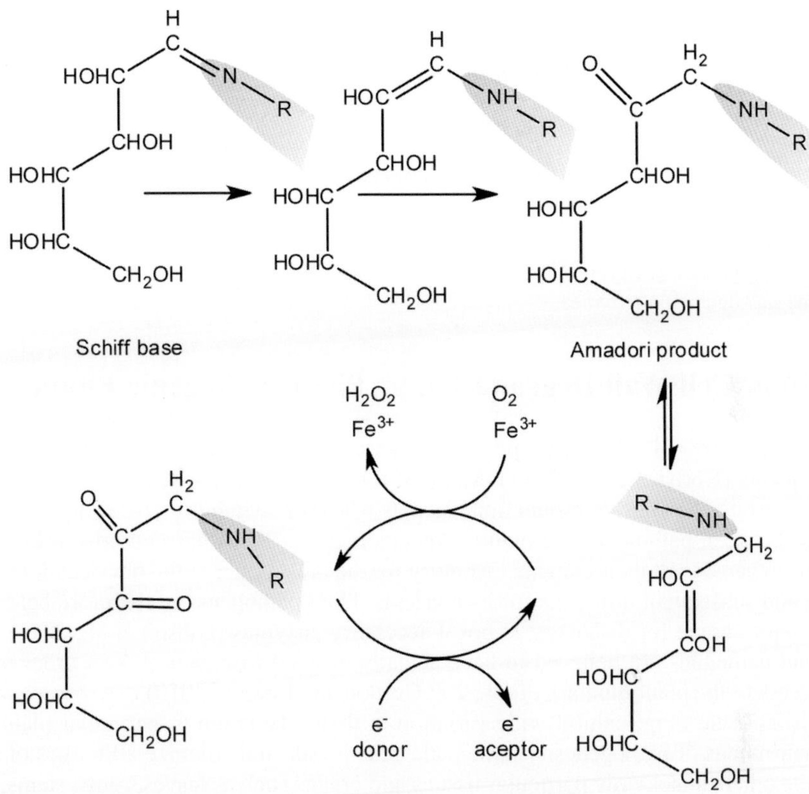

Figure 2.5. Mechanism of peptide catalyzed hydrogen peroxide formation. The peptide moiety is indicated by a shaded background. (Adapted from Baldrian and Valásková, 2008.)

2.2.3 *Soft Rot Fungi*

Soft rot degradation is a typical feature of ascomycetes and can be distinguished into two types: type 1, wherein the fungus forms longitudinal cavities within the secondary wall, and type 2, where the entire secondary wall is eroded. Soft rot gained its name from the appearance of a soft, decayed surface of wood after degradation by fungi. It has previously been claimed that a high moisture is obligatory for soft rot, but it now known that it can also occur in dry environments. In the latter case, its macroscopical appearance is indistinguishable to that of brown rot.

The major difference to white rot is that soft rot does not attack the middle lamella and does not involve ligninases. The only enzymes used are cellulases and hemicellulases, and soft rot is therefore characterized by an extensive loss in the carbohydrate polymers and consequently a significant reduction in the strength of the decayed wood. The rate by which soft rot fungi decay wood seems to be strongly dependent on the content and spatial distribution of lignin, guaiacyl components of softwood lignin being apparently the biggest obstacle (Eriksson et al., 1990).

The respective cellulolytic and hemicellulolytic enzymes of soft rotters are essentially of the same type as described above for white rot fungi and shall not be repeated here.

Table 2.2. Plant pathogenic lifestyles and lignocelluloses degradation.

	Necrotrophs	Biotrophs
Effect on host	Killing before invasion	Intimate contact with living plant
Parasitism	Nonobligate	Obligate
Survival and reproduction	In dead and in living plant	Only in living plant
Role of CWDEs	Invasion, colonization, nutrient acquisition	Invasion only

Adapted from Gibson et al. (2011).
CWDE, cell wall-degrading enzymes.

2.3 Plant Cell Wall Degradation by Plant Pathogenic Fungi

Approximately 10% of all fungal species known can cause diseases in more than 10,000 different plants (Horbach et al., 2011). Many of these fungi cause diseases in crops that are suitable substrates for biofuel production, such as wheat or switchgrass (see Chapter 1). The arms race between pathogens and plants (Anderson et al., 2010) has caused—among other tools—the expansion of their enzyme inventory for the breakdown of the physical defenses of the plant and subsequent utilization of its nutrients. Plant pathogens are therefore believed to offer a unique source for discovery of novel accessory enzymes (Gibson et al., 2011). Also, many plant pathogens are believed to have actually evolved from saprobes in order to have earlier access to the plant biomass (Table 2.2; Gordon and Leveau, 2010).

Plant pathogenic fungi exhibit wide variation in their adaptation to particular plant hosts and in their modes of pathogenesis: some pathogens invade and colonize all tissues of a host plant while others attack only particular tissues and organs such as leaves, roots, stems, floral structures and seeds, or reproduction organs (Gibson et al., 2011). They can thus be classified as obligate parasites that remain associated with their host plant for their entire life cycles or as facultative parasites that depend on the plant for only a limited period of time. Many of the latter are also saprophytes, which can proliferate both on the living as well as the dead plant.

Fungal pathogens also differ with respect to their host range, but they all release different kinds of enzymes during the infection process that enable a rapid breakdown of the cell wall for its penetration and digestion, in order to guarantee the pathogen's survival and reproduction. Recent papers, reviewed by Gibson et al. (2011), suggest a positive correlation between the presence and expression of certain plant cell wall-degrading enzymes (CWDEs) and virulence. King et al. (2011) reported that the production of CWDEs by plant pathogenic fungi is influenced by the respective plants (e.g., monocots vs. dicots), which is due to the respective differences in cell wall polymer composition. To protect themselves against the fungal attack, plants also produce proteins that inhibit fungal CWDEs (particularly pectinases and xylanases), and it is believed that this interaction has been the force driving evolution to develop unique enzymes by the plant pathogens, thus causing a large amplification of genes for CWDEs in these fungi (Gibson et al., 2011). However, no ligninases are present. Consistent with the above cited gene amplifications, the "secretome" of *F. graminearum* grown on hop cell walls produced 24 different classes of CWDEs and up to eight enzymes per class (Phalip et al., 2005). About 45% of these proteins were cellulases, xylanases, and enzymes for arabinose-, mannose-, galactose-, and fucose-polymer hydrolysis. Consistent with these data, Cuomo et al. (2007) also detected ten cellulose-degrading enzymes and six xylan-degrading enzymes among the proteins secreted during infection of barley by *F. graminearum*, and similar data were also reported for *Mycosphaerella graminicola* and *Botrytis cinerea* (reviewed by Gibson et al., 2011). In this regards, the report by King et al. (2011) that xylanase activity was the greatest contributor to the hydrolytic efficiency of the enzymes from plant pathogenic fungi on plant biomass is intriguing.

On the other hand, attempts to provide direct evidence for the roles of these enzymes in pathogenesis has been difficult: although some studies have been published where individual genes were knocked out, the effects were most often only marginal. This seems to be expectable, however, in view of the large gene families, in which a knock out of one allele is easily compensated by the proteins encoded by the remaining intact genes. As an example, a strain of *Magnaporthe grisea* in which two xylanase genes had been knocked out secreted three other xylanases that were not produced by the parent and thus fully retained its xylanase activity (Wu et al., 2006).

2.4 Anaerobic Fungi

Anaerobic fungi, first described by Orpin (1975), represent a special group of microorganisms almost exclusively inhabiting the gastrointestinal tract of ruminants and most nonruminant herbivores where they are involved in plant cell wall digestion (Bauchop, 1989). Seventeen distinct anaerobic fungi belonging to five different genera have so far been described from at least 50 different herbivorous animals (Ljungdahl, 2008; Griffith et al., 2010; Figure 2.6). Liggenstoffer et al. (2010) performed a metagenomic analysis of the phylogenetic diversity and community structure of members of the gut anaerobic fungi (phylum *Neocallimastigomycota*) in 30 different herbivore species that belong to ten different mammalian and reptilian families. About 36% of them corresponded to species from the genus *Piromyces* and were encountered

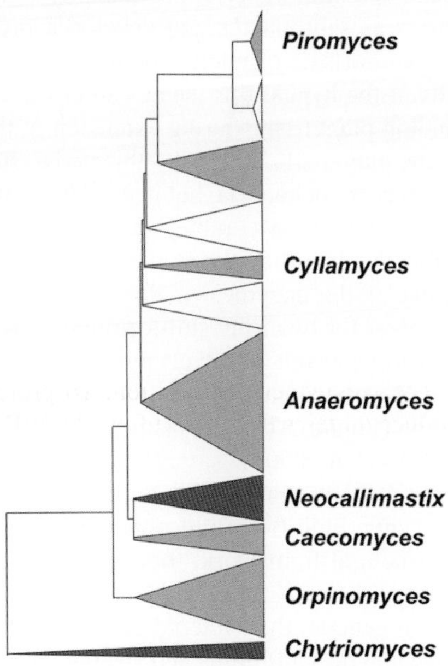

Figure 2.6. Phylogeny of the anaerobic fungi, based on the internal transcribed spacer sequences of the rRNA cluster. (Tree structure adapted from Griffith et al., 2010.)

in 28 of the 30 samples, whereas the genera *Cyllamyces* and *Orpinomyces* were the least abundant representing only 0.7%, and 1.1% of the total sequences obtained, respectively. Interestingly, 38.3% of the sequences formed eight phylogenetically distinct novel anaerobic fungal lineages, of which some were widely distributed, whereas others were animal specific. Thus, many species of anaerobic cellulose-degrading fungi still need to be identified.

Reports about the detection of anaerobic fungi in other habitats are almost nonexisting, but Lockhart et al. (2006), using DNA-based methods for detection of uncultured fungi, demonstrated the presence of anaerobic fungi in landfill sites. Because of the difficulties of separating each microbial group in the rumen, of measuring fungal biomass, and because of the complex nature of the rumen ecosystem, the precise role and overall contribution of these organisms to the degradation and fermentation of plant cell wall material is not understood yet. In spite of delicate interrelationships among the microorganisms in the rumen ecosystem, bacteria are believed to play a major role because they are predominant in number and exhibit the largest metabolic diversity (Russell and Rychlik, 2001). Also protozoa have been shown to digest from 25% to 30% of total fiber. However, Lee et al. (2000) found that fungal activity was responsible for most of the cell wall degradation, whereas cell wall degradation by anaerobic bacterial fraction was significantly less, and the protozoal fraction failed to grow under the conditions used in their study. In general, the protozoa inhibited cellulolysis of cell wall material by both bacteria and fungi, while the bacteria and fungi revealed a synergistic interaction. However, one must bear in mind that the interactions between different groups of microorganisms can range from synergism to antagonism and depend on the species involved and the type of substrate used, and a generalization from only a few species is not justified. Bauchop (1979) reported that fungi play a major role when very poor-quality forages are fed. However, ruminal fungi are inhibited by bacterially produced, bacteriocin-like substances and are typically present when large amounts of cereal grain are fed (Russell et al., 2009).

The higher fraction of plant cell wall material degraded by fungi can be explained by the fact that anaerobic fungi grow as saprophytes and develop a mycelium that penetrates the plant particle (Orpin, 1977), an intrinsic characteristic of fungal growth. Also, fungi release their enzymes predominantly at the hyphal tips (see Chapter 8), thus enabling the mycelium to degrade the local surrounding plant tissue during extension of their hyphae. This results in an increased surface within the plant particle that is subjected to fungal degradation.

Anaerobic fungi do not possess mitochondria, but instead have hydrogenosomes, organelles that have evolved from the mitochondria (Embley et al., 2003). They do not contain cytochromes and perform neither the tricarboxylic acid cycle nor electron-transport-linked oxidative phosphorylation. Most of the enzymes of their relatively simple metabolism have homologs in mitochondria except for two iron–sulfur proteins: pyruvate:ferredoxin oxidoreductase and hydrogenase. Hydrogenosomes convert pyruvate or malate under anoxic conditions to hydrogen, acetate, and carbon dioxide with the concomitant production of ATP by substrate level phosphorylation (Figure 2.7). This last, energy-generating step is catalyzed by succinyl-CoA synthetase (Dacks et al., 2006).

The way by which anaerobic fungi degrade the cellulose is principally the same as that used by soft rot fungi (i.e., using endo-β-1,4-glucanases and cellobiohydrolases), yet they are imbedded in a delicate structural framework: they form a multienzyme complex known as cellulosome, first described in the early 1980s in thermophilic anaerobe, *C. thermocellum* (cited in Bayer et al., 2008). In general, the bacterial cellulosome is composed of two major types of subunits: (i) the noncatalytic scaffoldins and (ii) the catalytically active components (=enzymes). The assembly of the cellulosome is forced by a high-affinity recognition between cohesin modules of the scaffoldin subunit and dockerin modules in the enzyme proteins

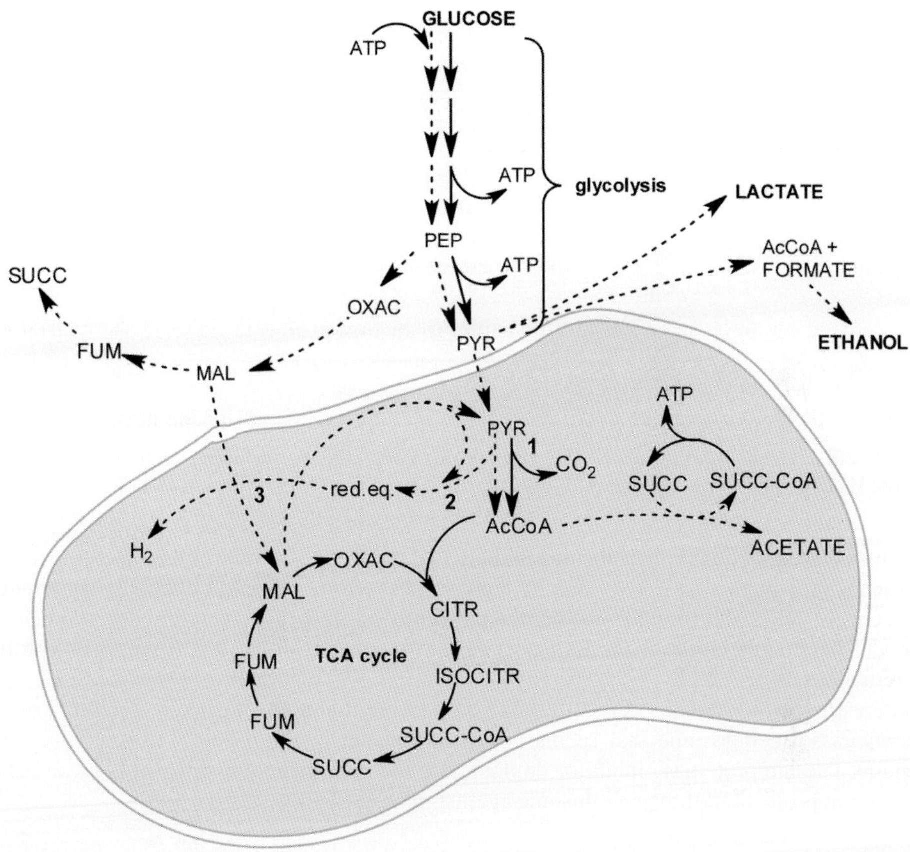

Figure 2.7. Generalized pathway of carbohydrate degradation in anaerobic fungi in comparison with aerobic fungi. Full lines indicate reactions in aerobic fungi, and dotted lines those occurring in anaerobic fungi. AcCoA, acetyl-CoA; CITR, citrate; FUM, fumarate; H2ASE, hydrogenase; ISOCITR, isocitrate; MAL, malate; 2OG, 2-oxoglutarate; OXAC, oxaloacetate; PEP, phosphoenolpyruvate; PFO, pyruvate:ferredoxin oxidoreductase; PDH, pyruvate dehydrogenase; PYR, pyruvate; red. eq., reducing equivalents; SUCC, succinate; SUCC-CoA, succinyl-CoA. The numbers specify selected enzymes involved: 1, pyruvate decarboxylase; 2, pyruvate:ferredoxin oxidoreductase; 3, hydrogenase. (Adapted from Dacks et al., 2006.)

(Figure 2.8). Scaffoldins often contain multiple cohesin modules and thus enable the assembly of several different enzymes into the cellulosome complex (Ding et al., 2003). Cellulosomes also contain carbohydrate-binding modules, either in the cellulosomal scaffoldin or in the enzyme subunits, which function to specifically bind to the cellulose.

The fungal cellulosome is much less well characterized. However, a number of notable differences are already known: first, its major product of cellulose digestion is glucose and not cellobiose as in bacteria (Dijkerman et al., 1997). Second, there seem to be significant differences in the cellulosomal architecture: the cellulases and xylanases of *Neocallimastix patriciarum* contain, besides the catalytic domains, a duplicated C-terminal domain that does not bind to cellulose, but to polypeptides of 97 and 116 kDa that are components of a multienzyme cellulose-binding complex (Gilbert et al., 1992). Nagy et al. (2007) supposed

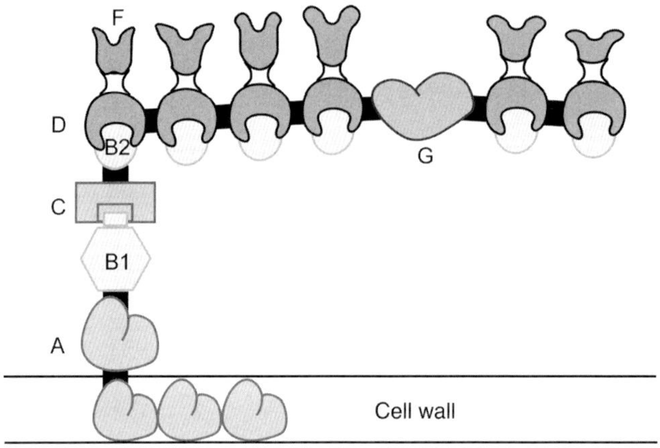

Figure 2.8. Generalized structure of a bacterial cellulosome: A, anchoring protein; B, cohesion; C, type II dockerin; D, type I dockerin; E, various CAZymes; F, cellulose-binding module.

that this C-terminal domain is a docking domain that mediates assembly of the anaerobic fungal multienzyme complex. However, despite considerable effort, the respective fungal scaffoldin and cohesin genes and proteins have not been identified, thus preventing the development of a model for the operation of the fungal system in the same way as has been done with the bacterial system.

Interestingly, Steenbakkers et al. (2008) have recently identified a protein ("celpin") in the *Piromyces* sp. cellulosome that seems not to be present in the anaerobic bacteria. Celpin contains a serine proteinase inhibitor domain, and the authors assumed that it is presumably involved in protection of the cellulosome against plant proteinases.

Chapter 3

The Tools—Part 1: Enzymology of Cellulose Degradation

3.1 General Properties and Classification of Enzymes That Hydrolyze Polysaccharides

Enzymes that are capable of cleaving the glycosidic bonds in oligo- or polysaccharides (including cellulose and hemicelluloses) are generally summarized under the term "glycoside hydrolases (GHs)." The hydrolysis of the glycosidic linkage leads to the formation of a sugar hemiacetal or hemiketal and the corresponding free aglycone. There are several means for categorization of these enzymes: the IUB (International Union of Biochemistry) enzyme nomenclature classifies them—as all enzymes in general—according to their EC (Enzyme Commission) number. This is a numerical classification scheme for enzymes based on the reactions they catalyze. Unfortunately, the EC classification does not distinguish between genetically and structurally different enzymes as long as they catalyze the same reaction, which is particularly unsatisfying in the case of GHs as I will explain in more detail later.

In the case of GHs, there are several additional differences between enzymes belonging to the same EC group, which are therefore traditionally used for further classification. One such difference is based on the topology of the action on the macromolecular substrate: exo-GHs cleave their substrate on one of the end of the polymer (most frequently, but not always, at the nonreducing end), whereas endo-GHs cleave within a chain (Figure 3.1a). In terms of the biological benefit for this, this principle aids to an efficient attack on any macromolecule because the endo-acting enzyme provides an increasing number of oligomers that can be attacked by the exo-enzyme. It is therefore not surprising that endo- and exo-GHs have been shown to act in synergy (called "exo-endo-synergism") and their simultaneous presence thus increases the rate of hydrolysis (Wood and McCrae, 1972).

Another related principle is the grouping of GHs into glycanases and glycosidases. The term "glycanases" is used for enzymes that preferentially act on polymeric or high-molecular weight substrates (which is then replaced by specifically naming the polymer and the linkage specificity; e.g., β-1,4-mannanase), whereas the latter term describes enzymes that are active

Fungi and Lignocellulosic Biomass, First Edition. Christian P. Kubicek.

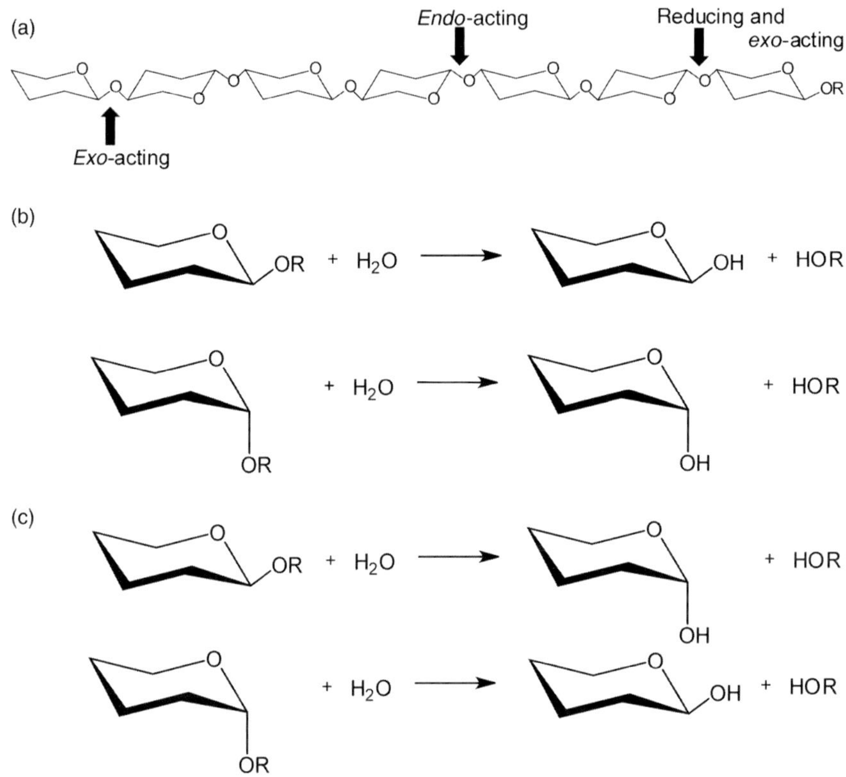

Figure 3.1. Basic mechanisms of oligo- and polysaccharide cleavage by glycosyl hydrolases: (a) endo- and exo-type hydrolysis, (b) retaining mechanism, and (c) inverting mechanism.

on oligomers. While basically simple, this distinction becomes difficult when distinguishing exoglycanases from glycosidases, however, because exoglycanases will act on the same soluble oligomers and from the same end as the glycosidases. Reese et al. (1968) has therefore proposed a method to distinguish exoglycanases from glycosidases: when acting on the same substrate (e.g., a cellooligodextrin), the affinity of an exoglycanase will increase with the polymerization length (i.e., the K_m will decrease), whereas the opposite will be observed with β-glycosidases. Unfortunately, this principle has been ignored in many cases where GHs were described, which led to some confusion about their identity in the literature.

Another important way to distinguish GHs is the reaction mechanism, that is, whether the position of the hydroxylic group at the C-1 atom of the sugar moiety in the glycosidic bond is inverted and retained (Figures 3.1b and c). The case of an inverting hydrolysis generally occurs via only one step, that is, as single-displacement mechanism, which involves an oxocarbenium ion-like transition states (Figure 3.2a), and catalysis proceeds by the canonical acid/base mechanism using (mostly) an E or D, which are typically located 6–11 Å apart (McCarter and Withers, 1994), as amino acid participants.

A hydrolysis in which retention of the configuration is maintained (also known as "Koshland retaining mechanism"; Koshland, 1953) mostly involves a two-step, double-displacement reaction that proceeds via a covalent glycosyl-enzyme intermediate (Figure 3.2b). Also this

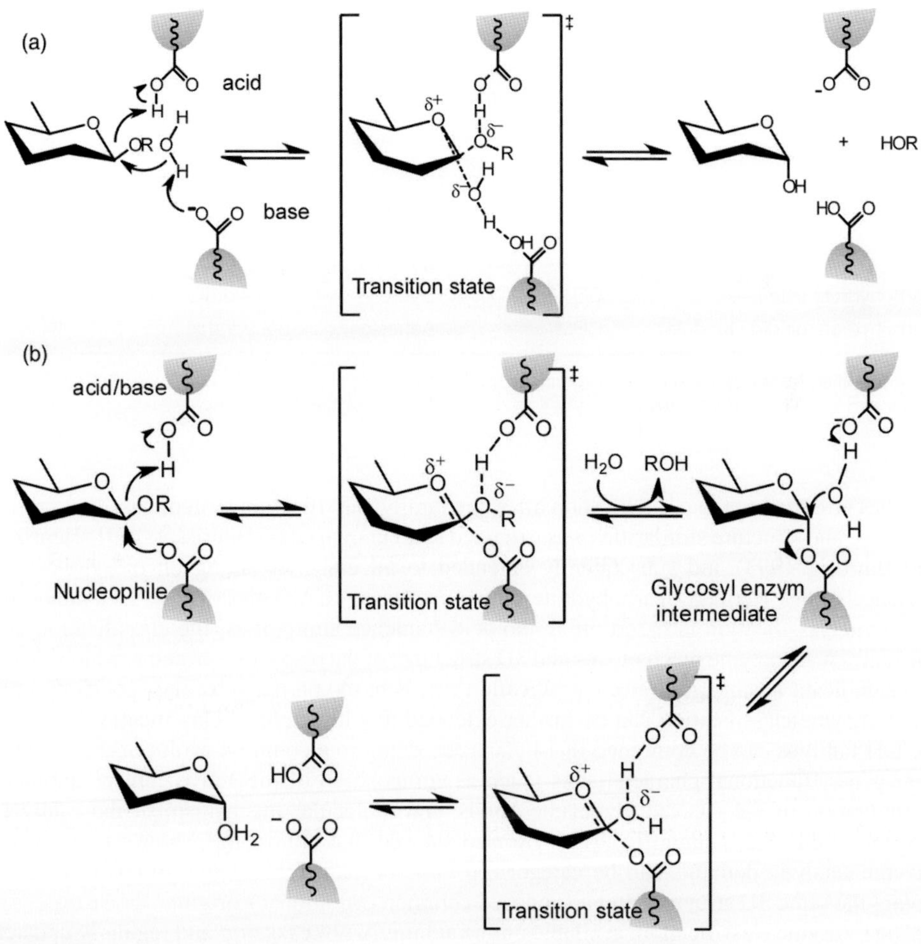

Figure 3.2. Reaction mechanisms of glycoside hydrolases. (a) Hydrolysis of a glycoside by a single-displacement mechanism; (b) hydrolysis by the Koshland retaining mechanism.

mechanism uses an oxocarbenium ion-like transition state and occurs with acid/base catalysis provided by an E or D residue. However, here these two amino acids are located much closer (5.5 Å) than in the inverting mechanism. The two steps occur as follows: in the first, one of the two amino acid residues performs a nucleophilic attack and displaces the aglycone to form a glycosyl enzyme intermediate. In the second step, the covalent bond between the enzyme and the glycosyl chain is hydrolyzed, which is aided by the residue that acts as the Lewis base and which deprotonates the water molecule (Williams, 2011).

Almost all of the GHs and particularly the cellulases and hemicellulases are distributed among various gene families that are genetically and structurally strongly different. In order to pay attention to this fact, Bernard Henrissat has nearly 20 years ago started to establish a classification concept that is based on amino acid similarity (and thus, although not initially pursued by him, phylogenetic relationship) in the respective enzymes. The first proof-of-principle was obtained by demonstrating the classification of cellulases into several distinct

Table 3.1. Enzymes and associated modules currently covered by CAZy.

CAZy	Abbreviation	Function	Current Number[a]
Glycoside hydrolases	GH	Hydrolysis and/or rearrangement of glycosidic bonds	125
Glycosyl transferases	GT	Formation of glycosidic bonds	94
Polysaccharide lyases	PL	Nonhydrolytic cleavage of glycosidic bonds	22
Carbohydrate esterases	CD	Hydrolysis of carbohydrate esters	16
Carbohydrate binding modules	CBM	Adhesion to carbohydrates	64

Data modified from http://www.cazy.org/Home.html (Cantarel et al., 2009).
[a]Checked last November 3, 2011.

families (Henrissat et al., 1989). Soon after, the family classification system based on protein sequence and structure similarities was extended to all known GHs (Henrissat, 1991; Henrissat and Bairoch, 1993) and subsequently extended to all enzymes acting on or synthesizing polysaccharides (termed "carbohydrate active enzymes," CAZymes; Table 3.1). Since the CAZyme classification is based on amino acid sequence similarities, the classification also correlates with enzyme mechanisms and 3D structures of the respective proteins, which clearly is a significant advantage over a classification that is based on the substrate specificity only. The CAZyme classification can be further extended to a hierarchical classification by which the GH families can be combined in 14 clans according to a common evolutionary origin of their genes, functional characteristics (such as composition of the active center), anomeric configuration of the cleaved glycosidic bonds, and molecular mechanism of the catalyzed reaction (Table 3.2; Naumoff, 2011). He also showed that almost the whole variety of the enzyme catalytic domains can be categorized into six main folds, large groups of proteins having the same 3D structure and a supposed common evolutionary origin.

The CAZyme classification, available online at http://www.cazy.org/ and regularly updated, is an indispensable tool for researchers in this field. A curatorium chaired by Harry Brumer from the Royal Institute of Technology at Stockholm, Sweden, also maintains a website describing the properties of the various carbohydrate active enzymes in detail (CAZypedia; http://www.cazypedia.org/index.php). To facilitate research and development of enzymes for the conversion of cell wall polysaccharides into fermentable sugars, Murphy et al. (2011a) have manually curated all the GH families that can be found in fungi. A total of 453 characterized GHs from 131 different fungal (mostly ascomycete) species were retrieved and shown to comprise 44 of the 115 CAZy GH families. The annotated genes and proteins were compiled in a searchable, online database (mycoCLAP; characterized lignocellulose-active proteins of fungal origin; http://mycoCLAP.fungalgenomics.ca/), which also includes information about available biochemical properties (temperature and pH optima, specific activity, kinetic parameters, and substrate specificities). A summary of GH families present in fungi, as far as they are related to lignocelluloses degradation, is given in Table 3.3.

Despite the invaluable information stored in the CAZy database, publicly available software tools utilizing this information for annotation of newly sequenced genomes by CAZy families are so far limited. A valuable attempt to alleviate this situation has been presented by Park et al. (2010), who elaborated two annotation approaches: (i) a similarity search against the

Table 3.2. Glycosyl hydrolase clans.

Clan	GH Families	Structure	Main Enzyme Activities
GH-A	1, 2, 5, 10, 17, 26, 30, 35, 39, 42, 50, 51, 53, 59, 72, 79, 86, 113	$(\beta/\alpha)_8$	Various
GH-B	7, 16	β-jelly roll	Cellulases, endo-β-1,3/1,4-glucanases
GH-C	11, 12	β-jelly roll	Xylanases, endo-β-1,3-glucanases
GH-D	27, 31, 36	$(\beta/\alpha)_8$	α-Galactosidases, α-xylosidases
GH-E	33, 34, 83, 93	Sixfold β-propeller	Sialidase, neuraminidase, α-1,5-endoarabinase
GH-F	43, 62	Fivefold β-propeller	α-L-arabinofuranosidases
GH-G	37, 63	$(\alpha/\alpha)_6$	α-Glucosidase, trehalase
GH-H	13, 70, 77	$(\beta/\alpha)_8$	Amylases and related α-glucosidases
GH-I	24, 46, 80	α+β	Lysozyme, chitosanase
GH-J	32, 68	Fivefold β-propeller	Invertases, inulinases
GH-K	18, 20, 85	$(\beta/\alpha)_8$	Chitinases, *N*-acetyl-β-glucosaminidases
GH-L	15, 65, 125	$(\alpha/\alpha)_6$	Glucoamylase, trehalose phosphorylase, α-1,6-mannosidase
GH-M	8, 48	$(\alpha/\alpha)_6$	Bacterial endo-processive cellulases and chitinases
GH-N	28, 49	β-helix	Pectinase, α-1,6-glucanase

entire nonredundant sequences of the CAZy database and (ii) an automatic annotation using links or correspondences between the CAZy families and protein family domains (CAZymes Analysis Toolkit; it is available at http://cricket.ornl.gov/cgi-bin/cat.cgi).

In the further course of this and the following chapters dealing with the lignocelluloses GHs, I will use a combined approach starting with a grouping according to the substrate that is hydrolyzed (cf. Table 3.3) and using the CAZyme concept as a subsequent classification criterion for each of these groups. As far as it is known and relevant to the understanding, I will also provide information on the protein structure, substrate specificity, and reaction mechanism for these protein families.

3.2 Fungal Cellulolytic Enzymes

As explained in Chapter 2, the canonical view of hydrolysis of cellulose involves the action of two types of cellulases in an exo-/endo-synergy, followed by a β-glucosidase that hydrolyzes the soluble cellodextrin oligomers to glucose. However, it has been recognized for some time now that the differentiation of cellulases into endoglucanases and cellobiohydrolases is an oversimplification: cellulases have evolved to a continuum of overlapping modes of actions ranging from totally random endoglucanases through processive endoglucanases to strictly exo-acting highly processive cellobiohydrolases (Teeri, 1997; Kurasin and Väljamäe, 2011). The exact roles of individual enzymes with different degrees of processivity and endo-activity in cellulose degradation are not known. The genome sequences of more than

Table 3.3. GH families related to lignocellulose degradation that have been characterized from fungi.

GH Family	No.[a]	Enzymatic Activities
GH1	7	β-glucosidase (7)
GH2	5	β-mannosidase (2), chitosanase (1), exo-glucosaminidase (1), β-galactosidase (1)
GH3	30	β-glucosidase (22), β-xylosidase (8)
GH5	45	Endoglucanase (22), exo-1,3-β-glucanase (12), β-mannanase (8), galactanase (2), endo-1,6-β-glucanase (1)
GH6	12	Cellobiohydrolase (11), endoglucanase (1)
GH7	29	Cellobiohydrolase (18), endoglucanase (10), xylanase (1)
GH10	19	Xylanase (19)
GH11	44	Xylanase (44)
GH12	24	Endoglucanase (20), xyloglucanase (3), licheninase (1)
GH26	3	β-mannanase (3)
GH27	6	α-galactosidase (6)
GH28	54	Endo-polygalacturonase (40), exo-polygalacturonase (9), endo-rhamnogalacturonase (3), exo-rhamnogalacturonase (1), xylogalacturonase (1)
GH31	10	α-glucosidase (8), α-xylosidase (1), invertase (1)
GH35	1	β-galactosidase (1)
GH36	7	α-galactosidase (7)
GH43	6	Endo-1,5-α-arabinanase (3), α-l-arabinofuranosidase (2), β-xylosidase (1)
GH45	8	Endoglucanase (8)
GH47	5	α-1,2-mannosidase (5)
GH51	5	α-l-arabinofuranosidase (5)
GH53	6	Arabinogalactanase (6)
GH54	9	α-l-arabinofuranosidase (9)
GH61	3	Cellulase-enhancing protein (3)
GH62	2	Arabinoxylan arabinofuranosidase (2)
GH67	4	α-glucuronidase (4)
GH74	6	Xyloglucanase (3), oligoxyloglucan cellobiohydrolase (2), endoglucanase (1)
GH78	3	α-rhamnosidase (3)
GH93	2	Exo-arabinanase (2)

Data modified from Murphy et al. (2011a).
[a]Total number of respective enzymes characterized from a fungal source; the numbers in the third column specify the numbers per enzyme activity, if more than one.

40 asco- and basidiomycetes that were available when this review has been written shows that these enzymes are confined to a relatively low number of GH families (Table 3.4): strictly processive "exocellulases" (=cellobiohydrolases) are found in GH families 6 and 7, and are usually present in the form of only 1–2 isoenzymes, whereas "endocellulases" (=endo-β-1,4-glucanases) are distributed throughout a larger number of GH families (GH families 5, 7, 12, and 45). However, when looking at these numbers, one must bear in mind that some

Table 3.4. Distribution of cellulase GHs in fungi.

		GH Family	Cellulases					β-glucosidases		PMO
Phyllum	Class	Species	5	6	7	12	45	1	3	61
Pezizomycota	Eurotiomycetes	*A. nidulans*	15	2	3	1	1	3	20	9
Pezizomycota	Eurotiomycetes	*A. niger*	10	2	2	4	0	3	17	7
Pezizomycota	Eurotiomycetes	*Penicillium chrysogenum*	13	1	2	3	0	3	17	4
Pezizomycota	Leotiomycetes	*Sclerotinia sclerotiorum*	5	1	3	2	2	1	12	9
Pezizomycota	Pezizomycetes	*Tuber melanosporum*	6	0	0	1	1	2	6	4
Pezizomycota	Leotiomycetes	*Blumeria graminis var. hordei*	0	0	0	0	0	0	0	2
Pezizomycota	Leotiomycetes	*Botrytis cinerea*	5	1	3	1	2	5	14	10
Pezizomycota	Sordariomycetes	*Fusarium graminearum*	15	1	2	4	1	3	22	15
Pezizomycota	Sordariomycetes	*Fusarium graminearum*	3	1	2	2	1	3	17	13
Pezizomycota	Sordariomycetes	*Nectria haematococca*	18	1	3	6	1	5	38	12
Pezizomycota	Sordariomycetes	*Neurospora crassa*	7	3	5	1	1	1	9	14
Pezizomycota	Sordariomycetes	*Magnaporthe grisea*	13	3	6	3	1	2	19	17
Pezizomycota	Sordariomycetes	*Podospora anserina*	15	4	6	2	2	1	11	33
Pezizomycota	Sordariomycetes	*T. atroviride*	14	1	2	3	1	4	14	3
Pezizomycota	Sordariomycetes	*T. reesei*	11	1	2	2	1	2	13	3
Pezizomycota	Sordariomycetes	*T. virens*	16	1	2	4	2	2	17	3
Pezizomycota	Dothideomycetae	*Mycospherella graminearum*	0	0	1	1	1	NI	NI	2
Pezizomycota	Dothideomycetae	*Stagonospora nodorum*	3	4	5	4	3	NI	16	30
Basidiomycota	Ustilaginomycetes	*Ustilago maydis*	0	0	0	0	3	0	3	0
Basidiomycota	Pucciniomycetes	*Puccinia graminis var. tritici*	9	0	7	2	0	0	2	3
Basidiomycota	Agaricomycetes	*Schizophyllum commune*	18	1	2	1	1	3	12	22
Basidiomycota	Agaricomycetes	*Laccaria bicolor*	22	0	0	3	0	0	2	8
Basidiomycota	Agaricomycetes	*Postia placenta*	36	0	0	4	0	4	9	4
Basidiomycota	Agaricomycetes	*P. chrysosporium*	20	1	9	2	0	2	11	15
Basidiomycota	Agaricomycetes	*Serpula lacrymans*	20	1	0	1	NI	NI	10	5
Basidiomycota	Agaricomycetes	*C. cinerea*	26	5	7	1	NI	NI	7	33

Data taken from Kubicek et al. (2011), Eastwood et al. (2011), and Goodwin et al. (2011).
NI means "no information" was available; PMO, polysaccharide monooxygenases, previously believed to be class GH61 cellulase-enhancing proteins.

of these GH families (particularly GH5) also contain a number of enzymes with other substrate specificities (such as endomannanases, α-1,6-galactanase, β-1,3-mannanase, xyloglucanases), whereas other families comprise cellulases only. β-Glucosidases are predominantly found in the GH1 and GH3 families, which however also contain other glycosidases such as β-galactosidase, β-mannosidases, and others.

In addition, it became recently very clear that there are further proteins (called "cellulase enhancing proteins") that are now known to strongly and synergistically raise the activity of the cellulases but do not exhibit any enzymatic activity on cellulose themselves (i.e., the nonhydrolytic "endoglucanase" CEL61, and the expansin-like protein swollenin; Saloheimo et al., 2002; Harris et al., 2010). They may thus fulfill the role of the missing link in the early theory of cellulase action, the Cx-C1 model (see Chapter 2), and be the "swelling factor," C1, a nonhydrolytic component that functions to make the substrate more accessible to Cx. Therefore, these proteins will also be described in this chapter.

Further, cellulases and related GHs very often display a modular structure that consists of a catalytic domain and a polysaccharide-binding domain connected by a loop ("hinge") region, which has been first discovered during the investigation of cellobiohydrolases I and II of *Trichoderma reesei* (Teeri et al., 1987). Because these polysaccharide-binding domains occur in most of the fungal cellulases, I decided to describe them first.

3.2.1 Cellulose-Binding Domains

Noncatalytic polysaccharide-binding modules of GHs were originally defined as cellulose-binding domains (CBDs) because the first examples investigated bound tightly to crystalline cellulose (reviewed by Boraston et al., 2004). Since thereafter, several polysaccharide-binding modules were found, which bind to polysaccharides other than cellulose; however, the more general term carbohydrate-binding module is used to reflect this diversity in specificity (Boraston et al., 1999). In analogy to the GH classification, CBMs are also divided into families based on amino acid sequence similarity, of which today 63 families have been defined and included in the CAZy database (http://www.cazy.org/Carbohydrate-Binding-Modules.html). They display a broad spectrum of ligand specificity, including very specific CBM-ligand interactions but some of them also bind to a broader range of different carbohydrates. Most of them are found in GHs that act on insoluble polysaccharides, indicating the necessity of locating the respective enzymes to such insoluble substrates. Interestingly, the largest number of CBM families is found only in bacterial enzymes, whereas only about 25% of the currently known CBMs are present in fungal enzymes. Table 3.5 lists those CBM families and the carbohydrates they bind to, which have been found in fungi and that act on polysaccharides found in lignocellulose biomass.

Regarding the structure, the most important conformational element of most CBMs is the β-sheet. The folds and architecture displayed by these β-sheets have led to a classification into seven families, of which the β-sandwich is the most recurrent fold (Boraston et al., 2004; Hashimoto, 2006).

Unfortunately, the grouping of CBM families based on the conservation of the protein fold is not predictive of function, as specific amino acids or binding-site topographies are not conserved. Consequently, another classification of CBMs based on structural and functional similarities has been proposed: "surface-binding" CBMs (type A), "glycan-chain-binding" CBMs (type B), and "small-sugar-binding" CBMs (type C) (Boraston et al., 2004; see also Table 3.5). Type A CBMs consist of the canonical β-sandwich structure, which is composed

Table 3.5. Fungal carbohydrate binding modules.

CBM Family	Structure	Length [aa]	Binding to	Found in
1	Antiparallel β-sheet	40	Cellulose, chitin	Cellulases, β-mannanases, α-arabinofuranosidases, acetyl xylan esterases
6	β-sandwich	120	Amorphous cellulose, β-xylan	Xylanases, α-arabinofuranosidases
13	β-trefoil	150	Plant-lectin-like galactose binding	α-galactosidases and α-L-arabinofuranosidases
29	β-sandwich	NI	β-mannan, glucomannan	Only found in enzymes from *Piromyces* spp.
32	β-sandwich	120	Galactose, lactose	Galactose oxidase (*Fusarium* sp.), similarity to CBM6
35	β-sandwich	130	β-galactan (mannan? arabinan?)	Galactan-1,3-β-galactosidase (*P. chrysosporium*)
42	β-trefoil	160	Arabinose	α-arabinofuranosidases
63	Double-psi-β-barrel	NI	Cellulose	Endo-β-1,4-glucanase (EglD) from *A. nidulans*

Data extracted from the CAZy database (http://www.cazy.org); CBMs classification based on fold is based on the reviews by Boraston et al. (2004) and Hashimoto (2006).

of two β-sheets, each consisting of three to six antiparallel β-strands, and (mostly) carries at least one structural metal ion. In fungi, most A types are represented by CBM1 domains (Table 3.5). The CBM of *T. reesei* CEL7A (see later) has a pronounced shape resembling a wedge, with a flat, hydrophobic binding surface characterized by three conserved Y residues. They are spaced at regular intervals in a line along this surface, nearly matching the length and spacing of a cellobiose molecule (Linder et al., 1995). In CBM1 domains of other enzymes, Y is sometimes replaced by W, and rarely by F. Two internal disulfide bonds stabilize the secondary structures of the overall fold. The upper surface of this wedge contains a groove running from the leading edge to the upper surface or "top" side of the protein, which is lined with several functional groups capable of hydrogen bonding, as well as significant hydrophobic patches (Mulakala and Reilly, 2005; Figure 3.3a). The specificity for binding is achieved by the location of aromatic amino acid side chains and the loop structures that shape the binding sites to mirror the conformation of the ligand (Boraston et al., 2004). Lehtio et al. (2003) showed that *T. reesei* CBM1 binds to the hydrophobic 110 face of *Vallonia* cellulose.

In contrast, the binding site architecture of type B CBMs, which bind amorphous cellulose or xylan, is arranged as a cleft in which aromatic residues interact with free single polysaccharide chains. Type B CBMs also recognize noncellulosic substrates like β-1,3-glucans, mixed β-(1,3)(1,4)-glucans, β-1,4-mannan, glucomannan, and galactomannan. The aromatic amino acid side chains in the ligand-binding sites of type B CBMs often sandwich with the sugar unit in the polysaccharide by stacking against the "b" and "a" face of the pyranose rings (Boraston et al., 2004).

Type C CBMs bind mono-, di-, or trisaccharides and are also termed "lectin-like" CBMs (Guillén et al., 2010).

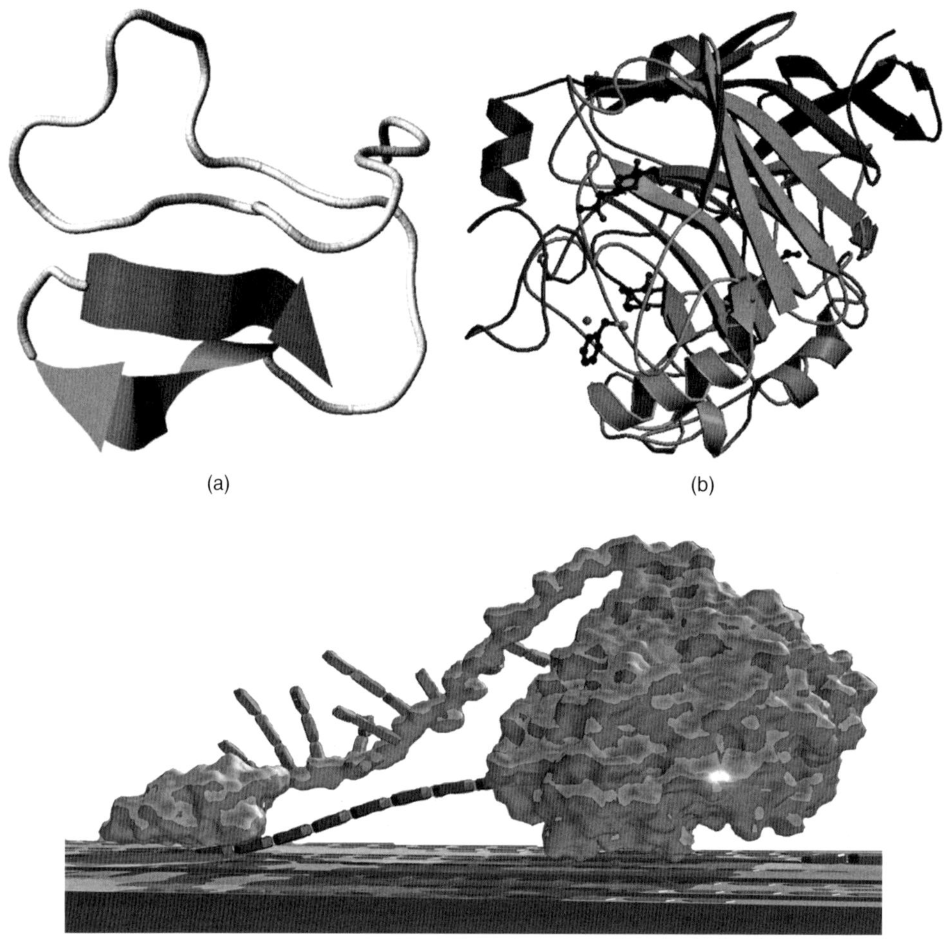

Figure 3.3. Structure of cellobiohydrolase CEL7A from *T. reesei*: (a) the CBM1 cellulose-binding domain (accession number 1CBH), (b) structure of the catalytic domain of CEL7A (accession number 1CEL), and (c) side view of the complex of CBH I docked onto the surface of the model cellulose microfibril. (I acknowledge the US Department of Energy Genomic Science program and the website http://genomicscience.energy.gov for the kind gift of this figure.)

Calcium is known to play a significant role in the interaction of lectins with their target ligands, either by maintaining the binding site in the correct conformation or via direct coordination with the carbohydrate itself. Indeed, xylan recognition by a bacterial CBM35 (from *Cellvibrio* spp., in Abf62A) is Ca^{2+}-dependent, but it has not been investigated as yet whether this also accounts for the CBM35 modules in ascomycetous β-mannanases, α-galactosidases, and α-arabinofuranosidases.

CBMs are generally believed to have three roles, which aid to the function of their cognate catalytic modules: (i) providing a proximity effect, (ii) a targeting function (i.e., bringing the enzyme and particularly the catalytic module to the substrate), and (iii) a disruptive function. With regards to the last point, binding of CBM1 domains from a *T. pseudokoningii* and *Penicillium janthinellum* cellobiohydrolase 1 to cellulose was shown to cause structural changes and release of short fibers, and the authors concluded that the CBM1 protein acts by disrupting hydrogen bonds between cellulose chains (Gao et al., 2001; Wang et al., 2008). Mulakala and Reilly (2005) suggested that the CBM1 of *T. reesei* CEL7A domain wedges itself under a free reducing chain end on the crystalline cellulose surface and feeds it to the active site tunnel of the catalytic domain. However, Igarashi et al. (2009), using high-speed atomic force microscopy to study real-time sliding of CEL7A molecules on crystalline cellulose, showed that the catalytic domain without the CBD moved with a velocity similar to that of the intact Cel7A enzyme with a speed of 3.5 nm/s, and this sliding was absent in proteins with loss-of-catalysis mutations in the catalytic domain. Consequently, the CBM1 module is dispensable for lifting and feeding of the cellulose chain. They therefore concluded that the role of the cellulose-binding CBM1 is merely to increase the enzyme concentration on the crystalline substrate.

For a deeper and more comprehensive review about all (not only fungal) CBMs, I refer the reader to the articles by Boraston et al. (2004), Shoseyov et al. (2006), Hashimoto (2006) and Guillén et al. (2010).

3.2.2 Cellobiohydrolases (EC 3.2.1.91)

GH7 Cellobiohydrolase Cel7A/CBH1

As mentioned earlier, fungal cellobiohydrolases are found in the glycosyl hydrolase families GH6 and GH7. Thereby, cellobiohydrolase I (CBH1, which according to its categorization into GH7 is now called CEL7A) is the archetypus of fungal cellulases, with which most of the mechanistic and structural studies have been performed, and its characterization from different fungi outnumbers that of all other cellulases (Table 3.6). CEL7A from *T. reesei* was the first cellulase protein that was purified and characterized, whose gene was cloned, and whose 3D

Table 3.6. Characterized cellobiohydrolase I proteins from fungi.

Species	Protein ID (Genbank)	Reference
Pezizomycota		
A. aculeatus	BAA25183	Takada et al., 1998
A. niger	AAF04491	Gielkens et al., 1999
A. nidulans	AAM54069	Lockington et al., 2002
Penicillium chrysogenum	AAV65115, AAX84833	Hou et al., 2007
Penicillium janthinellum	CAA41780	Koch et al., 1993
Humicola grisea var. thermoidea	BAA74517	Takashima et al., 1998
Thermoascus aurantiacus	AAL16941, AAL83303	Hong et al., 2003
T. koningii	CAA49596	Wey et al., 1994
T. reesei	CAH10320	Fagerstam and Pettersson, 1980
Melanocarpus albomyces	CAD56667	Haakana et al., 2004
Cochliobolus carbonum	AAC49089	Sposato et al., 1995
Cryphonectria parasitica	AAB00479	Wang and Nuss, 1995
Basidiomycota		
Irpex lacteus	BAA76363	Hamada et al., 1999
P. chrysosporium	AAB46373, CAA80253	Vanden Wymelenberg et al., 1993

Data retrived by searching MycoClap (Murphy et al., 2011) for "cellobiohydrolase 1."

structure had been elucidated. This all is due to the fact that this enzyme makes up more than 60% of the total protein secreted by *T. reesei* during growth under cellulase-inducing conditions (see Chapter 7), and it is also the major cellulase protein secreted by other fungi (with the exception of brown rot fungi, see Chapter 2). Orthologs of CEL7A have been found in all ascomycete and white-rot basidiomycete genomes, and a phylogenetic analysis reflects the species phylogeny, thus implying that CEL7A is an essential component of fungi.

The CEL7A protein consists of an N-terminal catalytic domain, a C-terminal CBM1 carbohydrate-binding domain, and an unstructured "hinge" domain that links these both (Figures 3.3b and c). The latter linker is rich in S and T residues, which are also highly *O*-glycosylated. Beckham et al. (2010) performed simulations that suggested that the linker is an intrinsically disordered protein of 16 Å extension and thus considerably expands the operating range of Cel7A. The 3D structures of the catalytic domain of CEL7A from *T. reesei* (Divne et al., 1994), *Phanerochaete chrysosporium* (Muñoz et al., 2001), from the thermophilic fungus *Talaromyces emersonii* (Grassick et al., 2004), and from *Melanocarpus albomyces* (Parkkinen et al., 2008) have been solved. They are built up around a β-jelly roll folded framework, in which two large antiparallel β-sheets pack face-to-face to form a highly curved β-sandwich. This β-sandwich is further extended along both edges by several of the loops that connect the β-strands, resulting in a long (about 50 Å) substrate-binding surface that runs perpendicular to the β-strands of the inner, concave β-sheet. A few further short α-helical segments occur in some of the loops at the periphery of the structure (Figure 3.3b; Stahlberg, 2011).

The catalytic domain of CEL7A contains this approximately 50 Å-long active-site tunnel that is decorated with amino acid side chains that contains ten subsites (−7 to +3) for the glycosyl units of a cellulose chain (Divne et al., 1994). The cellulose chain enters from the −7 subsite and is thread through the tunnel in a manner that twists the cellulose chain almost upside down. Once in the tunnel, the cellulose chain is then processively cleaved, two glucosyl residues at a time. The tunnel contains a total of seven glucosyl-binding sites, four of which are formed by W residues, which are major determinants for the formation of the −7, −4, −2, and +1 glucosyl-binding sites. The glucosyl rings seem to slide over the indole surface, and thus likely interact only rather weakly. This could be important for hydrolysis because strong interactions throughout the tunnel would hamper the advancement of the chain in the direction of the active site (Divne et al., 1994). All glucosyl units in the eleven different binding sites are in phase and aligned such that the glucosyl units at the active site, located at −1, are oriented for catalysis. The proposed nucleophile and acid/base catalysts, E212 and E217 (Divne et al., 1994; Ståhlberg et al., 1996), are positioned roughly on opposite sides of the glycosidic linkage that is to be cleaved, with their carboxylate groups about 6 Å apart. The sites for product binding are located at +1/+2, indicating that hydrolysis of the glycosyl-enzyme intermediate may proceed without prior release of the cellobiose product and suggests a product ejection mechanism during processive hydrolysis of cellulose (Ubhayasekera et al., 2005).

As also indicated by the spacing of E212 and E217 (see earlier), CEL7A is a retaining glycosidase that acts by a double-displacement mechanism (Figure 3.2b). These two E-residues perfectly mirror the typical -$\underline{E}_1$-X-D-X-X-$\underline{E}_2$- consensus motif, in which E1 acts as the catalytic nucleophile and E2 as general acid/base. Interestingly, the CEL7A enzymes act from the reducing ends of the cellulose chains, which is in conflict with the general IUBMB definition of cellobiohydrolases (EC 3.2.1.91), which implies that they would act from the nonreducing ends of cellulose.

The progress of action of CEL7A has been summarized by Levine et al. (2010): the enzyme first adsorbs to the crystalline regions of the cellulose surface via the CBM domain, then diffuses over the surface. Eventually, it binds a nonreducing end of a cellulose chain, catalyzes the hydrolysis event, generating cellobiose, and finally desorbs from the cellulose surface or starts to bind to the next nonreducing end. Fox et al. (2011) recently reported that the rate of CEL7A-catalyzed hydrolysis of crystalline cellulose is limited by the rate of binding of the enzyme to the cellulose chains, which in turn was equivalent to the rate of initial production of hydrolysis products and dependent on the micromorphology of the cellulose surface. This rate was enhanced in the presence of endoglucanases (see later).

At the time of this writing, the interplay between the catalytic and the CBM1 domain is still a matter of intense modeling and debate. Zhao et al. (2008) considered that the CBM1 module would prevent the catalytic domain from diffusing away from the surface of the substrate, and thus maintain it there in a high local concentration. Bu et al. (2009) showed that CBM1 alone exhibits regions of stability on the hydrophobic face of cellulose at every 5 and 10 Å, which would correspond to a glucose unit and a cellobiose unit, respectively. They observed that in the presence of hydrolyzed cellulose chain ends, CBM1 exerted a thermodynamic driving force to translate away from the free cellulose chain ends and concluded that it thus may be a driving force on the enzyme during processive hydrolysis of cellulose. Zhao et al. (2008) suggested that the linker segment might store energy, in the manner of a compressed spring, perhaps forcing the chain further into the active site after each bond scission and product escape or driving the CBM1 to advance along the substrate chain, pulling the catalytic domain along after it (Zhao et al., 2008). Indeed, the water bound by the heavy glycosylation of the linker chain could create a gel-like zone between the catalytic and the cellulose-binding modules, which would inhibit their relative motions and push the smaller CBM1 forward along the chain, away from the heavier catalytic domain, thus promoting processivity (Zhong et al., 2009).

GH6 Cellobiohydrolase CEL6A/CBHII

The second cellobiohydrolase that can be found in all (with the exception of brown rot) fungi and bacteria belongs to glycosyl hydrolase family GH6. This enzyme has previously been published as CBH II, and its enzymatic properties have been characterized in some detail from *T. reesei*, *H. insolens,* and *Coprinopsis cinerea* (see Koivula et al., 2002; Varrot et al., 2003; Liu et al., 2010, and references therein). CEL6A acts by removing cellobiose from the nonreducing end and inverting the anomeric stereochemistry. Just like CEL7A, CEL6A consist of a catalytic core protein, which is linked to a class I CBM via a hinge domain. However, in contrast to CEL7A, the CBM of CEL6A occurs at the N-terminus of the protein and has been duplicated (Teeri et al., 1987). The crystal structures of CEL6A have been obtained (Rouvinen et al., 1990): the catalytic core forms an α/β barrel folds, which, in deviation from the classical $(\beta/\alpha)_8$ "TIM" barrel (Wierenga, 2001), has only seven β-strands that form the central β-barrel (Figure 3.4). The catalytic core of *T. reesei* contains two target sites for *N*-glycosylation. Of these, N310 and N289 contain 70% and 82% *N*-glycosylation structures (see Chapter 8), respectively, whereas the remaining some 20% were occupied by single GlcNAc residues (Hui et al., 2002).

The active centre resides in a tunnel formed by two surface loops (Figure 3.4). Crystals of *H. insolens* Cel6A in complex with cellobiose revealed six binding subsites in the active center tunnel, a single glucose moiety being bound in the −2 subsite, and cellotetraose in the +1 to +4 subsites (Varrot et al., 2003). The β-1,4-glycosidic bond is cleaved by acid catalysis using an aspartic acid, D221, as the most likely proton donor, and another aspartate, D175, which likely ensures its protonation and stabilizes charged reaction intermediates. The catalytic base has not yet been identified experimentally: possible candidates are not within the distance

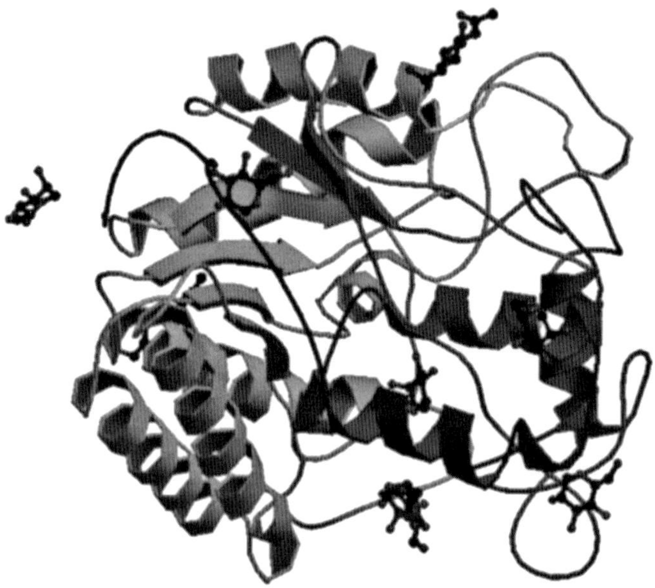

Figure 3.4. Three-dimensional structure of cellobiohydrolase II CEL6A from *T. reesei* (1CB2).

of a hydrogen bond to a water molecule that could act as the nucleophile in this inverting mechanism. Thus, the current interpretation proposes that the water is deprotonated through a "solvent wire" through to one of the conserved D residues near the active center (Piens and Davies, 2011).

The crystal structure of CBHII also shows a tyrosine residue, Y169, located close enough to the hydrolytic target bond to be involved in catalysis. Exchange of this residue to a phenylalanine (Y169F) increased the association constants of the mutant enzyme for cellobiose and cellotriose threefold and simultaneously reduced the catalytic constants toward the same substrates fourfold (Koivula et al., 1996). The data suggest that Y169 interacts with a glucose ring in the substrate at the second subsite, thereby distorting the glucose ring into a more reactive conformation. In addition, Y169 may affect the protonation state of the active site carboxylates, D175 and D221 (Koivula et al., 1996).

Interestingly, Varrot et al. (2003) showed that the catalytic core of *H. insolens* Cel6A undergoes several conformational changes upon substrate binding, the most significant of which is a closing of the two active site loops (residues 174–196 and 397–435) with main chain movements of up to 4.5 Å.

3.2.3 Endo-β-1,4-Glucanases (EC 3.2.1.4)

GH5 Endo-β-1,4-Glucanases

Glycosyl hydrolase family GH5 is one of the largest of all GH families and contains enzymes acting on a wide variety of substrates. Most of them are endo-β-1,4-glucanases and endo-β-1,4-mannanase, but other activities include endo-β-1,6-galactanase, endo-β-1,3-mannanase, endo-β-1,4-xylanase, as well as high specificity xyloglucanases. St. John et al. (2010) have recently shown that many members of GH5 actually should be classified as glycosyl hydrolase family 30, but the fungal endo-β-1,4-glucanases remained unaffected from this shift. Unfortunately, the 3D structure of these cellulases has only been solved for bacterial enzymes (e.g., *Clostridium thermocellum* endoglucanase CelC and *Bacillus agaradhaerens* Cel5A; Dominguez et al., 1995; Davies et al., 1998). They are members of Clan GH-A and thus have a classical $(\alpha/\beta)_8$-TIM barrel fold, with two E residues forming the active center. GH5 enzymes are retaining enzymes and act by the Koshland double-displacement mechanism and the two catalytic residues (catalytic nucleophile and general acid/base) are known to be E's found at the C-terminal ends of β-strands four (acid/base) and seven (nucleophile; Davies, 2011).

As for fungi, endo-β-1,4-glucanases from GH5 have been purified and characterized from *T. reesei* (previously called EGIII), *Aspergillus* spp., and also anaerobic fungi (Saloheimo et al., 1988; Eberhard et al., 2000; Hara et al., 2003). However, none of these has been investigated in more detail.

GH7 Endo-β-1,4-Glucanases

Endo-β-1,4-glucanases from glycosyl hydrolase family GH7 are produced by most fungi in major amounts and have therefore been characterized in some detail from several sources. Being a member of GH7, they are very similar in structure to CEL7A CBH1 (displaying a β-jelly roll-derived framework) and in the basic mechanism of action, that is, they attack at the reducing end by a retaining mechanism. All these features have been explained in detail at cellobiohydrolase CEL7A earlier (Section 3.2.2) and will thus not be repeated here. However, the major difference between the cellobiohydrolase and endoglucanase enzymes of GH7 is that the latter have an open substrate binding cleft instead of a tunnel, which enables the attack in the middle of the cellulose molecule and thus endo-action. Similar findings have also been reported for GH6 enzymes, which also (although not in fungi) contain endo-β-1,4-glucanases. When the structure of the first GH6 endo-β-1,4-glucanase was solved, the active center was observed in a long open groove and provided the first hint that endo or exo activity could be modulated through display of the active center either in an open grove (for endo) or loop-enclosed tunnel (for exo), respectively. Proof for this was obtained by showing that the exo activity was changed to endo when the extended loops of a cellobiohydrolase from the prokaryote *Thermomonospora fusca* were truncated (Meinke et al., 1995; Figure 3.5). Since the loops that form the active center of the cellobiohydrolases are flexible and show multiple conformations, the enzyme may switch between exo and endo activity depending on the conformational changes of that loop. There is thus an ongoing debate whether true exo- and endo-cellulases actually exist (cf. Stahlberg, 2011).

As with other cellulase, CEL7B occurs in several isoforms that exhibit different isoelectric points. In CEL7B, this has been studied in some detail: *T. reesei* CEL7B was shown to occur in at least 14 different glycoforms (García et al., 2001). The major isoform contained only a single *N*-linked GlcNAc-, and a single probably *O*-linked mannose residue. A minor population of the enzyme contained $Man_{5-7}GlcNAc_2$ antennae, and all the others contained a negatively charged phosphate ester on the *N*-glycans. Eriksson et al. (2004), using another preparation of CEL7B, arrived at basically similar results although they varied in detail, indicating that these modifications are dependent on biochemical processes that vary with the culture conditions. Of the five potential sites found in the wild-type enzyme, they found only N56 and N182 to be *N*-glycosylated. $GlcNAc_2Man_5$ was identified as the predominant *N*-glycan, although lesser amounts of $GlcNAc_2Man_7$ and glycans carrying a mannophosphodiester bond were

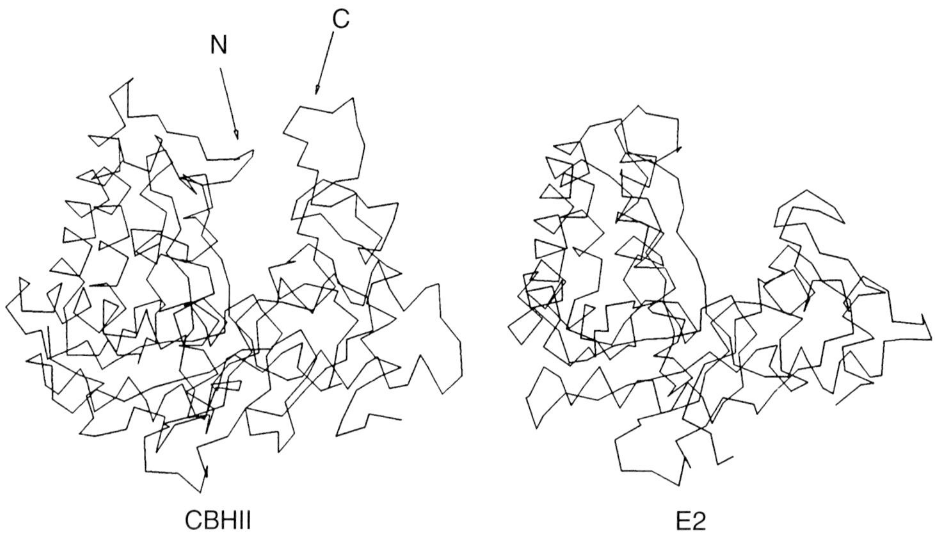

Figure 3.5. α-Carbon skeletons of the *T. reesei* cellobiohydrolase CEL6A and *T. fusca* endoglucanase E2 catalytic domains. The views are chosen to illustrate differences in the accessibilities of the two active sites. C and N, respectively, indicate the carboxyl- and amino-proximal loops that cover the active site of CBH II. (© 1995 The American Society for Biochemistry and Molecular Biology.)

also detected. In addition, they detected partial deamidation of N259 and a partially occupied *O*-glycosylation sites.

GH12 Endo-β-1,4-Glucanases

Members of GH family 12 (GH 12) are distributed throughout the bacterial and fungal kingdoms and comprise small (20–25 kDa) proteins that—in contrast to most other cellulases—lack a CBD. They are therefore unable to bind to crystalline cellulose and hydrolyze only amorphous cellulose (Henriksson et al., 1999). They perform their hydrolysis via a double-displacement reaction and a glycosyl-enzyme intermediate that results in retention of the anomeric configuration in the product (Schülein, 1997; Birsan et al., 1998). A phylogenetic analysis shows a division into five subfamilies of which two exclusively comprise fungal enzymes: 12–1 (fungal group I), 12–2 (fungal group II), 12–3 (*Streptomyces* group including *Rhodothermus marinus*), 12–4 (*Thermophiles* group), and 12–5 (*Erwinia carotovora*) (Figure 3.6; Goedegebuur et al., 2002). Some fungi (e.g., *Gliocladium roseum*) appeared to have duplicated their *cel12* genes in subfamily 12–1.

The enzymes from subfamily 12–2 showed the presence of an additional domain with unknown function, and also their catalytic domain was different from the other members. In addition, while most of the GH12 glycosyl hydrolase proteins comprise mainly endo-β-1,4-glucanase activities, the enzyme from *Aspergillus niger* and *Malbrachea cinnamomea* that are members of subfamily 12–2 were reported to exhibit xyloglucanase activity (Schülein et al., 2002; Powlowski et al., 2009). Powlowski et al. (2009) detected a short deletion and insertion following D112 and I129, respectively, in the *A. niger* protein AnXEG12A that are conserved among the subfamily 12–2 sequences but not subfamily 12–1 sequences. A

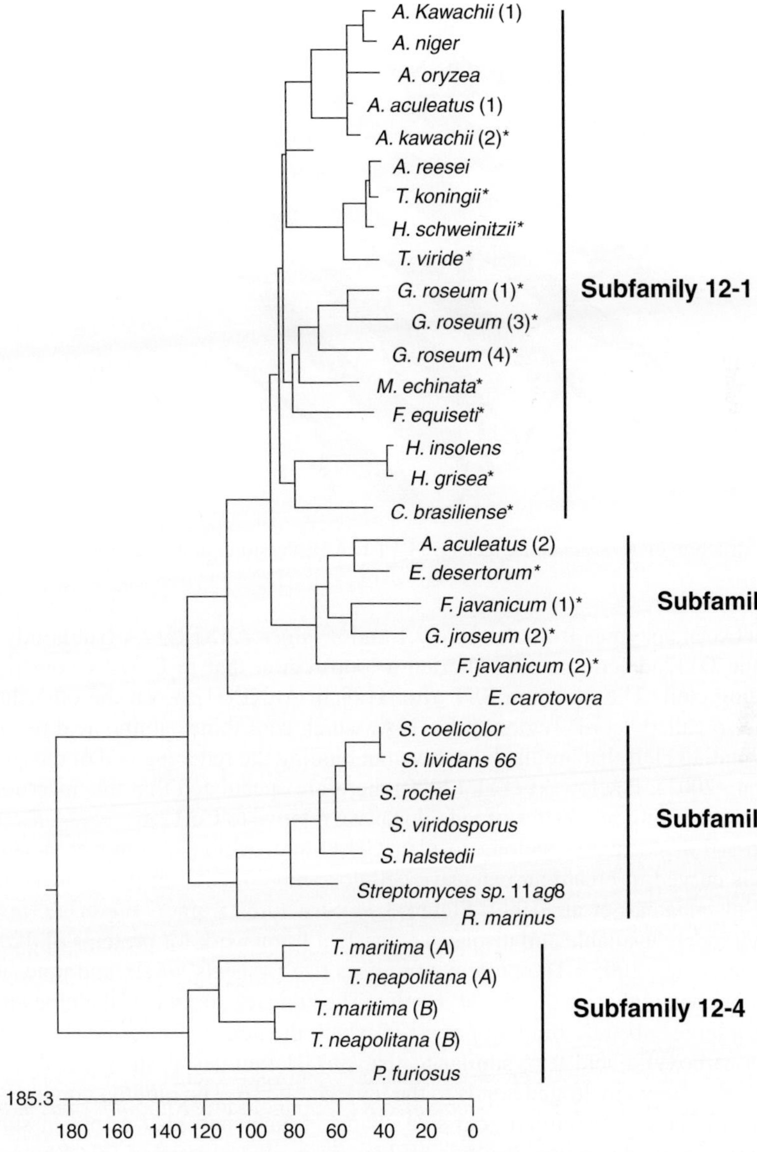

Figure 3.6. Phylogenetic tree of all known GH12 endoglucanases. Deposition numbers: *A. aculeatus* (1) (P22669), *A. aculeatus* (2) (O94218), *A. kawachii* (1) (Q12679), *A. kawachii* (2) (AF435072), *A. niger* (O74705), *A. oryzea* (O13454), *C. brasiliense* (AF434180), *E. carotovora* (P16630), *E. desertorum* (AF434181), *Fusarium equiseti* (AF434182), *F. javanicum* (1) (AF434183), *F. javanicum* (2) (AF434184), *G. roseum* (1) (AF435063), *G. roseum* (2) (AF435064), *G. roseum* (3) (AF435065), *G. roseum* (4) (AF435066), *Humicola grisea* (AF435071), *H. insolens* (A22907), *Hy. schweinitzii* (AF435068), *M. echinata* (AF435067), *P. furiosus* (AD54602.1), *R. marinus* (O33897), *S. coelicolor* (CAB61599.1), *S. halstedii* (O08468), *S. lividans* 66 (Q54331), *S. rochei* (Q59963), *S. viridosporus* (AAD25090.1), *Streptomyces* sp. 11ag8 (AF233376), *T. koningii* (AF435069), *T. maritima* A (Q60032), *T. maritima* B (Q60033), *T. naepolitana* A (P96491), *T. naepolitana* B (P96492), *T. reesei* (O00095), *T. viride* (AF435070). (Reprinted from Goedegebuur et al., 2002, with permission from Springer.)

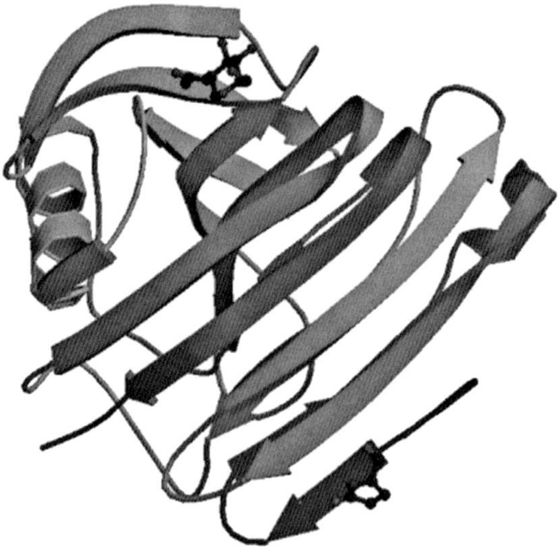

Figure 3.7. Structure of *T. reesei* endoglucanase CEL12A (accession number 1H8V).

comparison between enzymes from subfamily 1 and *A. niger* AnXEG12A (subfamily 12–2) showed that the D112 deletion would shorten a loop region that in Cel12A constricts the substrate-binding cleft. The insertion SST after I129 in AnXEG12A, on the other hand, is adjacent to the so-called "cord" region of Cel12A, which contributes amino acid residues to the substrate-binding cleft that are likely involved in binding the reducing end of the substrate (Sandgren et al., 2001). Powlowski et al. (2009) therefore speculated that this insertion may alter the substrate-binding properties of xyloglucanase relative to Cel12A.

GH12 endo-β-1,4-glucanases belong to clan GH-C and exhibit a compact β-sandwich structure that is curved to create an extensive cellulose-binding site on the concave face of the β-sheet (Sulzenbacher et al., 1999; Figure 3.7). Structures from *T. reesei, H. insolens,* and *T. citrinoviride* are available and display a consistent framework for proteins of the GH12 family (Sandgren et al., 2005). The protein comprises two β-sheets, of six and nine strands, packed on top of one another, and a single α-helix. The concave surface of the nine-stranded β-sheet forms a large substrate-binding groove in which the active-site residues are located. It comprises a carboxylic acid trio, similar to that of GH families 7, in which the strictly conserved D99 also forms hydrogen bonds to the invariant E116. The binding crevice is lined with both aromatic and polar amino acid side chains, which may play a role in substrate binding. The enzyme contains one disulfide bridge and is glycosylated at D164 by a single *N*-acetyl glucosamine residue.

GH45 Endo-β-1,4-Glucanases

The GHs belonging to GH45 have so far only been described as endo-β-1,4-glucanases, and previously been considered to belong to cellulase family K. They are known from bacteria and fungi only.

The first GH45 endoglucanase described was endoglucanase V from *H. insolens* (Davies et al., 1993) and *T. reesei* EGV (Saloheimo et al., 1994). The protein is unusually small (242

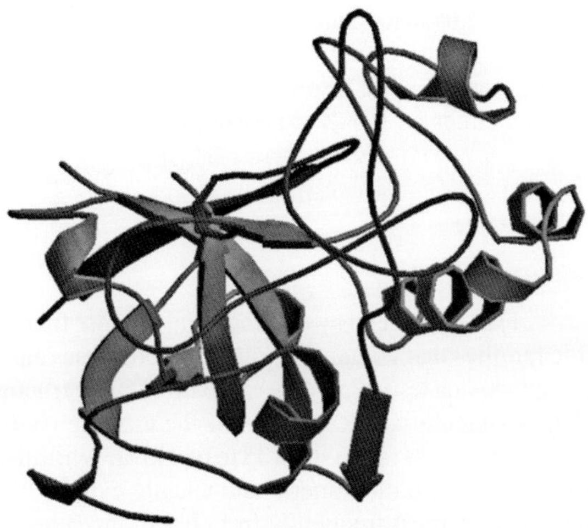

Figure 3.8. Structure of *T. reesei* endoglucanase CEL45A (accession number 2ENG).

amino acids), yet—in contrast to the small GH12 cellulases—contains a CBD. The catalytic and the CBDs are separated by only 36 amino acid-long linker, the smallest one known for cellulases. EGV consists of a six-stranded β-barrel domain with long interconnecting loops (Figure 3.8). A 40 Å groove exists along the surface of the enzyme, and this contains the catalytic residues, D10 and D121, which sit to either side of the substrate-binding groove in an ideal conformation for facilitating cleavage by inversion, their carboxyl groups being separated by approximately 8.5 Å (Davies et al., 1995). A disordered loop is located above the active center that becomes ordered upon the binding of cellooligosaccharides (Davies et al., 1995). D121 (located in an HxD motif) acts as the general acid and D10 most likely as the general base (located in a YxD motif; Davies et al., 1995). Unlike other cellulases, their pH optimum lies around neutrality, and they have thus been strongly investigated for their application in the textile/detergent industries (Schülein et al., 1998).

Seven subsites were detected, but no sugar seems to bind to the -1 subsite at the point of cleavage, and the geometry of the cleavage site suggests that the enzyme would favor the binding of the transition state (an elongated glycosidic bond) rather than the substrate. Upon substrate binding, the above described loop structure reorganizes (called "lid flipping"), which causes an increase in the hydrophobic environment of the catalytic proton donor, enclosing the active site at the point of cleavage and bringing a third aspartate (D114) in close proximity to the substrate (Davies et al., 1995).

The structure of GH45 endo-β-1,4-glucanases has also been found in so-called "Barwin endocellulases," plant defense proteins of unknown function (Ludvigsen and Poulsen, 1992). This similarity is due to the fact that both display a general, yet distant similarity in their β-barrel structure with the domain 1 of plant expansins (for review, read Sampedro and Cosgrove, 2005). Expansins are plant proteins that form a distinct protein family with high sequence identity (so-called α-expansins). They have been proposed to disrupt hydrogen bonding between cellulose microfibrils or between cellulose and other cell wall polysaccharides without having hydrolytic activity (McQueen-Mason and Cosgrove, 1994; Whitney et al., 2000). In this way, they are thought to allow the sliding of cellulose fibers and enlargement of the cell wall. Expansin domain 1 proteins also share a number of conserved cysteine residues with the GH45 family proteins. It is interesting that several residues that make up the catalytic site of GH45 endo-β-1,4-glucanases are also conserved in expansin, including the catalytic acid N, because until now neither α- nor β-expansin protein have been reported to display hydrolytic activity.

3.2.4 β-1,4-Glucosidases

β-glucosidases (EC 3.2.1.21) are produced by all fungi, and they are found in GH families 1 and 3. Both of them are large families that contain various β-glycosidases and that also show broad substrate specificities. β-glucosidases isolated so far exhibit high structural variability, partly reflecting the intracellular/extracellular localization of the enzyme (Baldrian and Valásková, 2008). The detected molecular masses range from 35 to 640 kDa. While the small enzymes with molecular masses up to 100 kDa are monomeric and usually extracellular, homo-oligomeric enzymes have also been isolated, most frequently from basidiomycetes.

Most of the GH1 β-glucosidases are intracellular enzymes. They act by the retaining Koshland double-displacement mechanism and are able to cleave soluble β-linked oligosaccharides from chain lengths up to nine glucose residues, as well as aglycone-linked β-glucosides. Many of them exhibit both β-glucosidase and β-galactosidase activity in the same protein. They are competitively inhibited by the product glucose and by β-glucono- and cellobionolactone, which could arise as a product of the action of cellobiose dehydrogenase (see Chapters 2 and 5). Enzyme catalytic details have been published for the enzyme from almond and *Agrobacterium tumefaciens* but not yet for any fungal enzyme (Withers, 2011a). Their 3D structure has only recently been reported (for the two GH1 β-glucosidases from *P. chrysosporium* and for the single enzyme from *T. reesei;* Nijikken et al., 2007; Jeng et al., 2011). They belong to clan GH-A and are organized in a classical α/β_8-TIM barrel fold, of which each contains a slot-like active site cleft and a more variable outer opening, related to its function in processing different lengths of β-1,4-linked glucose derivatives. While the two essential E residues for hydrolysis are spatially conserved in the active site, the residues around the aglycone-binding site are not. One of the isoenzymes of *P. chrysosporium* (BGL1A) has a unique aglycone specificity compared to other structurally known GH1 enzymes (i.e., activity toward aryl-β-D-glucopyranosides), which is lacking from the other GH1 β-glucosidases and which correlates with a unique subsite at +1 (Nijikken et al., 2007).

The second GH family that contains β-glucosidase enzymes is GH3. This is a very large family with most enzymes originating from microorganisms. The family 3 β-D-glucosidases are broad specificity exo-hydrolases that remove single glucosyl residues from the nonreducing ends of oligo- and polysaccharides using the retaining Koshland double-displacement mechanism. They are active on a wide range of substrates, including β-D-glucans, β-D-oligoglucosides and aryl-β-D-glucosides, β-1,3-D-glucans, β-1,4-D-glucans, β-1,3/1,4-D-glucans, β-1,6-D-glucans, and some β-D-oligoxyloglucosides (Hrmova et al., 1998). Although the GH3 β-glucosidases comprise the main extracellular fungal β-glucosidases, none of them has yet been structurally characterized. Such data are only available for the protein from barley (Varghese et al., 1999), which shows a two-domain, globular protein. The first part comprises—like the GH1 family β-glucosidases—a $(\beta/\alpha)_8$-TIM barrel domain, whereas the second one is arranged in a six-stranded β-sandwich, with three α-helices flanking the sheet on either side. The two domains are connected by a 16-amino acid helix-like linker. The broad substrate specificity is likely caused by the fact that the glucosyl residue occupying binding subsite −1 is tightly locked into a relatively fixed position, whereas the position of the glucosyl residue at subsite +1 is rather flexible. The active site is therefore largely independent of substrate conformation and will consequently accommodate a range of substrates (Hrmova et al., 2002).

While the fungal GH3 β-glucosidases are generally extracellular enzymes and secreted into the medium, many of them (notoriously that of *T. reesei*; Kubicek, 1981) remain tightly bound to the cell wall and are therefore not found in the extracellular fluid. One could speculate that this location, close to the cell membrane, would provide an ecological advantage because the products of hydrolysis are formed close to the respective mono- and disaccharide transporters and do not diffuse in the medium. Rath et al. (1995) identified a cell wall heteroglycan, composed of mannose, galactose, glucose, and glucuronic acid, that is responsible for the binding of β-glucosidase to the *T. reesei* cell walls.

3.3 Nonenzymatic Proteins Involved in Cellulose Hydrolysis

3.3.1 GH61 Proteins

The GH61 family comprises enzymes that were originally classified as endo-β-1,4-glucanases based on a very weak activity in one family member from *T. reesei* (Karlsson et al., 2001). However, more recent elucidation of the 3D structures and properties of GH61 proteins from *T. reesei*, *Thielavia terrestris,* and *Thermomyces aurantiacus* (Karkehabadi et al., 2008; Harris et al., 2010) confirmed that they lack enzymatic activity, but—in contrast—are able to enhance the degradation of cellulose by other cellulases in the presence of metal ions.

A genome-wide inventory of GH61 gene families shows that the respective members have become dramatically amplified in several asco- and basidiomycetes such as *C. cinerea*, *Podospora anserina*, *Chaetomium globosum*, and *Stagonospora nodorum*, in which more than 25 GH61-encoding proteins are present. On the other hand, no GH61-encoding genes have been detected in the genome sequences of yeasts or in noncellulolytic filamentous fungi such as *Rhizopus oryzae*, *Ustilago maydis*, or *Coccidioides immitis*. A phylogenetic analysis of the GH61 amino acid sequences from ascomycetes and basidiomycetes is discordant with the species tree and they are rather distributed throughout all branches (Figure 3.9), suggesting that they have either become duplicated and diverged in an already very early evolutionary period (the split between asco- and basidiomycetes is estimated to have happened 600 million years ago; Berney and Pawlowski, 2006) or are subject to frequent horizontal gene transfer between asco- and basidiomycetes (Harris et al., 2010). The fact that this diversity has been maintained until today further implies the operation of a significant selective pressure, which in turn lends to hypothesize an important ecological role of the GH61 family enzymes. Harris et al. (2010) also noted that the evolution of GH61 genes has gone in different direction in different species: *C. cinerea,* for example, has a much larger number of closely related paralogs that must be the product of relatively recent duplication event. In contrast, the phylogenetic distribution of GH61 proteins from *Chaetomium globosum* and *Phaeosphaeria nodorum* in the phylogenetic tree is much more dispersed and indicates a long-time ongoing diversification of their respective genes. Interestingly, gene loss seems to have occurred in some GH61 branches in some fungi like *Aspergillus* and *Trichoderma* spp. (Harris et al., 2010).

The structure of the GH61 proteins consists of a compact single-domain β-sandwich structure that is formed by two sheets in a variation of a fibronectin type III fold (Goll et al., 1998; Figure 3.10). Consistent with the absence of enzymatic activity, the protein is devoid of surface

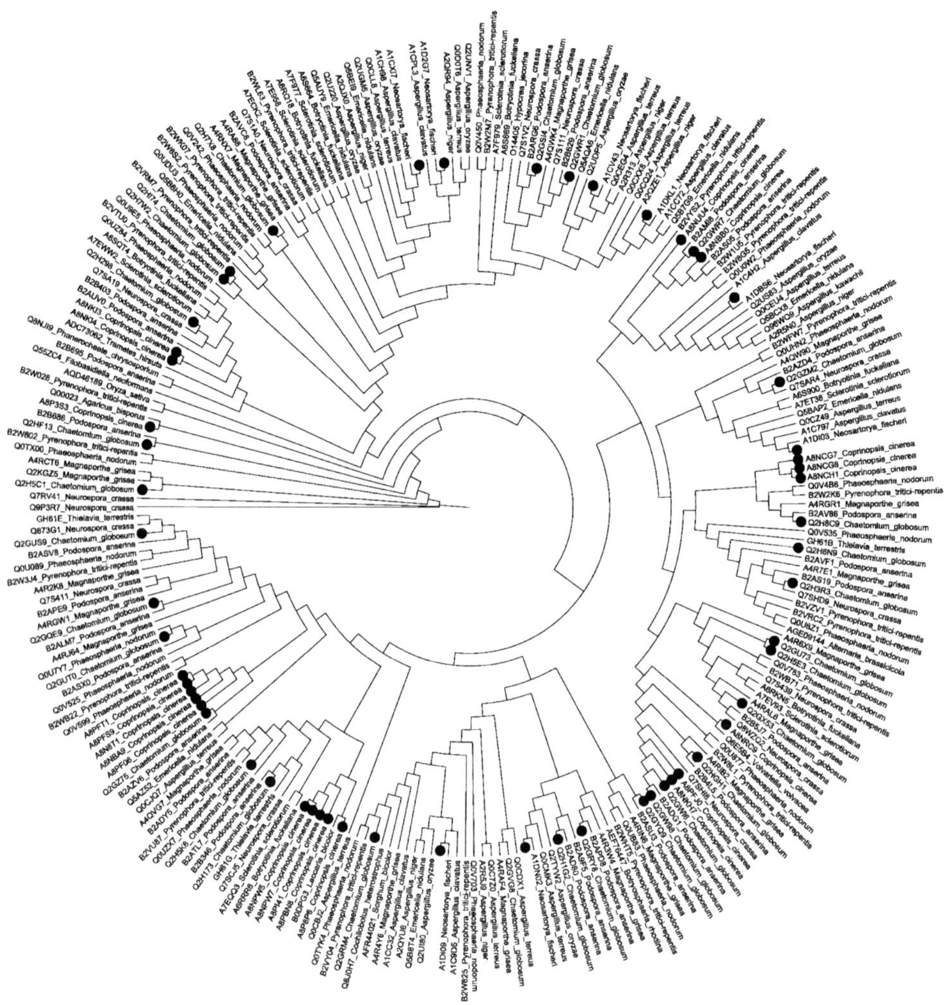

Figure 3.9. Phylogeny of GH61 proteins available in the Uniprot and GeneSeqP databases. The accession number and identified species are shown for each entry. (Reprinted with permission from Harris et al., 2010. Stimulation of Lignocellulosic Biomass Hydrolysis by Proteins of Glycoside Hydrolase Family 61: Structure and Function of a Large, Enigmatic Family. Biochemistry. Copyright 2012 American Chemical Society.)

crevices that could act as binding pockets for a substrate. Also, the conserved catalytic acidic D or E residues that are present in almost all known GHs are absent (Harris et al., 2010).

When added in concentrations between 5% and 20% of the total protein, GH61 proteins cause a dramatic increase in cellulose hydrolysis by cellulase mixtures such as the one from *T. reesei*. The effect was dependent on the presence of a divalent metal ion (Harris et al., 2010). There are three highly conserved H residues at the surface near the N-terminus of GH61A, two of which bind a metal ion that by default appears to be zinc.

Most recently, Phillips et al. (2011) identified the GH61 proteins of *N. crassa* to actually be carbohydrate monooxygenases (in fact those monooxygenases that were required for cellulose

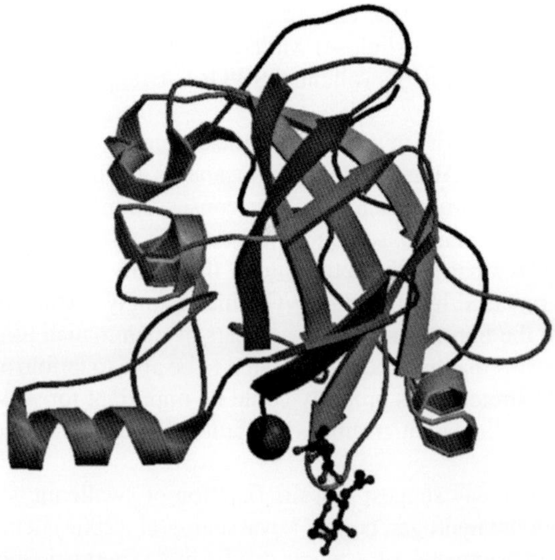

Figure 3.10. Structure of *T. reesei* cellulase-enhancing protein CEL61B (accession number 2VTC).

oxidation in the presence of cellobiose dehydrogenase), which have different regiospecifities on the cellulose chain, resulting in oxidized products modified at either the reducing or nonreducing end of the glucan chain. In contrast to previous models where oxidative enzymes were thought to produce reactive oxygen species that randomly attacked the substrate, the GH61 enzymes catalyze the direct oxidation of cellulose. The electron transfer necessary for their action was shown to come from cellobiose dehydrogenase (see also Chapter 2), thus linking this enzyme to cellulose oxidation. This is intriguing, because not all fungi—notably *T. reesei*—have such an enzyme, but their cellulases were nevertheless stimulated by the GH61 proteins (Harris et al., 2010). *T. reesei* has been shown to be able to oxidize cellulose (Szakmary et al., 1991) and thus must possess an alternative system for electron transfer to its GH61 proteins.

Interestingly, there is also a structural similarity between *T. reesei* GH61B and the chitin-binding protein CBP21 from *Serratia marcescens*, a protein that belongs to the CBM33 family and stimulates the chitin-degrading activity of chitinases while having no chitinase activity itself (Vaaje-Kolstad et al., 2005; Karkehabadi et al., 2008; Harris et al., 2010). This similarity is supported by the findings that the above mentioned H residues for metal ion binding are conserved in both proteins, and mutation of these residues significantly impacts the ability of both proteins to stimulate chitinase and cellulase activity by CBP21 and GH61B, respectively (Vaaje-Kolstad et al., 2003). A CBP21 homolog from *Thermobifida fusca* was able to also stimulate the hydrolysis of filter paper by cellulases (Moser et al., 2008). A phylogenetic analysis that combines both the GH61 and the CBM33 proteins is still lacking, but could test the hypothesis whether they form an ancient superfamily of proteins destined to enhance the degradation of polysaccharide substrates by monooxygenation of the carbohydrates.

3.3.2 Swollenin

Saloheimo et al. (2002) first described a gene and the encoded protein from *T. reesei* that has an C-terminal expansin-like domain with homology to the group 1 grass pollen allergens (pfam 01357) and an N-terminal CBD. This protein was able to loosen the structure of *Vallonia* cell walls (that consist mainly of cellulose) while retaining its integrity and not releasing any soluble sugars and was thus called swollenin (SWO1). This property is also reminiscent of expansins (*vide supra*).

The limited sequence similarity of swollenin to the expansins is in the same range of identity as the similarity between the GH family 45 endoglucanases and individual expansins (Brotman et al., 2008). However, there is almost no sequence conservation detectable between EGL5 and SWOI, and it is thus weaker than conservation between CEL45A and expansins. The sequence motif HFD, which forms a part of the active site of the family 45 hydrolases, is conserved in the expansins and is replaced by HLD in SWOI. Interestingly, swollenin also exhibits some sequence similarity to the fibronectin III-type repeats of mammalian titin proteins. The latter form β-sandwich domains that have been suggested to be able to unfold and refold easily, thus enabling the protein to stretch. This property could be important for swollenin, if its function is to allow slippage of cellulose microfibrils in plant cell walls as suggested for expansions (Saloheimo et al., 2002).

All the available data today suggest that the function of swollenin is to bind to cellulosic compounds and loosen the hydrogen bonds. Levasseur et al. (2006) demonstrated that fusion of swollenin to feruloyl esterase A of *A. niger* results in a synergistic increase of ferulic acid release. They speculate that the CBD of SWOI may increase the local concentration of the fused enzyme close to the substrate and facilitate the lateral diffusion of the FAEA along the surface of the cellulose microfibrils and consequently increase the final hydrolysis yields.

The genome of *T. reesei* also contains a gene encoding a second swollenin (SWO2), which is also expressed during growth on cellulose (C.P. Kubicek, unpublished data), but which has not been investigated yet. Swollenin genes are also present in the genomes of other *Trichoderma* spp. such as *T. asperellum*, a well-known biocontrol agent and inducer of plant defense responses (Brotman et al., 2008), where it was shown to be expressed during colonization of the plant rhizosphere. Root colonization rates were reduced in transformants silenced in swollenin gene expression. Swollenin was also capable of stimulating local defense responses in cucumber roots and leaves and to afford local protection toward infection by plant pathogens, but mutation studies showed that this was due to the presence of the CBD in SWO1 that is apparently recognized by the plant as a microbe-associated molecular pattern. Interestingly, swollenins appear to be absent from other fungi except for *A. fumigatus* and its close relative *Neosartorya fischeri*. The reason for this has so far remained unexplored.

Chapter 4
The Tools—Part 2: Enzymology of Hemicellulose Degradation

As described in Chapter 1, the hemicelluloses comprise xyloglucans, xylans, galactomannans, and pectins. While the linkage and sugars in the core chains are clearly different between these four major polysaccharides, the side chain substituents often comprise the same sugar and the same linkage, and therefore, the same enzymes are involved in their cleavage in different polymers. Table 4.1 presents an overview about the enzymes required for complete depolymerization of these four major hemicelluloses. In Chapter 5, I will first deal with the enzymes that are necessary for the degradation of the backbone and subsequently describe the others under "accessory enzymes."

4.1 Xyloglucan Hydrolysis

Fungi use both unspecific as well as specific groups of endo-β-(1→4)-glucanase for the hydrolysis of the backbone chain of xyloglucan. Endo-β-1,4-glucanases with xyloglucanase activity can be found in GH5, GH12, GH16, and GH44. GH5 and GH12 enzymes have already been discussed in Chapter 3. Because most xyloglucans are composed of XXXG-type of building units, distant subsites are needed to bind xyloglucan. Genes encoding GH44 proteins appear to be absent from fungal genomes, and although genes encoding GH16 enzymes are abundant in fungal genomes, there is only a single report on characterization of a GH16 enzyme from fungi (Hartl et al., 2011). It is a GPI-anchored endo-β-1,3-glucanase (ENG2) of *Aspergillus fumigatus*. This enzyme acts preferentially on soluble β-1,3-glucans, and also hydrolyzes carboxymethyl cellulose (CMC). Unfortunately, the authors did not test the hydrolysis of xyloglucan. However, the much better studied GH16 enzymes from bacteria can act on a variety of β-glycans (Eklöf and Planas, 2011), and it is thus likely that this will also be valid for the fungal enzymes.

Fungi also contain specific xyloglucanases, which completely lack or display only very low activities toward CMC and β-glucan and which belong to the GH family 74. They

Fungi and Lignocellulosic Biomass, First Edition. Christian P. Kubicek.

Table 4.1. Enzymes required for hydrolysis of different hemicellulose.

		Hemicellulose Polymer			
Enzyme	CAZy Fam.	Xyloglucan	Xylan	Galactomannan	Pectin
Acetylxylan/feruloyl esterase	CE1,2,3,5		+		
Arabinoxylan arabinofuranohydrolase	GH62		+		
Endo-/exo-(rhamno)galacturonases	GH28				+
Endoarabinanase	GH43				+
Exoarabinanase	GH93				+
Feruloyl esterase	CE1				+
4-*O*-methyl-glucuronoyl methylesterase	CE15		+		
Pectin lyase	PL1				+
Pectin methyl esterase	CE8				+
Rhamnogalacturonan acetyl esterase	CE12				+
Rhamnogalacturonan lyase	PL4,11				+
Unsaturated rhamnogalacturonan hydrolase	GH105				+
Unsaturated glucuronyl hydrolase	GH88				+
Xyloglucan β-1,4-endoglucanase	GH12,74	+			
α-1,4-galactosidase	GH27,36	+	+	+	
α-arabinofuranosidase	GH51,54	+	+		+
α-glucuronidase	GH67,115		+		
α-rhamnosidase	GH78				+
α-xylosidase	GH31	+			
β-1,4-endogalactanase	GH53				+
β-1,4-endomannanase	GH5,26			+	
β-1,4-endoxylanase	GH10,11		+		
β-1,4-galactosidase	GH2,35	+	+	+	+
β-1,4-mannosidase	GH2			+	
β-1,4-xylosidase	GH3,43		+		+

Modified from van den Brink and de Vries (2003) and complemented by own data.

comprise both exoenzymes, that is, which release oligosaccharides with two glucose units from the nonreducing end of xyloglucan oligosaccharides (oligoxyloglucan reducing end-specific cellobiohydrolase, EC 3.2.1.150; Bauer et al., 2005; Desmet et al., 2007; Yaoi et al., 2009), as well as enzymes with endo-type activity on xyloglucan (EC 3.2.1.151; Hasper et al., 2002; Yaoi et al., 2007). They are active on only xyloglucan oligosaccharides and cannot act on xyloglucan polysaccharides. Enzymes from GH74 utilize a single-displacement, anomeric configuration-inverting mechanism involving the direct attack of water on the sugar ring

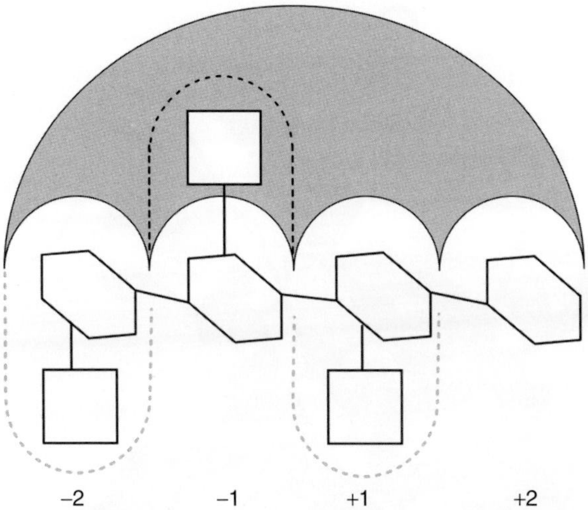

Figure 4.1. Schematic representation of subsites in the active site (−2/+2) of CEL74A from *T. reesei*. (Modified from Desmet et al., 2007.)

(Zechel and Withers, 2000). Most GH74 xyloglucanases hydrolyze the glycosidic linkage of unbranched glucose residues, but a few members accommodate side chain xylose residues at subsite-1 of the active site including *A. nidulans* OREX (Bauer et al., 2005) and *Trichoderma reesei* (Desmet et al., 2007; Figure 4.1).

The protein structure of GH74 xyloglucanase has been obtained for an exo- and an endo-acting xyloglucanase from *Geotrichum* sp. M128 (i.e., oligoxyloglucan reducing end-specific cellobiohydrolase OXG-RCBH, Yaoi et al., 2007, and XEG, Yaoi et al., 2009). Both consists of two seven-bladed β-propeller domains, and—although their sequence identity of these two β-propellers is only 18%—they share the same 3D structure and may be the result of a gene duplication event (Figure 4.2). The N-terminal domain is orientated at an angle of about 90° relative to the C-terminal domain, and the two domains are held in shape by H-bonding and hydrophobic interactions over a contact area of about 7500 Å. The N- and C-terminal domains are linked by three loop segments, one in the N-terminus, the second in the middle, and the third in the C-terminus of the sequence.

As can be expected, the substrate-binding region occurs in an open cleft, which is located at the intersection of the N- and C-terminal domains, the loops thereby forming the surface. Most of the amino acid residues that interact with the substrate are conserved between the exo- and endo-acting enzymes. However, there are two notable differences at the subsite positions −1 and +2, which may be relevant to the different mode of attack: first, the exo-xyloglucanase OXG-RCBH has a loop around the +2 site that blocks one end of the active site cleft, whereas this loop is lacking in XEG and allows binding to the middle of the main chain of the substrate. Second, at the −1 site in OXG-RCBH, an N488 interacts with the xylose side chain of the substrate, whereas this site is occupied by a bulky side chain (Y457) in XEG. Yaoi et al. (2009) showed that a Y457G mutation led to unspecific cleavage at various sites and concluded that the presence of this bulky Y457 may determine the substrate specificity of XEG.

Although the xyloglucanase CEL74A from *H. jecorina* and OXG-RCBH are both exo-acting enzymes that attack from the nonreducing end, they differ in their substrate specificity:

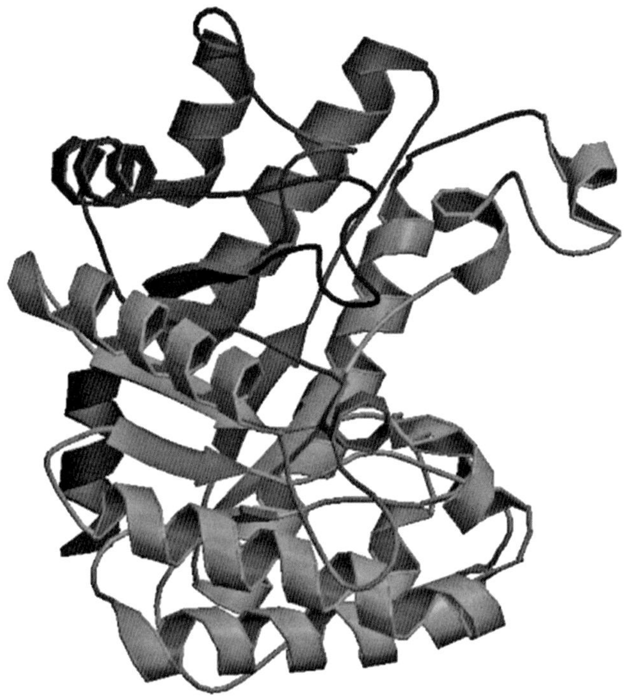

Figure 4.2. Structure of *T. reesei* CEL74A (accession number 3A0F).

CEL74A releases XGXX units by cleaving at substituted Glc residues, whereas OXG-RCBH hydrolyzes xyloglucan oligosaccharides at substituted glucose residues and releases only XG from XXXGXXXG. Desmet et al. (2007) compared the length of the exo-loop, that is typical for exo-type GH74 xyloglucanases, in CEL74A and OXG-RCBH. They found that the exo-loop in CEL74A is considerably shorter, and speculated that this would increase access of an oligosaccharide to the putative subsites +3/+4, and thus be the reason for the difference between the two enzymes. Enzymes that hydrolyze xyloglucan at branched glucose residues such as OXG-RCBH obviously need a −1 subsite that is relatively spacious.

Grishutin et al. (2004) also isolated a specific xyloglucanases from *A. japonicus* (32 kDa, pI 2.8), which exhibited endo-activity and high specific activity toward tamarind xyloglucan with almost no activity against CMC and barley β-glucan. The amino acid sequence of the respective protein did not fit to any other GH family.

4.2 Degradation of the Xylan Backbone

Endo-1,4-β-xylanase (1,4-β-D-xylan xylanohydrolase; EC 3.2.1.8) cleaves the glycosidic bonds in the xylan backbone, and thus reduces in the degree of polymerization of the substrate. Xylan is not attacked randomly, but the topology selected for hydrolysis depend on the nature of the substrate molecule, that is on the chain length, the degree of branching, and the presence of substituents. Most fungal xylanases have been identified as members of glycosyl hydrolase families GH10 and GH11. Other glycoside hydrolase families (GH5, GH7, GH8) have also been found to contain distinct catalytic domains that exhibit endo-1,4-β-xylanase activity (Collins et al., 2005), albeit so far not in fungi.

4.2.1 GH10 Xylanases

Family GH10 xylanases are abundant in fungi (Table 4.2), and it is interesting to note that a particularly high number of members is found in plant pathogenic or saprobic fungi. The first fungal GH10 xylanase was characterized from *T. reesei* (Törrönen et al., 1992, 1993) and shown to be a small (19 kDa) and acidic protein (IP 5.2). Yet orthologs of it have later on been investigated from several other fungi, including *Aspergillus* spp. (de Vries and Visser, 2001) and plant pathogenic fungi (for reviews, see de Vries and Visser, 2001; Ahmed et al., 2009). They hydrolyze glycosidic bonds by an acid–base-assisted catalysis via a double-displacement mechanism, leading to retention of anomeric configuration at the site of cleavage (Withers, 2011b).

Table 4.2. Xylan-degrading enzymes in fungi.

Enzyme	XYL	XYL	GR-XYL	XYD
GH family	10	11	30	54
Species				
A. nidulans	3	2	0	1
A. niger	1	4	1	1
P. chrysogenum	3	1	1	1
Sclerotinia sclerotiorum	2	3	NI	1
Tuber melanosporum	1	0	0	0
Blumeria graminis var. hordei	0	0	NI	0
Botrytis cinerea	2	2	NI	1
F. graminearum	5	3	0	1
F. graminearum	4	3	NI	1
Nectria haematococca	3	3	0	1
Neurospora crassa	4	2	1	1
Magnaporthe grisea	5	5	1	1
Podospora anserina	8	6	0	0
T. atroviride	1	4	2	2
T. reesei	2	4	2	2
T. virens	1	4	2	2
Mycospherella graminearum	2	1	NI	1
Stagonospora nodorum	7	7	NI	1
Ustilago maydis	2	1	NI	0
Puccinia graminis var. tritici	5	0	NI	0
Schizophyllum commune	5	1	NI	NI
Laccaria bicolor	0	0	8	0
Postia placenta	4	0	6	0
Phanerochaete chrysosporium	6	1	1	0
Serpula lacrymans	1	0	NI	NI
Coprinopsis cinerea	5	6	NI	NI

Data compiled as specified in legend to Table 3.4.
XYL, endo-β-1,4-xylanase; GR-XYL, methyl-glucuronyl-xylanase; XYD, β-xylosidase; NI, no information retrievable.

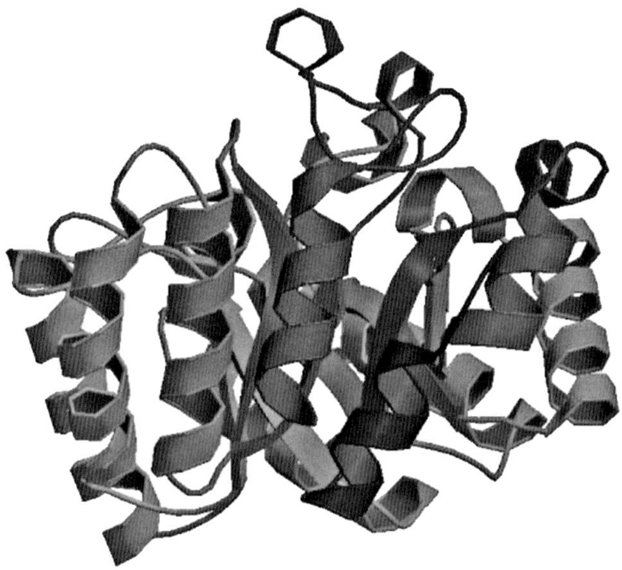

Figure 4.3. Structure of *T. aurantiacus* (XYN1 accession number 1K6A).

The crystal structures of GH10 xylanases from *Penicillium simplicissimum*, *Thermomyces aurantiacus,* and *A. nidulans* (Schmidt et al., 1998; Lo Leggio et al., 2001; Payan et al., 2004) show that GH10 enzymes fold into a $(\beta/\alpha)_8$-triosephosphate isomerase (TIM) barrel (Figure 4.3). Consistent with their "*endo*" mode of action, the substrate-binding cleft of xylanases extends along the length of the proteins and can accommodate from four to seven xylose residues. The crystal structure of the *P. simplicissimum* GH10 xylanase in complex with oligosaccharides and inhibitors revealed that short chain xylooligomers such as xylobiose and xylotriose bind to the active site cleft with their reducing end and hydrogen-bonded to the catalytic acid–base catalyst E132 (Lo Leggio et al., 2001). The trimeric hydrolysis product was always found to bind in the same way as xylotriose, whereas the monomer or dimer yielded only weak and diffuse electron density within the xylan-binding cleft and at the opposite side of the active center. Lo Leggio et al. (2001) interpreted these finding such that the two catalytic residues divide the binding cleft into a "substrate recognition area" (which ranges from the active site toward the nonreducing end of a bound xylan chain) that exhibits strong and specific xylan binding; and a "product release area" with considerably weaker and less-specific binding. In this model, the −1 subsite and the *O*-3 of xylose form a hydrogen bond with a highly conserved K and H, whereas *O*-2 interacts with an N and an E residue (the latter acts as the catalytic nucleophile). At the −2 subsite, xylose interacts with its *O*-2 by hydrogen bonding to an E and W residue, *O*-3 binding to an N, and the endocyclic oxygen atom to a K (Lo Leggio et al., 2001). All these amino acids are strongly conserved in GH10 xylanases. The aglycone region of the substrate-binding cleft of the enzyme appears to interact at subsite +1 with the xylan backbone via a highly conserved aromatic residue that stacks against the xylose.

GH10 xylanases are also able to attack decorated forms of xylan (Biely et al., 1997). However, in contrast to the GH11 xylanases, they tolerate only a low number of unsubstituted consecutive β-1,4-xylopyranosyl units in the main chain and show higher affinity for shorter linear β-1,4-xylooligosaccharides (see later). This ability to bind decorated substrates is conferred partly by the exposure of *O*-2 and *O*-3 groups of xylose at subsites +4, +3, +1, and −3, which may allow accommodation of the α-1,2-linked 4-*O*-methyl-D-glucuronic acid (MeGlcA) side chain in glucuronoxylan at these locations. Furthermore, the uronic acid makes hydrogen bonds and hydrophobic interactions with the enzyme at the +1 subsite, indicating that the sugar decorations in glucuronoxylan are targeted to this proximal aglycone-binding site. Finally, accommodation of 3′-linked L-arabinofuranoside decorations is observed in the −2 subsite and could, most likely, be tolerated when bound to xylosides in −3 and +4 (Pell et al., 2004; Fujimoto et al., 2004).

4.2.2 GH11 Xylanase

Although GH10 xylanases are abundant in fungi, the glycosyl hydrolase family GH11 xylanases comprise most of the xylanases that have been characterized from fungi and that were typically described as "small alkaline xylanases" because of their low M_r (20 kDa) and alkaline isoelectric point (9.0). The reason for this is that they make up for the major portion of the total secreted xylanase enzyme activity. The first fungal GH11 enzyme that has been characterized is XYN2 from *T. reesei* (Törrönen et al., 1992, 1994; Törrönen and Rouvinen, 1995): it exists as a jelly roll β-fold in which the main chain folds to form two mostly antiparallel β-sheets that are packed against each other in parallel. The β-sheet structure is twisted to form a large cleft on one side of the molecule that accommodates the active center (Figure 4.4). The structure has been confirmed by the analysis of several more GH11 xylanases from *T. longibrachiatum, T. harzianum, H. insolens, A. nidulans,* and several others (for details, see http://www.cazy.org/GH11_structure.html).

The catalytic nucleophile in the GH11 xylanase from *Cellulomonas fimi* was identified to be E78 and the general acid/base residue as E127 (MacLeod et al., 1994). They correspond to the two glutamic acid residues (E86 and E177) in the middle of the cleft of *T. reesei* XYN2, and the results with *C. fimi* can therefore likely be applied to the fungal GH11 xylanases as well.

GH11 xylanases are retaining enzymes and follow a classical Koshland double-displacement mechanism. From the 3D structure, putative binding sites for at least four xylose residues have been suggested (Törrönen et al., 1994). These subsites were investigated in detail by

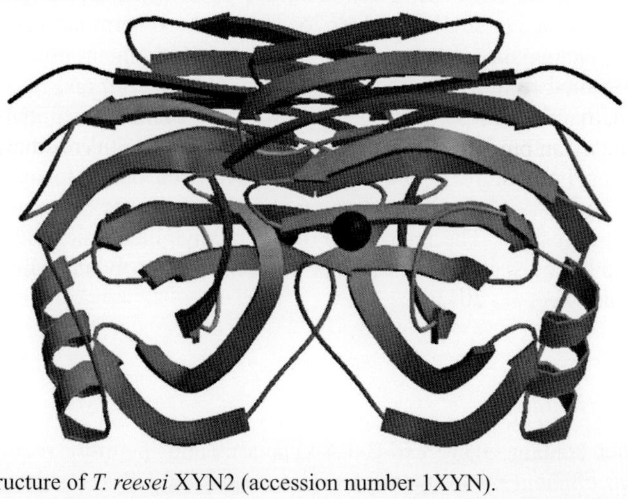

Figure 4.4. Structure of *T. reesei* XYN2 (accession number 1XYN).

site-directed mutagenesis of the *Bacillus coagulans* enzyme (Pollet et al., 2010). They detected several aromatic residues in the aglycone subsites that strongly interact with the substrate. Mutation analysis proved them to be indispensable for enzyme activity and specificity of the xylanase. Two residues (Y65 and W129), detected in subsite +2, are able to bind decorated substrates. Subsite +3 contributes significantly to substrate specificity. To study the binding of decorated substrates in more detail, Vardakou et al. (2008) crystallized a *Neocallimastix patriciarum* GH11 xylanase in complex with bound ferulic acid (FA)-1,5-arabinofuranose-α-1,3-xylotriose. The crystal structure revealed solvent exposure of the *O*-2 and *O*-3 groups of the backbone xylose at subsites −3 and +2 to allow accommodation of the α-1,2-linked 4-methyl-D-glucuronic acid and L-arabinofuranose side chains. In addition, they found that the ferulated arabinofuranose side chain make hydrogen bonds and hydrophobic interactions at the +2 subsite, indicating that the decoration may represent a specificity determinant at this aglycone subsite (Vardakou et al., 2008).

4.2.3 GH30 Glucuronoxylan Xylanohydrolases

The GH family 30 typically contains enzymes with β-glucosylceramidase, β-1,6-glucanase, and β-xylosidase activities, which are found in both pro- and eukaryotes. More recently, this family was revised—based on phylogenetics, primary amino acid sequence, and tertiary structure analysis—and now also accommodates xylanases, particularly glucuronoxylan xylanohydrolases that were previously classified as members of GH5 (St John et al., 2010). Genes encoding GH30 proteins are not abundant in fungi, with the exception of eight members in *Laccaria bicolor* and two in *T. reesei* (Table 4.2). The substrate specificity of one of the two members of *T. reesei*—XYN4—has been studied in detail (Saloheimo et al., 2003): it displays greater activity toward unsubstituted xylans or acetylated methylglucuronic acid-xylans than the GH10 and GH11 xylanases, and, most importantly, produces xylose as its main product and may thus be termed xylo-β-1,4-xylanase.

The kinetics and structure of GH30 xylanases from fungi have not yet been assessed. However, information from other proteins shows that the GH30 enzymes are retaining enzymes and believed to act by a Koshland double-displacement mechanism (Rempel and St John, 2011). They belong to the GH clan A and therefore possess the typical $(\alpha/\beta)_8$-TIM barrel fold with the two key active site E residues located at the C-terminal ends of β-strands 4 (acid/base) and 7 (nucleophile) (Kacher et al., 2008). A GH30 glucuronoxylan hydrolase was crystallized and studied from the bacterium *Erwinia chrysanthemi* and its substrate-binding pattern studied (Urbániková et al., 2011): recognition of the substrate involves binding of xylotriose to subsites −1, −2, and −3. The methylglucuronic acid residue, attached to the middle xylopyranosyl residue of xylotriose, forms hydrogen bonds to five amino acids of the active center and also performs ionic interaction between its carboxylate anion and a positively charged R residue. The interaction of the methylglucuronic acid residue with the enzyme appears to be indispensable for proper distortion of the xylan chain and its effective hydrolysis (Urbániková et al., 2011).

4.2.4 GH3 β-Xylosidases

Most fungi do not contain GH30 exo-β-1,4-xylanases and therefore require another exo-acting enzyme for efficient xylan hydrolysis. Not surprisingly, β-xylosidases have therefore been identified in numerous fungi, and they are distributed among three glycosyl hydrolase families: GH3, GH43, and GH54.

As already discussed in Chapter 3, GH3 is a large group that contains various β-glycosidases and also includes β-xylosidase. In many cases, these enzymes display a broad substrate specificity for monosaccharide residues, linkage position, and chain length of the substrate. A remarkable example for this is the well-characterized "bifunctional" β-D-xylopyranosidase/α-L-arabinofuranosidase of *T. reesei* (Herrmann et al., 1997). Its reaction mechanism has been investigated in detail in an ortholog from barley (Lee et al., 2003): it occurs via removing single glycosyl residues from the nonreducing ends of their substrates by a classical Koshland double-displacement mechanism. The β-anomeric configuration of the released xylose is retained. The *T. reesei* enzyme was suggested to actually be a β-D-xylan xylohydrolase, because it released xylose as an end product from debranched beech wood (Lenzing) xylan and 4-*O*-methylglucuronoxylan. It also hydrolyzed β-1,4-xylo-oligosaccharides (degree of polymerization 2–7), the apparent V_{max} increasing with increasing chain length, and exhibits α-L-arabinofuranosidase activity. This is in contrast to the GH3 β-xylosidase from *A. japonicus* that does not attack xylan (Semenova et al., 2009). The *T. reesei* enzyme hydrolyzes β-1,4-linkages preferentially to β-1,3-linkages, and β-1,2-linked xylo-oligosaccharides are not hydrolyzed at all. It can liberate terminal β-1,4-xylopyranose residues linked to a 2-*O* but not 3-*O* substituted xylopyranose residue. Rojas et al. (2005) studied the shape of this enzyme in solution by small-angle X-ray scattering: in contrast to the fold of other members of GH3 for which similar data were available, the shape of the β-xylosidase indicates the presence of a higher number (three) of distinct structural modules. The C-terminal part of the protein forms a third domain that adopts into a β-strand.

4.2.5 GH43 β-Xylosidases

GH43 glycoside hydrolases have been described to display α-L-arabinofuranosidase, endo-α-L-arabinanase, and β-D-xylosidase activity (Gilbert, 2011c). This glycosyl hydrolase family is also in fungi, and particularly strongly expanded in plant pathogenic taxa. These enzymes also display both α-L-arabinofuranosidase and β-D-xylosidase activity. Gilbert (2011c) speculated that the natural activity of these enzymes is conferred by the aglycone component of the substrate. In support of this, the *B. subtilis* arabinoxylan α-L-arabinofuranohydrolase specifically removes arabinofuranose side chains that are linked either α-1,2 or α-1,3 to backbone xylose residues, while the arabinoxylan arabinofuranohydrolase-D3 from *Bifidobacterium adolescentis* removes an α-1,3-linked arabinofuranose from xylans, where the xylose residue is substituted at both α-1,2 and α-1,3 with arabinose (van den Broek et al., 2005; Bourgois et al., 2007). GH43 β-xylosidases were only characterized from two fungi (*Cochliobolus carbonum*, Wegener et al., 1999; *P. herquei*, Ito et al., 2003). Interestingly, the enzyme is secreted into the medium but has no recognizable signal peptide and also does not seem to undergo any N-terminal processing events (Wegener et al., 1999).

The GH43 hydrolases use an inverting mechanism. The protein structure, as determined for a protein from *Cellvibrio japonicus*, is a "nonvelcroed" five-bladed-β-propeller that is based upon a fivefold repeat of blades, each composed of four-stranded β-sheets (Nurizzo et al., 2002). The substrate-binding surface is located in a long surface groove, with the active site located in a deep pocket (Figure 4.5). The orientation of the substrate appears to be very different between the enzymes, and it is likely that this contributes to their broad substrate specificity (Brüx et al., 2006; Vandermarliere et al., 2009).

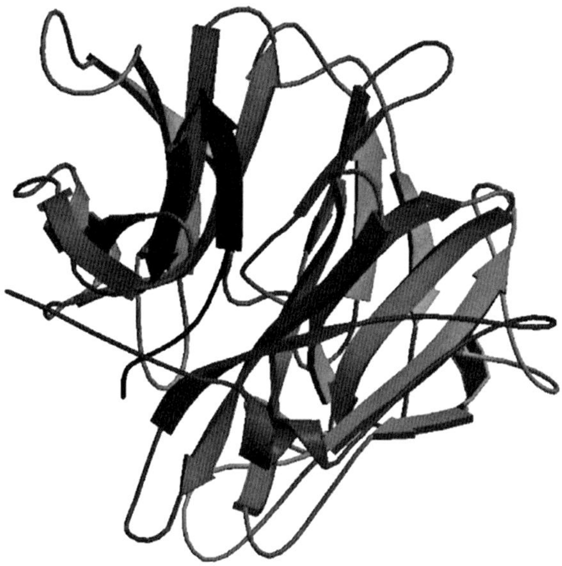

Figure 4.5. Structure of *C. japonicus* GH43 β-xylosidase (accession number 1GYD).

4.2.6 GH54 β-Xylosidases

The third glycoside hydrolase family that contains β-xylosidase enzymes is GH54, and these proteins also share β-xylosidase and α-L-arabinofuranosidase activity. Most members of GH54 are found in fungi and *Bacillus* spp. and have been widely studied in *Aspergillus* spp. (de Vries and Visser, 2001; de Vries, 2003 *Penicillium* spp., *Trichoderma* spp. (Seiboth and Metz, 2011), and *Fusarium oxysporum* (Chacon Martinez et al., 2004) with main emphasis on their α-L-arabinofuranosidase activity. Their properties will therefore be described in more detail in the paragraph about arabinofuranoside degradation.

GH54 enzymes use a retaining mechanism for catalysis. The 3D structure for the *A. kawachii* enzyme in complex with arabinose was solved and shown to exhibit a β-jelly roll fold that is slightly similar to other members of GH clan GH-B (Miyanaga et al., 2004). Interestingly, most of the members in GH54 have a carbohydrate-binding module of the CBM42 family at the C-terminus (see Chapter 3).

4.3 Degradation of the Galactomannan Backbone

The mannan-degrading enzymes comprise β-mannanase (1,4-β-D-mannan mannohydrolase, EC 3.2.1.78) and β-mannosidase (1,4-β-D-mannopyranoside hydrolase, EC 3.2.1.25) (Figure 4.6).

β-Mannanase, an endo-type enzyme, is responsible for the cleavage of β-1,4-linked internal linkages of the mannan backbone and produces new chain ends randomly. The degradation of galactomannan and galactoglucomannan by β-mannanase is greatly affected by the extent and pattern of substitution of the mannan backbone (Moreira and Filho, 2008). β-mannanases are found in the glycosyl hydrolase families GH 5, subfamilies 7 and 8 (Larsson et al., 2006), and

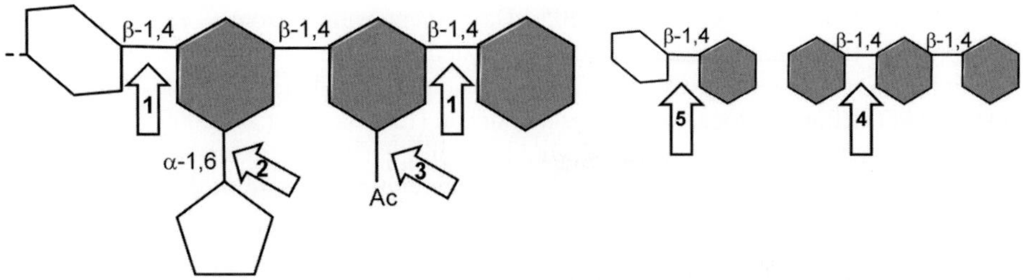

Figure 4.6. Enzymes necessary for the degradation of galactomannan. Arrows indicate the position of cleavage, and the respective enzymes are given by numbers: 1, β-mannanase; 2, α-galactosidase; 3, acetyl-mannan esterase; 4, β-mannosidase; 5, β-glucosidase.

GH 26. Members of both enzyme families display a double-displacement mechanism, with retention of anomeric configuration. GH5 comprises mannan-degrading enzymes from pro- and eukaryotic organisms, such as *B. circulans*, *Clostridium cellulolytic*, *T. reesei*, *A. aculeatus*, and *Agaricus bisporus*, and displays a typical $(\beta/\alpha)_8$-TIM barrel fold structure (Sabini et al., 2000). Interestingly, the β-mannanases from GH5 frequently contain a CBM1 cellulose-binding module: an example is the β-mannanase MAN5A from *T. reesei* that is composed of an N-terminal catalytic module and a C-terminal class 1 carbohydrate-binding module (CBM1), which binds to cellulose but not to mannan, and is necessary for the ability to hydrolyze mannan/cellulose complexes (Hägglund et al., 2003). Similar data were also obtained for a β-mannanase from *Phanerochaete chrysosporium* (Benech et al., 2008). In the anaerobic fungi, ManA from *Orpinomyces* sp. strain PC-2 is a polypeptide of 579 amino acid residues that, besides the catalytic domain, contains a CBM1 domain and a fungal-type dockerin (Ximenes et al., 2005). Expectedly, this CBM1 also bound strongly to acid-swollen cellulose, Avicel PH-105, bacterial crystalline cellulose, filter paper, dewaxed cotton, chitin, and insoluble oat spelt xylan but failed to bind to soluble oat spelt xylan, carboxymethylcellulose, arabinanan, pectic galactan, laminarin, or lichenan.

Although glycosyl family GH26 also contains enzymes with β-1,3-xylanase activity, enzymes belonging to this family that were characterized from fungi (e.g., from *Aspergillus*, *Humicola, N. crassa, P. chrysogenum,* and *Orpinomyces* sp.) were exclusively β-mannanases. None of their 3D structures has been assessed yet. However, on the basis of the information available for bacterial GH26 members, they belong to clan GH-A and have the canonical $(\alpha/\beta)_8$-TIM barrel fold with the two key active site E residues located at the C-terminal ends of β-strands 4 (acid/base) and 7 (nucleophile) (Gilbert, 2011c).

Hogg et al. (2003), on the basis of a detailed analysis and comparison of the *Cellvibrio japonicas* GH5 and GH26 β-mannanases, supposed that the primary function of the GH5 β-mannases would be to hydrolyze crystalline mannan and glucomannan, an integral component of the cell wall of angiosperms. In contrast, the primary substrates for the GH26 mannanases would be the soluble mannans or mannans that are loosely associated with the plant cell wall, such as galactomannans and mannooligosaccharides, and thus can rapidly generate large quantities of mannose as an energy and carbon source for the degrading organism.

β-1,4-mannosidase cleaves the β-1,4-linked mannosides and releases mannose from their nonreducing end (Moreira and Filho, 2008). Almost all known β-mannosidases belong to the glycosyl hydrolase families GH1 or GH2. Both families include enzymes that operate *via* a double-displacement mechanism, resulting in retention of the configuration at the anomeric carbon of the sugar ring. They also are often able to catalyze transglycosylation reaction

(Gilbert et al., 2008). An extracellular 105-kDa β-mannosidase from *T. reesei* was shown to hydrolyze β-1,4-mannooligosaccharides, exhibited decreased rates with increasing chain length (which would have been expected for a glycosidase), and liberated mannose from soluble and insoluble fractions of locust bean galactomannan (Eneyskaya et al., 2009). The presence of a galactose-substituted mannose residue adjacent to the terminal mannose residue reduces the activity of a β-mannosidase from *A. niger*, depending on the size of the oligosaccharide (Ademark et al., 1999).

The structure of the *T. reesei* β-mannosidase has been examined by X-ray small-angle scattering and exhibited excellent similarity with a β-galactosidases from the same glycosyl hydrolase family (GH2; Aparicio et al., 2002).

4.4 Degradation of Pectin

As I have outlined in Chapter 1, the main chain of pectin is composed of hairy and smooth regions. The structural differences between these two regions have significant implications for the enzymes necessary to degrade these regions: the backbone of the smooth region can easily be hydrolyzed by enzymes that can cleave the linkage between α-1,4-galacturonate residues, whereas the hydrolysis of the hairy regions needs additional accessory enzymes.

Since the building block of pectin is a sugar acid, the glycosidic bond can not only be hydrolyzed (using polygalacturonases, EC 3.2.1.15 and 3.2.1.67), but can also be cleaved in a nonhydrolytic manner by β-elimination (Figure 4.7) using pectin lyases (EC 4.2.2.10) and pectate lyases (EC 4.2.2.2).

4.4.1 Hydrolytic Pectin Degradation

Polygalacturonidases comprise endo- and exo-acting enzymes, both of which are accommodated in glycosyl hydrolase family GH28. They are present in all fungi but to varying numbers

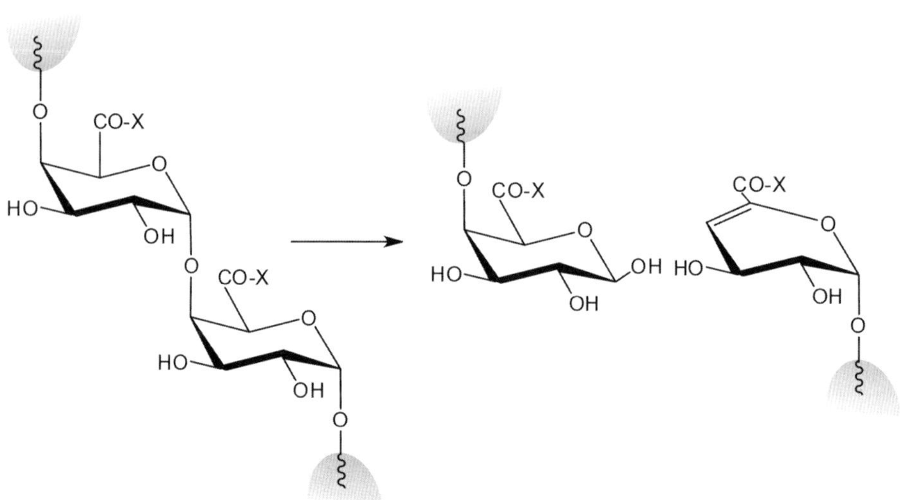

Figure 4.7. Reaction of pectin/pectate cleavage by pectin/pectate lyase.

Table 4.3. Pectinases (GH28) in fungi.

Enzyme	PECT	RHAM
GH family	28	78
A. nidulans	9	8
A. niger	21	8
P. chrysogenum	5	5
Tuber melanosporum	2	2
F. graminearum	6	7
Nectria haematococca	11	11
Neurospora crassa	2	0
Magnaporthe grisea	3	1
Podospora anserina	0	1
T. atroviride	**6**	**3**
T. reesei	**6**	**3**
T. virens	**4**	**1**
Mycospherella graminearum	2	0
Stagonospora nodorum	4	1
Ustilago maydis	1	NI
Puccinia graminis var. Tritici	0	NI
Schizophyllum commune	3	NI
Laccaria bicolor	6	0
Postia placenta	11	4
Phanerochaete chrysosporium	4	1
Serpula lacrymans	7	NI
Coprinopsis cinerea	3	NI

Data taken from the same source as in Table 3.4.
PECT, pectinase; RHAM, α-L-rhamnosidase; NI, no information retrievable.

(Table 4.3). They are best studied in *A. niger* and *Rhizopus oryzae* (because enzymes from these two fungi are used in the food and feed industry) and in some plant pathogenic fungi because of their role in plant pathogenesis (de Vries and Visser, 2001; Lara-Márquez et al., 2011). The genome of *A. niger* contains seven endopolygalacturonases, each of them exhibiting distinct kinetic properties, sensitivity to the presence of methylation groups in the pectin and mode of action. For instance, although the structures of polygalacturonases PgaI and PgaII are highly similar, only PgaI performs a processive mode of action because it contains a narrower substrate-binding cleft and contains an R-residue at position 96, which has a crucial role in the processive behavior because it is flexible and binds to the oxygen atoms in several well-defined galacturonic acid conformations. Thus, R96 may bind the polygalacturonic acid substrate and prevent its release, while at the same time being flexible enough to push the oligosaccharide chain toward the active site. Using a bacterial exo-polygalacturonase (*Yersinia enterocolitica*, YeGH28), Abbott and Boraston (2007) showed that the exo (processing) activity is due to inserted stretches of amino acid residues that transform the active site from the open-ended channel observed in the *endo*-polygalacturonases to a closed pocket that restricts the enzyme to the exclusive attack of the nonreducing end of oligogalacturonide substrates.

A. niger also contains four exopolygalacturonases (PgaX, PgxA, PgxB, and PgxC) that differ in their relative activities on homo- and xylogalacturonan. Thus, PgxB acts only on homogalacturonan, PgxC acts on both, and PgxA predominantly acts on xylogalacturonan. The latter may in fact represent a new exoxylogalacturanase (van den Brink and de Vries, 2003). A similar situation occurs in *R. oryzae*, because 15 of the 18 putative GH28 polygalacturonases showed different substrate specificity for polygalacturonan, trigalacturonan, and digalacturonan, and their activity against polygalacturonan ranged from less than 1 to 200 μmol/min/mg protein (Mertens and Bowman, 2011).

All glycosyl hydrolase family GH28 enzymes act by an inverting mechanism, using three D residues as the catalytic residues. Using polygalacturonase II from *A. niger*, van Santen et al. (1999) identified D201 as the proton donor and D180/D202 as the activators of the nucleophilic water molecule.

Rhamnogalacturonan hydrolases are also members of GH28 and cleave the α-1,2-glycosidic bonds formed between D-galacturonic acid and L-rhamnose residues in the hairy regions both by an exo- or endo-mechanism (van den Brink and de Vries, 2003). The structure of rhamnogalacturonase (RGase A) from *A. aculeatus* (Petersen et al., 1997) is a parallel β-helix composed of four parallel β-sheets extending along their longitudinal axes and displaying ten turns. A comparison of the structure of a bacterial polygalacturonase (from *E. carotofora*) with *Aspergillus* rhamnogalacturonases showed that the latter contain one additional turn at the C-terminus and an N-terminal extension (Pickersgill et al., 1998). The α-helix in the N-terminal extension and the additional polypeptide chain at the C-terminal end of the parallel β-helix form an antiparallel β-strand at the C-terminal end of parallel β-sheet 1 followed by an extension that partially covers parallel β-sheet 3. The loop structures are also different.

4.4.2 Pectin Degradation by β-Elimination

Polysaccharide lyases (PLs; EC 4.2.2.-) are a group of enzymes that cleave uronic acid-containing polysaccharide chains via a β-elimination mechanism to generate an unsaturated hexenuronic acid residue and a new reducing end (Figure 4.7). The CAZy database has classified these enzymes in families as well as subfamilies based on amino acid sequence similarities, to reflect their structural features and diversity (Lombard et al., 2010). Enzymes belonging to the same subfamily are believed to display the same substrate specificity. The PL database in CAZy lists 21 families, of which PL families 1, 3, and 9 contain fungal pectin/pectate lyases and family 11 contains a fungal rhamnogalacturonan lyase.

The homogalacturan backbone of pectin varies in its degree of methylation from the highly methylated and relatively hydrophobic pectin, to the fully demethylated and highly charged form called pectate. Consequently, these structures are named pectin and pectate. The methylated and demethylated regions of pectin are cleaved by pectin lyase and calcium-dependent pectate lyases, respectively. Both enzymes cleave via a β-elimination mechanism the α-1,4-linked D-galacturonic acid residues within the smooth regions of pectin (Lombard et al., 2010). A comparison between the structures of pectin and pectate lyases has indicated that both lyases most likely descended from a common ancestor enzyme (Mayans et al., 1997; Vitali et al., 1998). However, these two types of lyases display important differences in their active site, which are responsible for substrate binding and optimum pH (pectin lyases attack preferentially heavily methyl-esterified substrates and have their optimum pH around 5.5, whereas pectate lyases favor lower degrees of esterification, have their optimum pH around 8.5, and require Ca^{2+} for their activity; Mayans et al., 1997). Currently, all characterized pectin lyases belong to the PL1 family, while the fungal pectate lyases are found in PL1, PL3, and PL9 (Table 4.4). In the *Aspergilli*, where pectin and pectate lyases have been most comprehensively investigated, six pectin lyases and only a single pectate lyase have been identified in *A. niger* and partially characterized, whereas *A. nidulans* has only two pectin lyases, but eleven pectate lyases (of which yet only five have been investigated), suggesting significant differences in

Table 4.4. Fungal pectate lyases.

PL Family	Activity	EC Number	Structure	Subfamily	Found in
PL1	Pectate lyase	EC 4.2.2.2	Parallel β-helix	4,7	*A. niger, A. oryzae, Colletotrichum gloeosporioides*
	Pectin lyase	EC 4.2.2.10			
PL3	Pectate lyase	EC 4.2.2.2	Parallel β-helix	2	*A. nidulans, A. oryzae, Leptosphaeria maculans, Magnaporthe grisea, N. crassa*
PL4	Rhamnogalacturonan lyase	EC 4.2.2.-	β-sandwich + β-sheet	1,2,3,4	*Aspergillus* spp., *L. maculans, Podospora anserina*
PL9	Pectate lyase	EC 4.2.2.2	Parallel β-helix	1	*Aspergillus* spp.
	Exopolygalacturonate lyase	EC 4.2.2.9			
PL11	Rhamnogalacturonate lyase	EC 4.2.2.-	β-propeller	2	*Aspergillus* spp.
	Exo-unsaturated rhamnogalacturonate lyase	EC 4.2.2.-			

Data extracted from the CAZy database (www.cazy.org).

the inventory of these enzymes even between species of the same fungal genus (Coutinho et al., 2009).

The 3D structure of *A. niger* pectin lyase B (PLB) has been determined to be a large right-handed cylinder, termed a parallel β-helix (Figure 4.8; Vitali et al., 1998). This is a right-handed parallel β-helix architecture first seen in this group of enzymes. It consist of seven complete coils, each consisting of three β-strands and three turn region (named PB1, T1, PB2, T2, PB3, and T3 in their order of occurrence; Jenkins et al., 1998). The region PB2-T2-PB3 is the best conserved region in the lyases, showing a strongly conserved pleating and hydrogen bonding in the sheets conserved and also a conserved α-L-helix conformation.

Loops of various sizes and conformations protrude from the central helix of PLB, the largest one (53 amino acids) folding into a small domain consisting of three antiparallel β-strands, one turn of an α-helix, and one turn of a 3(10) helix. The active center has been deduced by comparison with the structure of *E. chrysanthemi* pectate lyase PelC (Yoder et al., 1993): the substrate-binding site in PLB is considerably less hydrophilic than in *E. chrysanthemi* and consists of an extensive network of highly conserved W and H residues. The site in PLB analogous to the Ca^{2+} site in PelC is filled with a positive charged amino acid (R) that is conserved in the pectin lyases and thus explains their lack of requirement for Ca^{2+}.

The substrate specificity of the *A. niger* pectate lyase A has been investigated (Benen et al., 2000): the bond cleavage frequencies for tetra-, penta-, and hexagalacturonides changed in a $CaCl_2$-dependent way, and the response toward Ca^{2+} exhibited a sigmoidal relationship.

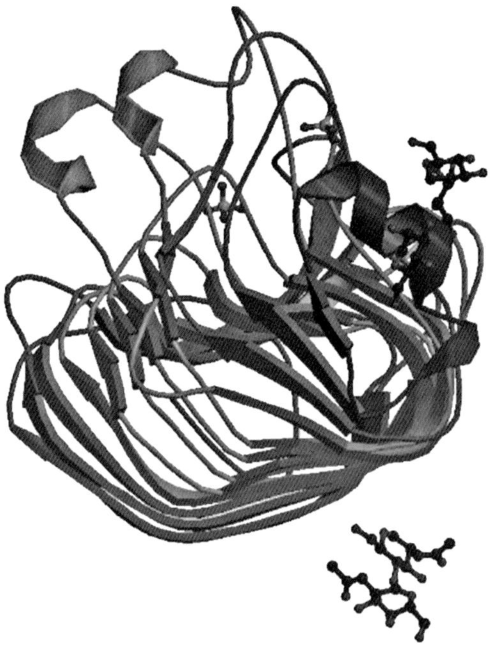

Figure 4.8. Structure of polygalacturonate hydrolase from *A. niger* (accession number 1NHC).

Rhamnogalacturonan lyases differ substantially in their structure from the pectin and pectate lyases and cleave within the hairy regions of pectin. This group of lyases belongs to two carbohydrate lyase families, PL4 and PL11, where PL4 lyases have a much lower optimum pH than PL11 lyases (Jensen et al., 2010). The PL4 rhamnogalacturonan lyase from *A. aculeatus* showed that cleavage preferably occurs at the fourth residue before the reducing end and is severely affected by the presence of acetyl groups in the backbone of rhamnogalacturonan (Mutter et al., 1998; de Vries et al., 2000), implying a cooperative action with rhamnogalacturonan acetylesterases (see later) for efficient degradation of rhamnogalacturonan (van den Brink and de Vries, 2003).

Only a single rhamnogalacturonan lyase has been identified in the Aspergilli. This enzyme has a higher molecular mass than the pectin and pectate lyases, and while it was also positively influenced by Ca^{2+}, it was able to perform catalysis also in the absence of Ca^{2+}. The activity of the enzyme was positively affected by the presence of galactose side chains and negatively affected by the presence of arabinose side chains and acetyl residues (Mutter et al., 1996).

4.5 Accessory Glycoside Hydrolases for Hemicelluloses Degradation

In contrast to the enzymes described in Sections 4.1–4.4, which act on the main chain of plant cell wall polysaccharides, accessory enzymes act on the side chains of these structures. Some of these enzymes act directly on the linkages between the backbone-chain residue and the substituent, whereas other enzymes cleave bonds that are internal or terminal linkages of side chains only.

4.5.1 Enzymes that Act on Arabinose-Containing Substituents

Endoarabinases (EC 3.2.1.99) belong to glycosyl hydrolase family GH93 and hydrolyze the α-1,5-linkages of arabinan polysaccharides that are present as side chains of pectin. They are both exo- and endo-acting enzymes that release arabinobiose from the nonreducing end of the α-1,5-L-arabinan chain and putatively are retaining enzymes based on studies on an endoarabinase from *P. chrysogenum* (Sakamoto et al., 2004). Endoarabinases appear to be absent from most basidiomycetes, but up to three GH93 members are present in some Sordariomycetes such as *Podospora anserinea* and *T. atroviride*.

Their mode of action was studied with the GH93 α-L-arabinohydrolases Abn2 from *Chrysosporium lucknowense* C1 (Kühnel et al., 2011): Abn2 binds two arabinose units at the subsites −1 and −2, and releases arabinobiose.

The arabinohydrolases from *F. graminearum* and *P. chrysogenum* have been crystallized and their structure been determined (Carapito et al., 2009; Sogabe et al., 2011): typically for GH-E proteins, they display a six-bladed β-propeller fold with a typical," in which the last four-stranded β-sheet is completed by the incorporation of a strand from the N-terminus. Two residues, E64 and Y66, contribute to the recognition of the nonreducing chain end of the polysaccharide and enclose the substrate-binding groove at one end. The *P. chrysogenum* enzyme is extraordinarily active, displaying k_{cat}/K_m values from 1.15×10^4 to $3.92 \times 10^4\ s^{-1}M^{-1}$ (in contrast to those of the *Fusarium* enzyme; 1.05×10^7 to $4.66 \times 10^7\ s^{-1}M^{-1}$), indicating that the former enzyme catalyzes a reaction almost every time it encounters a substrate molecule, whereas many substrate molecules seem to leave the active site of the *Fusarium* enzyme without catalytic reaction. In other words, the lower active enzyme has substrate-binding mode that is difficult to shift to the transition state (Sogabe et al., 2011). A comparison of the structure of the *Fusarium* and *Penicillium* enzymes revealed a quite similar overall structure but different amino acid residues that are involved in funneling substrates to the active site and/or to stabilize the transition state, which could be used to improve arabinohydrolases (Sogabe et al., 2011).

Single arabinose residues can be removed by α-L-arabinofuranosidases (EC 3.2.1.55) and arabinoxylan arabinofuranohydrolases (EC 3.2.1.99). These enzymes and their corresponding genes from several microorganisms have been studied and have been shown to differ strongly in substrate specificity. α-L-arabinofuranosidases can be found in two glycosyl hydrolase families: GH51 and GH62. The GH51 L-arabinofuranosidases belong to the GH-A clan and are particularly abundant in *Aspergillus* and *Penicillium* spp. The α-L-arabinofuranosidases from *H. insolens* and from the white-rot basidiomycete *Meripilus giganteus* have been cloned and characterized (Sørensen et al., 2006): they catalyze the removal of α-1,2- and α-1,3-L-arabinofuranosyl residues from singly substituted xylopyranosyl residues in arabinoxylan. Furthermore, they exhibit a synergistic interaction with GH43 α-L-arabinofuranosidases, probably because the GH51 enzymes can remove the single α-1,2-L-arabinofuranosyl residues that remain after the action of the GH43 enzymes. The GH51 α-L-arabinofuranosidases are retaining enzymes and follow a classical Koshland retaining mechanism (Shoham, 2011). No 3D structures are available for fungal enzymes yet; however, the bacterial enzymes form a trimer, which is composed of three dimers that consist of a catalytic $(\beta/\alpha)_8$ domain and a 12-stranded β-sandwich jelly roll domain, the former serving as the catalytic domain (Paës et al., 2008; Figure 4.9). The pentasaccharide is bound in a groove on the surface of the enzyme, with the monoarabinosyl branch entering a tight pocket harboring the catalytic dyad. In the absence of substrate, the enzyme adopts an open conformation, but in the substrate-bound form, the long loop that connects β-strand 2 to α-helix 2 closes the active site and interacts with the substrate through the amino acid residues H98 and W99 (Paës et al., 2008).

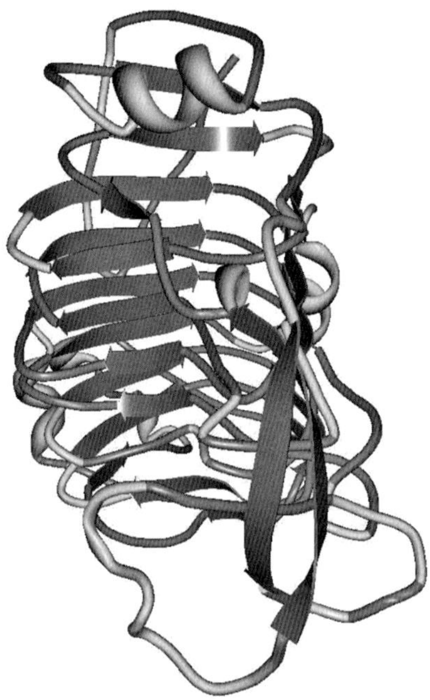

Figure 4.9. Structure of pectate lyase B from *A. niger*. Disulfide bonds are designated by thick, black lines (accession number 1QCX).

GH62 α-L-arabinofuranosidases contain enzymes of GH-clan F that specifically cleave either α-1,2- or α-1,3-L-arabinofuranose side chains in xylans, but do not act on the xylose moieties that are decorated by an arabinose side chain on *O*-2 and *O*-3. Also they lack nonspecific arabinofuranosidase activity and should thus correctly be termed arabinoxylan-arabinofuranhydrolases (de Vries and Visser, 2001). Several of these enzymes contain either a CBM1 (cellulose) or a CBM43 (xylan)-binding domain. The enzymes from *A. niger* and *A. oryzae* have been studied in some detail: the latter enzyme is active only on arabinoxylan, whereas the former also releases arabinose from L-arabinan and arabinogalactan. The GH62 enzymes are present in all ascomycetous genomes (with the exception of *N. crassa*), but absent from those of basidiomycetes. The reaction mechanism of these enzymes is unknown, as is their 3D structure. However, since they are members of GH hydrolase clan F, it can be anticipated that they consist of a fivefold β-propeller fold and perform a single displacement or inverting mechanism just like the GH43 β-xylosidases/α-L-arabinofuranosidases that belong to the same clade (*vide supra*).

4.5.2 Enzymes that Act on Galactose-Containing Substituents

β-galactan side chains in polysaccharides can be degraded by endo-β-1,4-galactanases (endo-β-1,4-galactanase; EC 3.2.1.89), endo-β-1,6-galactanases (EC 3.2.1.164), and exo-β-1,3-galactanases (EC 3.2.1.145).

Endo-β-1,4-galactanases belong to glycosyl hydrolase family GH53 and are members of clan GH-A. They are mainly involved in the breakdown of the side chains in pectin, act by a retaining mechanism, and most characterized members have been reported to be endo-acting, although processivity has been suggested in one case (Lo Leggio, 2011). Ryttersgaard et al. (2004) first reported the structure of an endo-β-1,4-galactanase from *A. aculateus*: typically for members of GH-A, they assemble to a $(\beta/\alpha)_8$-TIM barrel structure with fairly compact loop structure and a sequence under 400 residues in length (Lo Leggio, 2011). The catalytic residues are positioned at the C-terminal ends of β-strands 4 and 7 in the barrel. In the enzyme from *A. aculateus, Myceliophthora thermophila,* and *Humicola insolens,* the β/α loops 7 and 8 are linked by a disulfide bond (Ryttersgaard et al., 2004; Le Nours et al., 2003). The proteins further contain a conserved β-turn, which contributes to subsites −2 to +3. This β-turn does not occur in any other clan GH-A members and thus might be a structural signature for endo-β-1,4-galactanase specificity (Le Nours et al., 2009). Torpenholt et al. (2011) reported that loop 8 in the two fungal β-1,4-galactanases from *A. aculeatus* and *Meripilus giganteus* contains substrate-binding subsites −3 and −4, which—in contrast to the bacterial endo-β-1,4-galactanases—enable the fungal enzymes to bind 4-nitrophenyl-1,4-β-D-thiogalactobioside.

The first endo-β-1,6-galactanase was isolated from a commercial cellulase preparation (Onozuka) of "*T. viride*" (see Chapter 9) and putatively been attributed to GH5. It specifically acted on β-1,6-galactan side chains with a degree of polymerization higher than 3 and their acidic derivatives with 4-*O*-methyl-glucosyluronic or glucosyluronic groups at the nonreducing terminals (Okemoto et al., 2003). Takata et al. (2010) also partially characterized a GH5 endo-β-1,6-galactanase from *N. crassa* that specifically acted on β-(1→6)-galactosyl residues. Kotake et al. (2011) recently characterized an endo-β-1,3-galactanase from the winter mushroom *Flammulina velutipes*. The enzyme specifically hydrolyzed β-1,3-galactan, but did not act on β-1,3-glucan, β-1,3:1,4-glucan, xyloglucan, and agarose. It released various β-1,3-galactooligosaccharides together with D-galactose from β-1,3-galactohexaose in the early phase of the reaction, demonstrating that it acts on β-1,3-galactan in an endo-fashion. Interestingly, a phylogenetic analysis revealed that the *F. velutipes* enzyme is a member of a novel subgroup in glycosyl hydrolase family GH 16 rather than GH 5 and distinct from known glycoside hydrolases such as ENG2 and endo-β-1,3:1,4-glucanase from the GH16 family. This subgroup also contains putative uncharacterized proteins from other fungi such as *Aspergillus* pp., *N. crassa,* and *M. grisea,* and is thus probably present in several fungi (Kotake et al., 2011).

Exo-β-1,3-galactanases (galactan-β-1,3-galactosidase; EC 3.2.1.145) belong to glycosyl hydrolase family GH43 and hydrolyze β-1,3-galactan main chains, thereby bypassing the β-1,6-galactan side chains. The major activities reported for the GH43 family are α-L-arabinofuranosidases, endo-α-L-arabinanases and β-D-xylosidases (*vide supra*), and—although crystallization of the exo-β-1,3-galactanase from *P. chrysosporium* has been reported (Ishida et al., 2009)—no 3D structure has as yet been reported from fungi or bacteria. Yet, as members of clan GH-F, GH43 hydrolases typically display a fivefold β-propeller, and it would thus be logical to assume a similar basic structure for the exo-β-1,3-galactanases.

Fungal exo-β-1,3-galactanases have been isolated from *Irpex lacteus* and *Phanerochaete chrysosporium* (Ichinose et al., 2005; Kotake et al., 2011). The *I. lateus* enzyme specifically hydrolyzed β-1,3-galactooligosaccharides and showed high activity toward arabinogalactan-proteins as well as β-1,3-galactan. Product analysis revealed that the enzyme released β-1,3-galactobiose, β-1,6-galactotriose, and α-L-arabinofuranosyl-1,3-β-galactosyl-1,6-β-galactose together with galactose from β-1,3-galactans attached with and without β-1,6-galactosyl branches. This indicates that the enzyme acts on the β-1,3-galactan main chains and bypasses β-1,6-galactosyl substituents (Kotake et al., 2011). Likewise, the enzyme from *P. chrysosporium* showed reactivity only toward β-1,3-linked galactosyl-containing oligosaccharides and polysaccharides as substrates but did not hydrolyze β-1,4-linked galactooligosaccharides, β-1,6-linked galactooligosaccharides. The enzyme also did not hydrolyze β-1,3-galactosyl galactosaminide, β-1,3-galactosyl glucosaminide, or β-1,3-galactosyl arabinofuranoside, suggesting that it specifically cleaves the internal β-1,3-linkage of two galactosyl residues (Ichinose et al., 2005). The enzyme also produced galactose from β-1,3-galactan in an exo-acting manner, but did not exhibit activity toward *p*-nitrophenyl β-galactopyranoside.

Interestingly, this protein also showed the presence of a C-terminal CBM6 (see Chapter 3) domain, yet it did not bind to the known CBM6 substrates such as xylan, cellulose, and β-1,3-glucan, but only to β-1,3-galactan and was shown to interact with oligosaccharides containing a β-1,3-galactobiose moiety independently of the degree of polymerization (Ichinose et al., 2005).

The removal of single D-galactose residues from plant cell wall polysaccharides requires the action of α-galactosidases (EC 3.2.1.22) and β-galactosidases (EC 3.2.1.23). α-galactosidases are involved in the degradation of galacto(gluco)mannan, removing galactose from the mannose residues of the backbone, whereas β-galactosidases release terminal galactose residues from the galactan side chains of pectins. However, since terminal β-linked galactose residues also occur in certain galactoglucomannans (Sims et al., 1997), these two enzymes may play a role in the degradation of both of these polysaccharides.

α-Galactosidases are distributed among two glycosyl hydrolase families, GH27 and GH36, which are members of clan GH-D. Enzymes of the former are found in asco- and basidiomycetes, whereas the latter seems to occur only in ascomycetes (Table 4.5).

Family GH27 enzymes hydrolyze α-galactooligosaccharides using Koshland double-displacement mechanism and retaining the anomeric configuration. The conserved amino acids that functions as the catalytic nucleophile/general acid base in GH27 have been identified in *P. chrysosporium* (Hart et al., 2000) and *T. reesei* α-galactosidase (Golubev et al.,

Table 4.5. Fungal characterized α-D-galactosidases.

	Species	Protein ID (Genbank)	References
GH27	*A. niger*	CAA44950	de Vries et al., 1999 den Herder et al., 1992 Ademark et al., 2001
	P. purpurogenum	BAA22992	Shibuya et al., 1998
	P. simplicissimum	CAA08915	Luonteri et al., 1998
	T. reesei	CAA93244	Margolles-Clark et al., 1996
	A. nidulans	ABF50881	Bauer et al., 2006
GH36	*Mycocladus corymbifera*	AAF68953	Baik et al., 2000
	Penicillium sp. F63	ABC70181	Mi et al., 2007
	A. niger	CAB63901	Ademark et al., 2001 de Vries et al., 1999
	A. nidulans	ABF50891, EAA58775, CBF73961	Bauer et al., 2006
	Gibberella sp.	ACQ72829	Cao et al., 2009a
	Rhizopus sp. F78	ACM48349	Cao et al., 2009b
	T. reesei	CAA93245	Margolles-Clark et al., 1996

Data compiled by querying MycoCLAP (Murphy et al., 2011a) for GH27 and GH36 enzymes.

2004) and are also conserved among other GH27 members. Structural analysis of the *T. reesei* enzyme indicated that it composed of an N-terminal $(\alpha/\beta)_8$-TIM barrel domain and a C-terminal anti-parallel jelly roll domain, of which the former contains the enzyme catalytic center composed by loop residues at the ends of β strands 1–7. Interestingly, the *T. reesei* enzyme contains a 40 amino acid insertion in loop β4-α4.

Glycosyl hydrolase family GH36 α-galactosidases are very similar to the enzymes from GH27, as predicted by their common membership in Clan GH-D, as far as their structure and kinetic mechanism are concerned (Comfort et al., 2007). This is also reflected by the findings that the substrate specificities of two GH27 and one GH36 α-galactosidase from *T. reesei* are similar (Margolles-Clark et al., 1996). However, the GH36 enzymes can be distinguished by an additional large β-supersandwich N-terminal domain, which contributes a key substrate-binding amino acid residue (W65) to the active site, and a C-terminal β-sheet domain (Comfort et al., 2007).

Fungal β-galactosidases can be found in glycosyl hydrolase families GH2 and GH35. Enzymes of the GH2 family are essentially intracellular, and only the GH35 family members are secreted enzymes. The GH2 β-galactosidases have been thoroughly characterized from the bacterium *Escherichia coli* (encoded by the well-known *lacZ* gene; Savageau, 1999) and the lactose-utilizing yeast *Kluyveromyces marxianus* (Rubio-Texeira, 2006), but not from any multicellular fungus. Yet, their genes (mostly single alleles) are present in the genomes of several fungi. Most fungi possess only a single member of GH2, but expansion to up to four members can be observed in the Aspergilli or *Sclerotinia sclerotiorum*.

The extracellular GH35 β-galactosidases catalyze the hydrolysis of terminal nonreducing β-D-galactose residues in, for example, lactose (1,4-*O*-β-D-galactopyranosyl-D-glucose), but also other oligosaccharides that contain a β-1,3-, β-1,6-, or β-1,4-galactosidic linkages. They act by a Koshland retaining mechanism. Further, Gamauf et al. (2007) showed that the *T. reesei* GH35 β-galactosidase BGA1 can also release D-galactose residues from galactan polymers, thus acts as an exo-β-D-galactanase.

GH35 enzymes belong to Clan GH-A and are therefore composed of an $(\alpha/\beta)_8$-TIM barrel that forms the catalytic domain, in which two E-residues act as the general acid–base and nucleophilic catalysts. These residues are located in strands 4 and 7 of the barrel. The conformations of the loop regions are significantly different from those of the bacterial enzymes, whereas the superimposition of the active sites of the GH35 β-galactosidases shows a remarkable similarity. On the basis of the crystallographic models of β-galactosidases in the presence of galactose from a *Penicillium* sp. and *T. reesei*, a single galactose molecule is bound to the active site of the GH35 enzyme in the chair conformation and β-anomeric configuration (Kulminskaya et al., 2011).

Maksimainen et al. (2011) have described conformational changes in two loop regions of the active site of *T. reesei* BGA1, which implicates a conformational selection mechanism for the enzyme, whereby the open and closed conformation are both favorable in the native structure and the closed conformation becomes more favorable in the complex structures. When binding the substrate, the acid/base catalyst E200 undergoes two different conformations, which affect its pK_a value and thus the catalytic mechanism of the enzyme.

4.5.3 α-Xylosidases

Most α-glycosidases are found in the glycosyl hydrolase families GH13 and GH31, and indeed, the latter also contains enzymes that can be identified as α-D-xylosidases. However, none of them has so far been characterized at the kinetic or structural level. An extracellular α-D-xylosidase has been purified from *A. niger*, and although its nucleotide or amino acid sequence are not known, its size would be in accordance with a putative GH31 enzyme: it is an oligomeric protein consisting of four subunits, each of which have a molecular weight of 123 kDa (Matsushita et al., 1985). The enzyme was highly specific for *p*-nitrophenyl α-D-xylopyranoside and isoprimeverose (6-*O*-α-D-xylopyranosyl-D-glucopyranose), and hydrolyzed the α-D-xylopyranosyl residue of the nonreducing terminal of the backbone of oligoxyloglucans.

Members of GH31 are retaining enzymes that follow the classical Koshland double-displacement mechanism. The crystal structure of a GH31 α-xylosidase from *E. coli* (YicI) has been elucidated (Lovering et al., 2005): the catalytic domain features a $(\beta/\alpha)_8$-TIM barrel and belongs to GH clan GH-D, like the α-galactosidases comprising glycosyl hydrolase families GH27 and GH36.

4.5.4 α-Fucosidases

Fucose-α-1,2-galactose linkages are attached as branches to xyloglucans. α-L-fucosidases that are capable of cleaving this bond belong to glycosyl hydrolase family GH95, and a comparison of the genome inventory of fungi shows that they are particularly rich in *Trichoderma* spp. and some *Aspergillus* spp. (Table 4.6). However, the fungal enzymes have not been studied in sufficient detail. In plants, studies with the 1,2-α-L-fucosidases from *Arabidopsis thaliana* and *Lilium longiflorum* show that they liberate L-fucose from Xylα1-6Glcβ1-4(Xylα1-6)Glcβ1-4(Fucα1-2Galβ1-2Xylα1-6)Glcβ1-4Glc as well as 2′-fucosyllactose, but not from 3-fucosyllactose (Katayama, 2011). Hydrolysis catalyzed by GH95 members proceeds via an inverting mechanism. The 3D structure has been solved for the 1,2-α-L-fucosidase from *Bifidobacterium bifidum* and shown to adopt an $(\alpha/\alpha)_6$-barrel fold typical for the clan GH-L enzymes (Nagae et al., 2007).

4.5.5 α-Glucuronidases and Glucuronan Lyases

Glucuronic acid residues and their 4-*O*-methyl ethers are side chains of the xylan backbone and can be removed by α-glucuronidases (EC 3.2.1.131) that belong to GH67 and GH115.

Table 4.6. Number of α-L-fucosidases in fungi.

Species	Number
A. nidulans	3
A. niger	2
P. chrysogenum	1
Tuber melanosporum	0
F. graminearum	2
Nectria haematococca	3
Neurospora crassa	0
Magnaporthe grisea	1
Podospora anserina	0
T. atroviride	4
T. reesei	4
T. virens	4
Laccaria bicolor	1
Postia placenta	1
Phanerochaete chrysosporium	1

Data extracted as described in legend to Table 3.4.

In addition, longer glucuronan chains and glucuronan polymers (such as ulvan, a complex sulfated aldobiouronan from marine green algae) can be cleaved by endo-β-1,4-glucuronan lyases, which will therefore be also treated here (EC 4.2.2.14).

The GH67 enzymes target the glucuronic acid appended to the C2-OH of the xylose at the nonreducing end of xylooligosaccharides and display a preference for MeGlcA side chains. The length of the oligosaccharide does not influence the catalytic rate, indicating that the enzyme only interacts with the uronic acid and the linked xylose residues, but does not remove glucuronic acid from internal regions of xylan (Gilbert, 2011c). α-glucuronidases are inverting enzymes that hydrolyze their target glycoside bond through a single-displacement mechanism assisted by general acid and general base residues. Consequently, the glucuronic acid is formed in a β-configuration (Biely et al., 2000).

GH67 enzymes are present in all fungi, but their structure has so far only been analyzed from prokaryotes (*Cellvibrio japonicas, Geobacillus stearothermophilus*). They are composed of three distinct domains (Nurizzo et al., 2002; Golan et al., 2004): the N-terminal domain forms a two-layer β-sandwich, whereas the catalytic domain is a classical (β/α)8-TIM barrel whose catalytic center is located opposite to the N-terminal domain. The remaining, C-terminal domain consists mainly of α-helices. The active site comprises a deep, partially hydrophobic, pocket.

Glycoside hydrolases of GH115 catalyze the cleavage of 4-*O*-methyl D-glucuronic acid side chains from native xylan polysaccharides. In contrast to GH67 enzymes, which only cleave glucuronosyl linkages at the nonreducing ends of xylooligosaccharides, GH115 enzymes from *Thermomyces aurantiacus* and *Schizophyllum commune* remove glucuronic acid from the both terminal and internal regions of xylooligosaccharides and xylans (Khandke et al., 1989; Tenkanen and Siika-Aho, 2000; Ryabova et al., 2009). These enzymes release the β-anomer of MeGlcA as the first-formed product, thus suggesting a one-step, inverting mechanism (Kolenová et al., 2010).

At the time of this writing, the 3D structure of GH115 α-glucuronidases has not yet been assessed.

Glucuronan lyases have so far been only characterized from *T. reesei*, but orthologs are present in Eurotiomycetes and Sordariomycetes (Konno et al., 2009). The enzyme was purified from an unidentified *Trichoderma* sp. and was shown to be a monomer of about 27 kDa. The enzyme was active toward acetylated and nonacetylated glucuronans, leading to various (4,5)-unsaturated products as oligoglucuronans (Delattre et al., 2006). Konno et al. (2009) cloned the gene from *T. reesei*: since its amino acid sequence showed no similarity to the amino acid sequences of previously described functional proteins, it was considered as a novel PL family (PL20). The recombinantly produced enzyme catalyzed depolymerization of cellouronate endolytically by β-elimination and was highly specific for cellouronate (Konno et al., 2009). The enzyme displays a typical β-jelly roll fold, similar to glycosyl hydrolase family GH16 and the alginate lyases of PL family PL7. A calcium ion is bound to the site far from the cleft and appears to contribute to the stability (Konno et al., 2009).

4.5.6 Accessory Hydrolases for Pectin Degradation

α-L-rhamnosidases (EC 3.2.1.40) are found in glycosyl hydrolase family GH78 and catalyze the hydrolysis of rhamnosides by an inverting mechanism. Although genes encoding these enzymes are present in many fungal genomes, they have so far only been characterized from *Aspergillus* spp. (Manzanares et al., 1997, 2000, 2001; Koseki et al., 2008).

An α-L-rhamnosidase from *A. kawachi* (90 kDa) and two α-L-rhamnosidases (RhaA and RhaB) from *A. aculeatus* (92 and 85 kDa) hydrolyzed α-1,2- and α-1,6-glycosidic bonds (Koseki et al., 2008; Manzanares et al., 2001). None of the fungal enzymes has so far been structurally characterized, but the enzyme from a *Bacillus* sp. was shown to consist of an $(\alpha/\alpha)_8$ structure.

Unsaturated glucuronyl hydrolases (GH88) and unsaturated rhamnogalacturonan hydrolases (GH105) have recently been described from bacteria. They do not yet have an EC number. Although several putative orthologs have been identified in the fungal genomes, none of them has been characterized as yet.

4.6 Other Accessory Enzymes

4.6.1 Feruloyl and p-Coumaroyl Esterases

Cell wall polysaccharides contain a small amount of ester-linked hydroxycinnamic acid derivatives, such as FA and *p*-coumaric acid (see Chapter 1; Ishii, 1997). Feruloyl esterases (FAEs; EC 3.1.1.73), also known as FA esterases, cinnamic acid esterases, or cinnamoyl esterases, are a class of enzymes that are able to hydrolyze the ester linkages of FA and diferulic acid (diFA) that are present in plant cell walls (Williamson et al., 1998).

FAEs were initially classified on the basis of the primary amino acid sequence and substrate specificity exhibited against four model substrates (methyl caffeate, methyl *p*-coumarate, methyl ferulate, and methyl sinapate; Crepin et al., 2004) into type A, B, C, and D. Type A enzymes exhibit high sequence similarity with fungal lipases, although they lack lipase activity and are represented by FaeA enzymes from *A. niger*, *A. tubingensis,* and *A. awamori*. They display preference to substrates with methoxy substituents, such as ferulic and sinapinic acids. The catalytic triad, typical for lipases and which consists of the amino acids S, D, and H, is also present in the type A enzymes, and the spacing between these residues in the amino acid sequences is conserved.

Type B enzymes have been described from *N. crassa* (Fae-1) and *P. funiculosum* (FaeB) and act on substrates containing one or two hydroxyl substituents, such as *p*-coumaric or caffeic acids, respectively. These FAEs show poor hydrolytic activity toward acetylated substrates, such as α-naphthyl acetate and acetylated xylan, respectively (Koseki et al., 2009). They have similarity to carbohydrate esterases (CEs) of the CE family CE1.

The type C FAEs exhibit sequence similarity to fungal tannases and display a broad substrate specificity such as toward ferulic, *p*-coumaric, caffeic, and sinapinic acids. They were characterized from *A. niger* (FaeB), *Talaromyces stipitatus* (FaeC), *F. oxysporum* (FaeC), and *A. oryzae* (FaeC; for review, see Crepin et al., 2004; Koseki et al., 2009).

Type D enzymes have so far only been found in bacteria.

The positional specificity of the different types of fungal FAEs has been investigated by Puchart et al. (2007) using several feruloyated 4-nitrophenyl glycosides. Type A enzymes exhibit higher specificity toward 5-*O*-(*E*)-feruloyl-α-L-arabinofuranoside than toward 2- and 3-*O*-(*E*)-feruloyl-α-L-arabinofuranosides and 2-, 3-, and 4-*O*-(*E*)-feruloyl-β-D-xylopyranosides. Type B enzymes prefer 2- and 5-*O*-(*E*)-feruloyl-α-L-arabinofuranosides to other substrates. Type C enzymes readily hydrolyze all feruloylated xylopyranosides and arabinofuranosides, except for 3-*O*-E-feruloyl-β-D-xylopyranoside (Koseki et al., 2009).

These four current classes of FAE sequences have different evolutionary origin: Olivares-Hernández et al. (2010) performed a phylogenetic analysis of FAE sequences and found five, highly supported major clades (I–V): clade I contained the three biochemically characterized member's FAEs from *A. niger*, *A. awamori,* and *A. tubingensis,* which have earlier been attributed to type A. With the exception of a putative FAE sequence from *Laccaria bicolor* belonging to the *Agaricomycetes* class of basidiomycetes, all the clade I members were from *Eurotiomycetes*. Clade II contains sequences from both *Sordariomycetes* and *Eurotiomycetes*, but the clade does not reflect any phylogenetic structure, and therefore, the included proteins have been maintained only in some species and lost in most others. They correspond to type B. Clades III and IV are phylogenetically poorly resolved: clade III is composed of sequences from *Eurotiomycetes*, *Sordariomycetes,* and *Leotiomycetes,* and these genes display considerably longer branches than those of the genes in the other clades, suggesting a higher rate of evolution. Clade IV includes putative FAE from four taxonomic classes, with proteins from *Fusarium*, *Phaeosphaeria*, *Pyrenophora,* and *Magnaporthe* occupying a basal position. Interestingly, clade IV contains enzymes that were previously characterized as type B and type C. Clade V is basal to I–IV and contains sequences from different taxonomic classes. None of these have been biochemically characterized to date (Olivares-Hernández et al., 2010).

Being related to lipases, the structure of a type-A feruloyl esterase from *A. niger* (AnFaeA) was shown to display a typical α/β-hydrolase fold (Hermoso et al., 2004). The catalytic triad (S133-H247-D194) was identified, and the active-site cavity is—in analogy to lipases—confined by a lid and a loop that confers plasticity to the substrate-binding site. The lid exhibits a high ratio of polar residues, which in addition to a unique *N*-glycosylation site stabilize it in an open conformation, and confers the typical esterase character to the enzyme. A comparison with structurally related lipases reveals that subtle amino acid and conformational changes have endowed this protein with new enzymatic properties that point to a functional convergence after evolutionary divergence within the FAEs family (Hermoso et al., 2004). Indeed, Levasseur et al. (2006)—examining the evolution of the feruloyl esterase family A family—suggested that a functional shift followed by a duplication event has occurred within the ancestral lipase genes, thus resulting in the acquirement of a new function ("neofunctionalization"). Consistent with this hypothesis, they showed that certain amino acids were indeed under positive selection during the functional shift. They speculated that drastic environmental changes such as the colonization of land by terrestrial plants might have driven adaptation in Euascomycetes (*Aspergilli*) by functional diversification, thus conferring a selective advantage on this group.

4.6.2 Acetyl- and Methylesterases

Acetylxylan esterases (EC 3.1.1.72) are common components of microbial hemicellulolytic and cellulolytic enzyme systems, and are destined to de-esterify partially acetylated hardwood 4-*O*-methyl-D-glucuronoxylan and xylans of annual plants (see Chapter 1). These esterases are part of hemicellulolytic enzyme systems and found in 8 of 15 CE families (Table 4.7). Most of the CEs are serine-type esterases because they involve a S-residue at their active center. Only the acetylxylan esterases of CE family 4 do not possess the S-D-H catalytic triad and have been identified as aspartate deacetylases, which contain a D-residue and are activated by bivalent metal cations that are coordinated by two H-residues (Puchart et al., 2006). They have so far only been found in anaerobic fungi (Caufrier et al., 2003).

Members of the CE-2 family were first described in *N. patriciarum* as acetyl esterases (AcEs), which are active on synthetic aryl esters, but with low or no activity against acetylated birchwood xylan (Dalrymple et al., 1997). Their protein structure was resolved from bacterial

Table 4.7. CE xylan esterases in fungi.

			AS	NE	PE	MA	LP	NC	TR
CE1	(α/β/α)-sandwich	Acetylxylan esterase (EC 3.1.1.72),	+		+				
		feruloyl esterase (EC 3.1.1.73)	+					+	
CE2	NI	Acetylxylan esterase		+		+	+		
CE3	(α/β/α)-sandwich	Acetylxylan esterase	+	+		+	+	+	
CE5	(α/β/α)-sandwich	Acetylxylan esterase			+	+		+	+
CE8	β-helix	Pectin methylesterase	+		+	+		+	
CE12	(α/β/α)-sandwich	Rhamnogalacturonan acetylesterase (EC 3.1.1.-)	+		+	+	+		
CE15	NI	4-*O*-methyl-glucuronoyl methylesterase (EC 3.1.1.-)						+	+

Data extracted from the CAZY-database (www.cazy.org). EC number only given at first occurrence. + indicates characterization of the respective gene or protein.
AS, various *Aspergillus* spp.; NE, *Neocallimastix patriciarum*; PE, *P. chrysogenum*; MA, *Magnaporthe grisea*; LP, *Leptosphaeria maculans*; NC, *Neurospora crassa*; TR, *T. reesei*; NI, no information.

members (*Cellvibrio japonicus*), revealing the α/β hydrolase fold typical for esterases and lipases, and they were shown to act by deacetylation of the 6-position in aldohexoses (Montanier et al., 2009).

The CE5 family contains most of the fungal acetylxylan esterases that were studied in some detail, although their structures have not yet been reported. On the basis of their amino acid similarity, they resemble some cutinases, for which the structure has been determined as an α/β/α-sandwich (Nyon et al., 2009), and a similar structure may therefore be anticipated for them. Acetylxylan hydrolases have been characterized from Aspergilli, *Penicillium* spp., *T. reesei,* and also basidiomycetes (Tenkanen, 1998). Some of them (e.g., *T. reesei*) also carry a cellulose-binding domain of the CBM1 type that appears to have been arisen by a gene fusion event.

Members of CE6 share the primary structure of GDS(L)-like hydrolases (Upton and Buckley, 1995). A catalytic S serves as the nucleophile and is located in a highly conserved sequence G(D/Q)SX. The active-site E is part of a highly conserved sequence HQGE, situated near the so-called Block III in the CE6 family, but the role of the other conserved amino acids has so far not been identified (López-Cortés et al., 2007).

Li et al. (2008) identified a new group of carbohydrate AcEs that were phylogenetically unrelated to the CE-family members and that form the basis for a new, not yet defined CE family. The prototype, *T. reesei* AES1 was characterized in some detail: its deduced amino acid sequence displayed significant (>50% identity) homology to that of many putative fungal proteins, some of which had been previously roughly annotated as lipases/fatty acyltransferases. The orthologs of *A. clavatus* and *A. fumigatus* also had a CBM1 domain. Using 2-, 3-, and 4-*O*-acetyl 4-nitrophenyl-β-D-xylopyranosides as substrates, AES1 preferred acetyl groups in positions 3 and 4, while AXE1 (see earlier) preferred them in position 2.

4.6.3 Pectin Esterases

The acetyl and methyl residues in the smooth regions of pectins are removed by pectin esterases, which comprise enzymes that hydrolyze the esters present in the pectin backbone. Three classes of esterases have been identified so far: (i) the pectin methylesterases (EC 3.1.1.11), (ii) the pectin acetylesterases (EC 3.1.1.6), and (iii) the rhamnogalacturonan acetylesterases (RgaE; EC 3.1.1.).

Pectin methylesterases belong to family 8 (CE8) of the CEs and catalyze the specific demethylesterification of homogalacturonan within plant cell walls, thereby releasing methanol and protons (and creating negatively charged carboxyl groups). They act specifically on *O*6-methyl-esterified D-galactosiduronic acid units in pectin (Pelloux et al., 2007). Two PME 3D structures have been determined to date, that is from *E. chrysanthemi* (PemA) and carrot. A phylogenetic analysis of plant and microbial PMAs showed that the bacterial and fungal enzymes formed sister clades that appear to have evolved most recently from plant PMEs of clade X2 (Markovic and Janecek, 2004), and it is thus likely that the fungal enzymes share the same 3D structure, which is a right-handed parallel β-helix structural domain such as in pectate lyase (see earlier). On the basis of the putative active site found, PME is like an aspartyl esterase. Pectin methylesterase is unable to remove methyl residues from the nonreducing end of the pectin backbone and also cannot de-esterify a methyl-esterified galacturonic acid dimer (Kester et al., 2000).

A RgaE has been purified from *A. aculeatus* (Kauppinen et al., 1995). The enzyme was found to be essential for the action of rhamnogalacturonan hydrolases and shown to randomly remove the different types of acetyl residues present in it (de Vries and Visser, 2001). RgaE folds into an α/β/α-structure; its active site comprises an open cleft containing a typical S-H-D catalytic triad. However, the position of the three residues relative to the central parallel β-sheet and the lack of a nucleophilic elbow motif that is otherwise present in structures possessing the α/β/α hydrolase fold show that the structure of RgaE is likely no member of the α/β/α-hydrolase family, but represents the archetype and a new hydrolase family, the SGNH-hydrolase family, that includes the CE family 12 (a distinct family specifically created for it; Mølgaard et al., 2000). This hydrolase family is characterized by having four conserved blocks of residues, of which one (SGNH, which all contribute to catalysis) is fully conserved.

4.6.4 Glucuronoyl Esterases

Hemicellulose and lignin have been shown to be connected *via* ester linkages between D-glucuronic or MeGlcA of glucuronoxylans and lignin alcohols (see Chapter 1). Therefore, enzymes, which would hydrolyze these linkages (Figure 4.10), play an important role in lignocelluloses hydrolysis. A glucuronoyl esterase has only recently been identified and characterized from *Schizophyllum commune* (Špániková and Biely, 2006). This esterase hydrolyzes alkyl and arylalkyl esters of D-glucuronic acid and MeGlcA (Špániková and Biely, 2006; Špániková et al., 2007) and has been postulated to hydrolyze ester linkages between MeGlcA residues of glucuronoxylans and aromatic alcohols of lignin (see Figure 4.10). These substrates were not attacked by acetylxylan esterases, feruloyl esterase, and pectin methyl esterase, and in turn glucuronoyl esterase did not hydrolyze the substrates of other CEs (Špániková and Biely, 2006).

Orthologs of the *S. commune* glucuronoyl esterase were found in several but not all fungi and given a new CE family number, CE15 (Li et al., 2007; Duranová et al., 2008): one ortholog was found in the genomes of *T. reesei* (CIP2, a gene previously recognized to encode a cellulose-binding domain and to be upregulated under conditions of induction of cellulolytic and

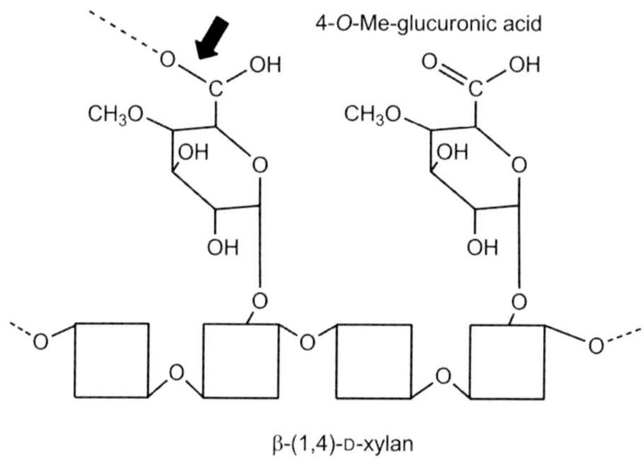

Figure 4.10. The ester linkage between 4-*O*-methyl-D-glucuronic acid residues of glucuronoxylans and aromatic alcohols of lignin as potential cleavage sites for glucuronoyl esterases.

hemicellulolytic enzymes; cf. Foreman et al., 2003), *Neurospora crassa*, *Magnaporthe grisea*, *A. fumigatus*, and *Phaeosphaeria nodorum;* two orthologs in *Phanerochaete chrysosporium* and *Chaetomium globosum;* and four orthologs in the genome of *Coprinus cinereus*. In contrast, neither *A. niger* nor *A. oryzae* have orthologs of *cip2* in their genomes (Duranová et al., 2008). A phylogenetic analysis of CE15 together with other CE family proteins from fungi reveals that CE15 members form a sister clade to the clade comprising all the other fungal feruloyl-, acetylxylan, and pectin methylesterases (Figure 4.11).

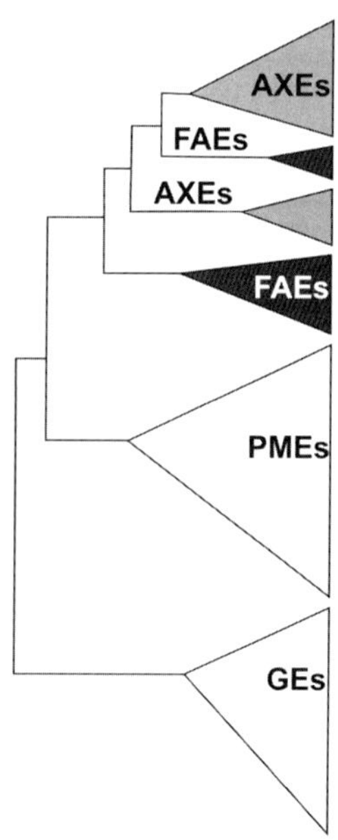

Figure 4.11. Simplified scheme of the phylogenetic relationship between glucuronoyl esterases and other carbohydrate esterases. Glucuronoyl esterase (GE), acetylxylan esterase (AXE), ferulic acid esterase (FAE), and pectin methylesterase (PME). (Data adapted from Duranová et al., 2009.)

Figure 4.12. Structure of *T. reesei* CIP2 (accession number 3PIC).

The structure of the catalytic domain (aa 90–460) of glucuronoyl esterase CIP2 from *T. reesei* was recently determined and represents the first structure of the newly established CE family 15 (Pokkuluri et al., 2011; Figure 4.12). The structure revealed the residues S278-H411-E301 that are present in a triad arrangement, as the active site. S278 is present in the novel consensus sequence GCSRXG reported earlier in the members of CE15 family. The active site is exposed on the surface of the protein that is apparently necessary for the ability of the enzyme to hydrolyze ester bonds of large substrates (Pokkuluri et al., 2011).

Chapter 5
The Tools—Part 3: Enzymology of Lignin Degradation

In contrast to the polysaccharides present in the plant cell wall, lignin is a polyphenylpropan polymer in which the linkage between its building blocks is the result of radical reactions and not due to removal of water. Consequently, lignin degradation does not require hydrolytic enzymes. Also on the contrary to polysaccharides, lignin cannot be used for as a carbon or energy source for most of the wood-decaying fungi, and its deconstruction therefore rather serves to obtain access to the energy-rich and thus more desirable cellulose and hemicelluloses in the plant biomass.

First observations on the degradation of lignin revealed that the degradation products had a higher oxygen content than the original lignin, which suggested the operation of oxygen-introducing mechanisms. This was in principal agreement with what was known about the biochemistry of catabolism of aromatic compounds, which as a first step involve an oxidation of the aromatic ring. The depolymerization of lignin essentially follows this rule; however, while catabolism of aromatic compounds within cells mostly makes use of mono- and dioxygenases, the oxidative attack on lignin is achieved by the action of peroxidases. These are enzymes that use hydrogen peroxide (H_2O_2) or organic hydroperoxides (R-OOH) as electron-accepting cosubstrates to oxidize a large variety of compounds (Hofrichter et al., 2010). The majority of peroxidases are heme proteins bearing an iron protoporphyrin IX in their active site. In addition, however, there are also a number of nonheme peroxidases known, which contain either certain metal ions (vanadium, manganese) or specific metal-free prosthetic groups like the thiol peroxidases or the alkyl hydroperoxidases. The PeroxiBase database (http://peroxibase.toulouse.inra.fr/; Passardi et al., 2007a) has been established to provide a tool for linking the peroxidase nucleotide sequences and protein structures with other databases and also contains useful information about the evolution of the peroxidases, their putative functions, and transcription regulation.

The huge group of heme-containing peroxidases can be diversified into two main superfamilies, that is, the peroxidase-cyclooxygenases (formerly called mammalian peroxidases) and the nonanimal peroxidases (plant peroxidases), which include most of the so far recognized and deposited heme-dependent peroxidase sequences. Phylogenetic analysis further subdivides

Fungi and Lignocellulosic Biomass, First Edition. Christian P. Kubicek.

the latter into three main classes that differ by organism and physiological function: class I contains intracellular nonanimal peroxidases, class II contains secreted fungal peroxidases, and class III represents the secreted plant peroxidases (Passardi et al., 2007b; Zamocky et al., 2009; Lundell et al., 2010). As for class II peroxidase, four groups are today known to be involved in lignin oxidation. They are all secreted enzymes and comprise (a) lignin peroxidase (LiP; EC 1.11.1.14; 1,2-bis(3,4-dimethoxyphenyl)propane-1,3-diol:hydrogen-peroxide oxidoreductase), (b) manganese peroxidase(MnP; EC 1.11.1.13), (c) so-called "versatile peroxidase" (VP; EC 1.11.1.16), and (d) several phenol oxidases. Yet this list is likely incomplete as research of the last years discovered several additional new cases of peroxidases that are capable of attacking lignin (Hofrichter et al., 2010). Knowledge about the above given enzymes has mostly been derived from studies in white rot fungi, but the enzymes can also be found in some litter-decomposing basidiomycetes (*Agaricus bisporus*, *Coprinus* spp.).

The above classification into four groups is also supported by phylogenetic analysis (Figure 5.1; Lundell et al., 2010), which grouped the fungal class II peroxidases into three main lineages, that is, group A, containing the short LiP genes, which further branches into several

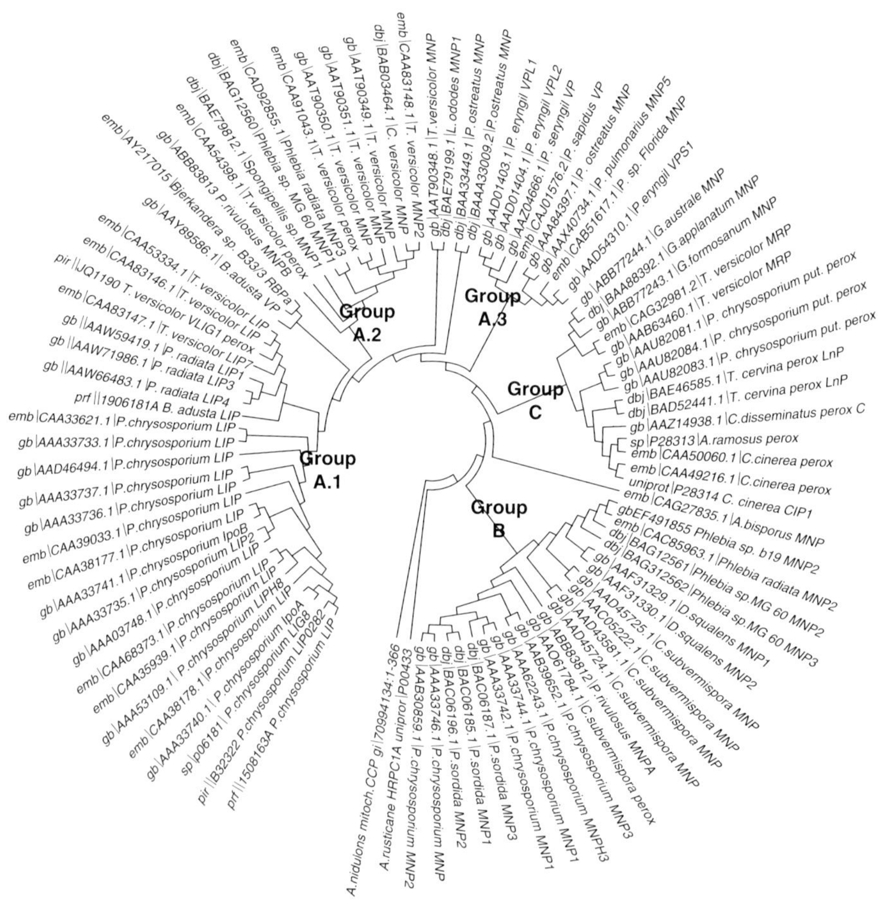

Figure 5.1. Phylogeny of fungal lignin and manganese peroxidases. (From Lundell et al. (2010), by permission from John Wiley & Sons, Inc.)

Table 5.1. Properties of fungal secretory heme peroxidases.

Enzyme	EC No.	Reaction	Substrates	Redox-Potential (V)
Lignin peroxidase	1.11.1.14	Oxidation	Nonphenolic aromatics	1.4–1.5
Manganese peroxidase	1.11.1.13	Oxidation	Manganese ions	1.0–1.2
Coprinopsis cinerea peroxidase	1.11.1.7	Oxidation	Phenolic compounds	0.9–1.1
Chloroperoxidase	1.11.1.10	Oxidation	Halogenides and phenolic compounds	NI
Aromatic peroxygenase	1.11.-	Peroxigenation	Aromatics and *n*-alkanes, ether cleavage	NI
Dye-decolorizing peroxidase	1.11.-	Oxidation	Dyes, phenol	1.2–1.5

Modified after Hofrichter et al. (2010).
NI, no information.

enzyme families and species-specific clades, and to the rather narrow group B that consists only of the MnPs. This grouping is also consistent with the biochemical functions of these enzymes functions (see Table 5.1). The phylogenetic tree also displayed a further group (group C) that includes enzymes with variant sequences and functions. It is also evident from the tree that the LiP group has apparently evolved by gene duplications and eventual horizontal gene transfer.

As explained in Chapter 2, lignin degradation depends on the availability of other fungal metabolites like phenolic and other aromatic compounds, smaller peptides, and oxalic acid, and, therefore in this chapter, I will also deal with the enzymes responsible for their production.

5.1 Lignin Peroxidase

LiP is, together MnP, the eldest and best known peroxidase that acts on lignin, and both were first described from the white rot fungus *Phanerochaete chrysosporium*. They are relatively nonspecific for their substrates and known to oxidize phenolic aromatic substrates and a variety of nonphenolic lignin model compounds, as well as a range of organic compounds with a redox potential up to 1.4 V in the presence of H_2O_2 (Valli et al., 1990). The enzyme activity of LiP can easily be measured with veratryl alcohol as a substrate, the most popular LiP substrate, which is oxidized to veratraldehyde, which can be followed by the increase in its absorbance at 310 nm. LiPs have often been reported to occur in the form of multiple isozymes, whose nature is both due to the presence of multiple genes as well as to posttranslational modifications (Table 5.2). They are (*N*- and *O*-linked) glycoproteins that contain 1 mol of iron protoporphyrin IX per mole of protein, exhibit an average molecular weights between 38 and 46 kDa, and possess rather low isoelectric points (3.2–4.0). The latter is due to their phosphorylation during the secretory process (Kuan and Tien, 1989).

LiP displays a globular structure, which is composed of eight major and eight minor α-helices and a small β-sheet, yielding a two-domain protein with a dimension of $50 \times 40 \times 40$ Å held together by four disulfide bridges (Figure 5.2). Simulation studies of LiP in aqueous solutions predict the deformation and enlargement of this channel to allow the movement of small substrates such as veratryl alcohol into it and toward the active site (Gerini et al., 2003).

Table 5.2. Lignin peroxidases characterized.

Fungus	Lignin Peroxidase Name
Bjerkandera adusta	BaLiP (LPO)
Ceriporiopsis subvermispora	CsuLiP01 (Csg1)
	CsuLiP02 (Csg2)
Perenniporia medulla-panis partial	PmpLiP
P. chrysosporium	PcLiPI_BKMF1767 (LIG3, GLG3)
	PcLiPA_RP78 (LPOA, Ligninase H8)
	PcLiPB_RP78 (LPOB, Ligninase B)
	PcLiPC_BKMF1767 (GLG5, LIP6)
	PcLiPD_BKMF1767 (GLG4, LiP2)
	PcLiPE_RP78 (LiP2, GLG2)
	PcLiPF (GLG6)
	PcLiPI_ME446 (LIG1)
	PcLiPJ_RP78 (lipJ)
	PcLiPC_RP78
	PcLiPD_RP78
	PcLiPE_BKMF1767 (lpo)
	PcLiPF_RP78
	PcLiPG_RF78
	PcLiPI_RP78
	PcLiPA_SC26 (rLDM(TM)6)
P. sordida partial	PsoLiP01
	PsoLiP02
Phlebia radiata	PrLiP01 (lip1)
	PrLiP03 (lip3, lgp3)
	PrLiP04 (lip4)
Pycnoporus coccineus partial	PcocLiP
	PYsLiP
Trametes versicolor (Coriolus versicolor)	TvLiP (LPGIV, LPG1, Ligninase C)
	TvLiP07 (LP7, LPGVI)
	TvLiPGIII

Data retrieved from PeroxiBase (http://peroxibase.toulouse.inra.fr).

The heme is located in the crevice between the two domains and is accessible from the outside via two small channels. Each domain further contains one calcium ion that is believed to be structurally important for maintaining the active site in an appropriate topology. The heme iron is predominantly in a high spin state, pentacoordinated with H176 at the proximal side as the fifth ligand and W339, which is stabilized by a hydrogen bond to H46 (Figure 5.3; Wong, 2009).

The peroxide-binding pocket is located on the distal side of the heme, with a channel extending to the exterior of the protein. It contains a D–H–Fe triad that is conserved in all peroxidases of the plant superfamily (Figure 5.3): the distal H47, which forms a hydrogen bond with D82, acts as a proton acceptor for the bound peroxide. In the proximal pocket, H176 coordinates with the heme iron and also forms hydrogen bond with D235, which helps to stabilize the Fe(IV)–O intermediate in compound I. In addition, W171 is believed to play an important role in the binding and oxidation of veratryl alcohol through long-range electron transfer (Doyle et al., 1998). R43 stabilizes the negative charges developed in the cleavage of peroxides and the ferryl oxygen of compound I. The catalytic cycle of LiP is shown in Figure 5.4.

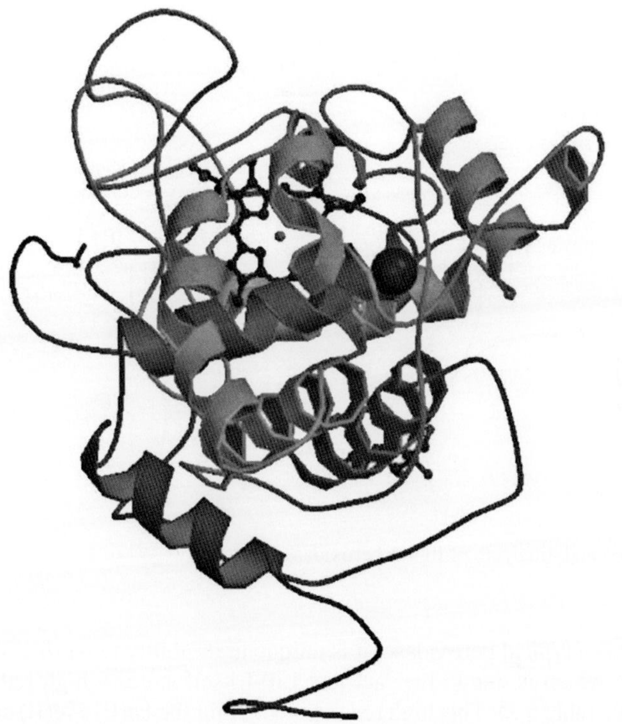

Figure 5.2. Three-dimensional structure of *P. chrysosporium* lignin peroxidase (accession number PDB#1lga).

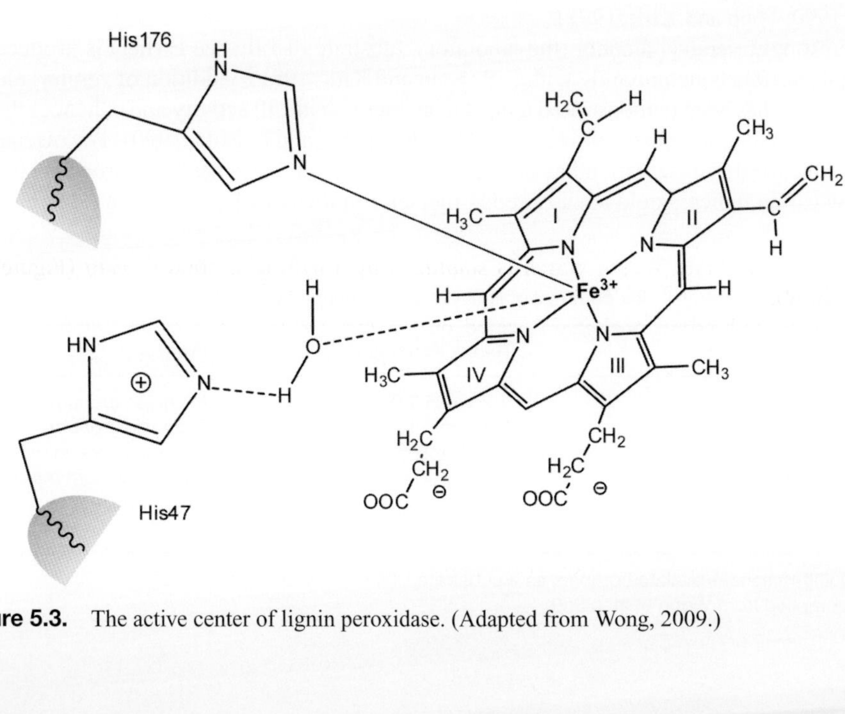

Figure 5.3. The active center of lignin peroxidase. (Adapted from Wong, 2009.)

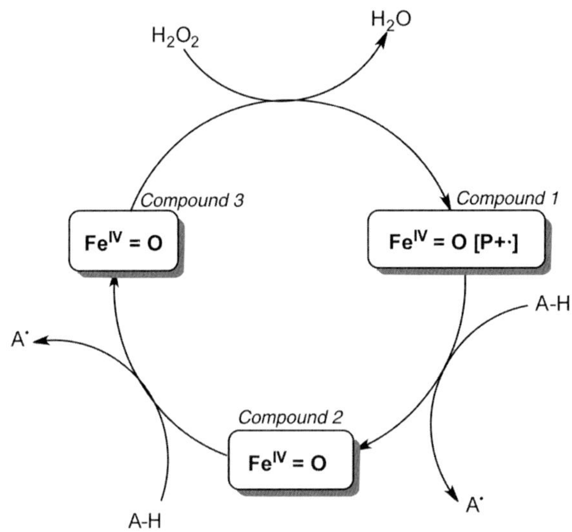

Figure 5.4. The catalytic cycle of lignin peroxidase.

Although LiP is a typical peroxidase, it is unique in its ability to oxidize substrates of high redox potentials, which is due to the fact that LiP-I itself stores a high redox potential (E'_0 ~1.2 V at pH 3.0; Table 5.3). This high redox potential for the Fe(III)/Fe(II) couple is reflected in the fact that the heme active sites are more electron deficient than other peroxidases. A comparison of LiP, MnP, and laccase in the oxidation of a homologous series of 12 methoxybenzenes (which ranged from 0.81 to 1.76 V at pH 3.0) showed that the reaction correlated both to the redox potentials of the enzyme as well as that of the compound substrate (Kersten et al., 1990; Popp and Kirk, 1991).

Interestingly, veratryl alcohol (the laboratory substrate of LiP; see earlier) is produced by *P. chrysosporium* concomitantly with LiP (Fenn and Kirk, 1981). Addition of veratryl alcohol to the fungus has been demonstrated to lead to an increase in LiP activity and enhances the rate of lignin degradation (Faison and Kirk, 1985; Schoemaker and Leisola, 1990). The mechanism responsible for this observation has not been discovered yet, but it has been proposed that the veratryl cation radical could act as a redox mediator in the oxidation of lignin (Harvey et al.,

Table 5.3. Comparative redox potential of lignin and other peroxidases.

Enzyme	Redox Potential	From
Lignin peroxidase I	1.2 V (pH 3.0)	*P. chrysosporium*
Manganese peroxidase[a]	0.8 V (pH 4.5)	*P. chrysosporium*
Laccase T1/T3	0.79 V (pH 5.5)	*Trametes versicolor*
	0.47 V (pH 6.0)	*Myceliophthora thermophila*
Peroxidase	0.95 V (pH 6.3)	Horseradish
Cytochrome C peroxidase	0.19 V (pH 7.0)	Mammalian

With Mn(III)-malonate/oxalate complex as a substrate.
[a]Data compiled from Wong et al. (2009).

1986), or, alternatively, veratryl alcohol could protect LiP from inactivation by an excess of H_2O_2 (Valli et al., 1990).

5.2 Manganese Peroxidase

MnPs utilize Mn(II) as the reducing substrate for hydrogen peroxide by oxidizing it to Mn(III), which in turn oxidizes monomeric phenols. The genes encoding these enzymes are strongly conserved within different fungal species. Their phylogeny suggests species-correlated evolution with recent gene duplications (Figure 5.1). MnPs are also able to generate oxidative Mn^{3+} chelates that can diffuse into the wood, and it has therefore been suggested that they may be of particular advantage to fungi that are specialized in selective decay of lignin (see Chapter 2).

MnPs are acidic glycoproteins with isoelectric points between 4 and 5 near 4.5 and an M_r of around 45,000 (Table 5.4). Similar to LiP, the paralogous MnPs are also subject to posttranslational modification, which gives rise to a high number of detected isoenzymes. MnP reacts with Mn(II) or preferentially with its chelated form, oxalate–Mn(II), which likely occurs *in vivo* in the environment and the active center contains one molecule of heme as iron protoporphyrin IX in the high-spin, pentacoordinate, ferric state with a His residue coordinated as the fifth ligand (Wong et al., 2009). The structure of MnP has first been discovered from *P. chrysosporium* (Sundaramoorthy et al., 1994): like LiP, it is composed of ten major and one minor helices, which form two domains with the heme located between them (Figure 5.5). However, MnP is distinguished from LiPs by containing one more disulfide bonds (five), and this additional bond (C341–C348) functions as the Mn(II)-binding site and pushes the C-terminal part of the enzyme away from the main body of the protein. It also coordinates to the carboxylate oxygens of E35, E39, and D179, the heme propionate oxygen, and two water oxygens. D–H–Fe triad is formed by H173-D242 and an oxyferry iron. Two heptacoordinate structural calcium ions are also present and function in the stabilization of the active site of the enzyme (Sundaramoorthy et al., 2005).

The catalytic cycle of MnP, which bears high resemblance to that of LiP, is shown in Figure 5.6. The Mn(III) thus formed dissociates from the enzyme and can be stabilized by forming a chelate with oxalate (that is abundantly secreted by white rot fungi; Wariishi et al., 1992; Kishi et al., 1994), which has a high redox potential of 0.8–0.9 V. Other α-hydroxy acids (such as malonate) can also function as chelators *in vivo*. This Mn(III)–oxalate complex then acts as an oxidant of phenolic substrates by performing $1e^-$ oxidation of the substrate to produce a phenoxy radical intermediate, which in turn stabilizes by undergoing various rearrangements to yield various breakdown products (Wong et al., 2009). Interestingly, MnP can even initiate its own H_2O_2 formation in the absence of exogenous H_2O_2, by oxidizing NADPH, glutathione, or dithiothreitol. It is believed that this oxidase activity would make H_2O_2 available to start the catalytic cycle of both MnP and LiP (Paszczynski et al., 1986).

While the chelated Mn(III) is unable to oxidize nonphenolic compounds, the radicals that are formed by the oxidation of Mn(III) are able to oxidize benzyl alcohols and other components with a diarylpropane structure. This reaction involves the oxidation of thiol radicals, which then withdraw hydrogen from the above named substrate to form benzylic radicals. The latter then react spontaneously to their final products.

MnP is also capable of lipid peroxidation, which may be important for the generation of oxalate: glyoxal—which can be converted to glyoxalate and oxalate by glyoxal oxidase (GLX; see later in this chapter)—was shown to be produced from linoleic acid by a MnP-dependent lipid

Table 5.4. Characterized manganese peroxidases.

Fungus	Manganese Peroxidase Name
Agaricus bisporus	AbMnP01
	AbMnP02
Amylostereum areolatum	AarMnP01
Amylostereum chailletii	AchMnP01
Amylostereum ferreum	AfeMnP01
Amylostereum laevigatum	AlMnP01_B1371
Amylostereum sp.	AMspMnP01
Bondarzewia montana	BmoMnP02_DAOM415
	BmoMnP01
Ceriporiopsis rivulosa	CriMnP01
Ceriporiopsis subvermispora	CsuMnP02a
	CsuMnP03
	CsuMnP04
	CsuMnP01
	CsuMnP02b
Coriolopsis gallica	CgalMnP01
Dichomitus squalens	DsMnP02
	DsMnP01
Echinodontium japonicum	EjMnP01
Echinodontium tinctorium	EtiMnP01_B1122
Echinodontium tsugicola	EtsMnP01
Fomitiporia mediterranea	FmeMnP01
Heterobasidion abietinum	HabMnP02
	HabMnP01_B1166
	HabMnP03
Heterobasidion annosum	HanMnP03_B1257
	HanMnP03_B163
	HanMnP01_B825
	HanMnP02_B1169
	HanMnP01_B1256
	HanMnP02_B825
	HanMnP04
	HanMnP03b
Heterobasidion araucaria	HauMnP01_B1080
	HauMnP02_B1080
Heterobasidion insulare	HiMnP02_B1159
	HiMnP01_B1279
	HiMnP03
Heterobasidion parviporum	HpaMnP01_B146
	HpaMnP03_B304
	HpaMnP02
Lentinula edodes	LedMnP01
P. chrysosporium	PcMnP01a
	PcMnP04
	PcMnP02_ATCC24725
	PcMnP03
	PcMnP01b
	PcMnP01_OGC101
	PcMnP02_OGC101
	PcMnP02_FPL5157

Table 5.4. (*Continued*)

Fungus	Manganese Peroxidase Name
P. flavidoalba	PflMnP02
	PflMnP01
P. sordida	PsoMnP01
	PsoMnP02
	PsoMnP03
Phlebia radiata	PrMnP03
	PrMnP02
Pleurotus ostreatus	PoMnP03
Pleurotus pulmonarius	PpulMnP03
	PpulMnP02
	PpulMnP05
	PpulMnP01
	PpulMnP04
Spongipellis sp.	SPOspMnP01
Trametes versicolor	TvMnP04
	TvMnP01_KN9522
Tylospora fibrillosa	TfMnP01

Data retrieved from PeroxiBase (http://peroxibase.toulouse.inra.fr).

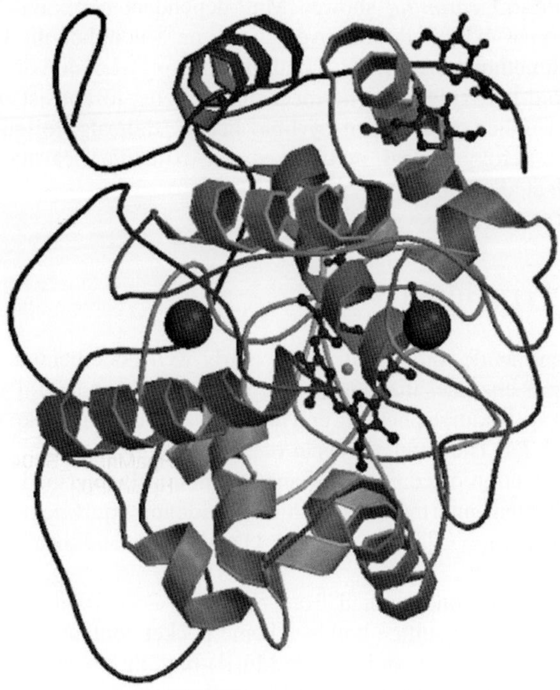

Figure 5.5. Structure of manganese peroxidase (accession number 1MNP).

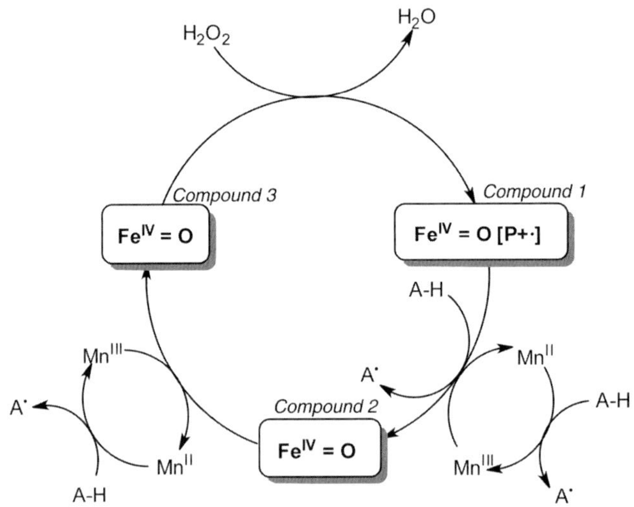

Figure 5.6. The catalytic cycle of manganese peroxidase.

peroxidation (Watanabe et al., 2001). This ability for lipid peroxidation may also be important for cleaving the Cα-Cβ and β-*O*-4 lignin structures: culture liquids of the litter-decomposing basidiomycete *Stropharia coronilla* showed MnP-dependent peroxidation of linoleic acid, and the purified enzyme exhibited oxidative activity on a nonphenolic β-*O*-4 lignin model compound [1-(3,4-dimethoxyphenyl)-2-(2-methoxyphenoxy)-1,3-dihydroxypropane] *in vitro* when coupled to lipid peroxidation. The mechanism seems to consist of an abstraction of hydrogen from the Cα benzylic carbon by lipid peroxy radicals, followed by the addition of oxygen to form a peroxy radical, and subsequent oxidative cleavage and nonenzymatic degradation (Kapich et al., 2005).

5.3 Versatile Peroxidase

In addition to LiP and MnP, some fungi (particularly well documented in *Bjerkandera* and *Pleurotus* spp.) possess enzymes that are specific for Mn(II) (just as MnP), but in the absence of Mn(II), it can also still oxidize phenolic and nonphenolic substrates like LiP. These enzymes are therefore called VPs. The catalytic cycle of VP is identical to that of LiP and involves the formation and reduction of compound I and compound II intermediates. Compound I is a two-oxidizing equivalent intermediate, with one-oxidizing equivalent resided in the ferryl state of the iron as [Fe(IV) = O] (low spin, $S = 1$) and the second localized as a porphyrin π radical cation (Wong, 2009).

VP has been crystallized and studied from *Pleurotus eryngii:* it consists of 12 helices, which are stabilized by four disulfide bonds, a heme pocket containing two H residues (H47 and H169), two Ca^{2+}-binding site, and a single Mn(II)-binding site (Figure 5.7). A propionate residue in heme ring III is responsible for Mn(II) fixation and electron transfer to compound I or compound II/Fe(IV) = O (Ruiz-Dueñas et al., 1999; Perez-Boada et al., 2005). A surface-located W164, which exists in the form of a neutral radical, has been identified to be involved in the oxidation of high redox potential aromatic substrates, such as veratryl alcohol via

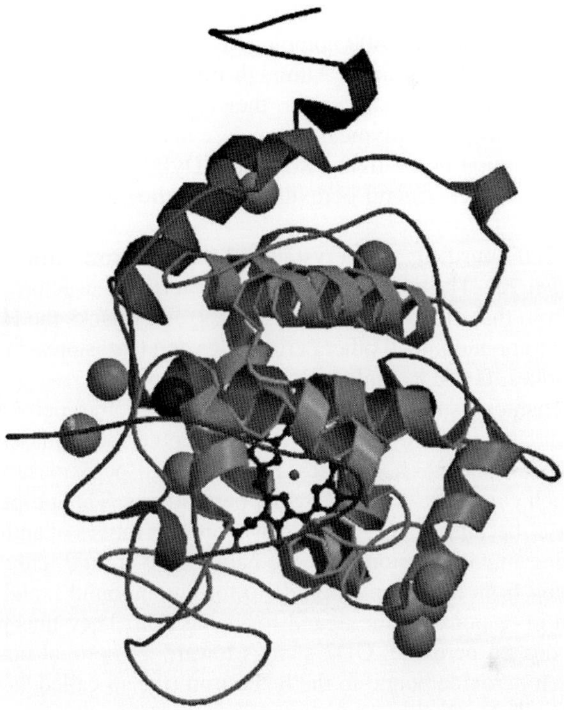

Figure 5.7. Structure of versatile peroxidase (accession number PDB#2boq).

long-range electron transfer pathway (Wong et al., 2009). Consequently, its exchange strongly reduces Mn(II)-independent oxidation of substrates, but leaves the Mn(II)-dependent activities unaffected (Perez-Boada et al., 2005; Tinoco et al., 2007). The detection of the W164 radical implies that alternative forms for compound I may exist, which could be involved in the ability of VPs to oxidize different aromatic substrates (Wong et al., 2009).

In recent years, a few other class II fungal hybrid-type peroxidases have been cloned and described. In *Phlebia radiata*, two divergent MnPs have been described: classical-long MnP type (*Pr*MnP2) and short-VP type comprising the conserved Mn^{2+}-binding site but lacking the exposed tryptophan (*Pr*MnP3; Hildén et al., 2005). Because of this, the latter enzyme is unable to oxidize veratryl alcohol but shares with VP the Mn^{2+}-independent ability to oxidize phenols, small dye compounds, and amines (Hofrichter et al., 2010).

Accumulating sequence data also predict the presence of similar, short-type hybrid MnP enzymes in *Trametes* spp. and *Pleurotus* spp. (Lundell et al., 2010), and in particular, in litter-decomposing basidiomycetes, such as *Agaricus bisporus*, *Agrocybe praecox*, and *Stropharia coronilla* (Hofrichter et al., 2010).

5.4 Dye-Oxidizing Peroxidase

Recently, a new heme-containing peroxidase enzyme family, the so-called dye peroxidases (DyP), was detected. This family appears to be abundantly present in fungi and bacteria

(Sugano, 2009). Thus far, 80 putative members of this family have been registered in the PeroxiBase database and more than 400 homologous proteins have been detected by Sugano and colleagues via PSI-BLAST search. Although only few studies have characterized the function and structure of these proteins so far, they appear to be bifunctional enzymes with hydrolase or oxygenase, as well as typical peroxidase activities (Sugano, 2009). Because of their widespread distribution in the living world, the DyP-type peroxidase family may have evolved earlier than the other general peroxidases, and they actually form a separate basal phylogenetic branch.

Sugano et al. (2007)purified and crystallized the enzyme from the basidiomycete *Thanatephorus cucumeris*. The protein had a unique tertiary structure with a distal heme region that differs from that of most other peroxidases, and it lacks the H residue that assists in the formation of compound I in all other peroxidases and that is present in LiP (Figure 5.8). Two amino acid residues, D and R, replace this H in DyP.

DyP has several characteristics that distinguish it from all other peroxidases, such as very wide substrate specificity, a lack of homology to most other peroxidases, and the ability to be active at low pH (around 3). It oxidizes all the typical peroxidase substrates, but also anthraquinones, which cannot be attacked by the peroxidases. This property gave rise to the name "dye peroxidase" because many synthetic dyes are derivatives of anthraquinone. Yoshida et al. (2011), studying mutant versions and the native DyP from *T. cucumeris*, showed that D171 accepts a proton from hydrogen peroxide to form compound I, and it can swing to the appropriate position in response to the ligand for heme iron. They thus proposed that when DyP reacts with hydrogen peroxide, OD2 swings toward an optimal position to accept the proton from hydrogen peroxide bound to the heme iron (the so-called "swing mechanism of compound I formation").

5.5 Laccases

Laccases (EC 1.10.3.2, benzenediol:oxygen oxidoreductase) belong to a family of multicopper enzymes, which include ascorbic oxidase and ceruloplasmin. Their name reminds the Japanese lacquer tree, *Rhus vernicifera,* where they were first detected (Wong et al., 2009). They catalyze the removal of a single electron from phenolic hydroxylic groups such as vanillyl glycol or syringaldehyde with the concomitant reduction of molecular oxygen to water and formation of a phenoxy radical (Figure 5.9). These radicals undergo further chemical reactions that ultimately lead to oxidized quinones and coupled oligomeric products. The analysis of metal ion contents of the laccase from *Trametes hirsuta* has shown the presence of copper and manganese in a 3:1 ratio (Giardina et al., 2010). However, some "nonblue" fungal laccases have also been described, for example, POXAW1w (a white laccase from *Pleurotus ostreatus*), which contains only one copper atom/molecule instead of the usual four, along with two zinc atoms and one iron atom in each protein molecule. Other so-called white laccases were purified from *Pycnoporus sanguineus*, *T. hirsuta*, and *Phlebia radiata* BP-11-2.

Laccases serve manyfold functions in fungi, including the synthesis of melanin and other pigments, and the formation of reproduction structures such as fruiting bodies and conidia (Baldrian, 2006; Giardina et al., 2010). However, their function in the decomposition of lignin, although demonstrated *in vitro*, is doubtful because laccases are present in nearly all fungi, even those unable to degrade lignin (e.g., several ascomycetes; Baldrian, 2006). On the other hand, some white rot fungi (e.g., *Pycnoporus cinnabarinus*) only produce laccases and do not have LiP or MnP orthologs, but still are efficient lignin degraders (Eggert et al., 1996).

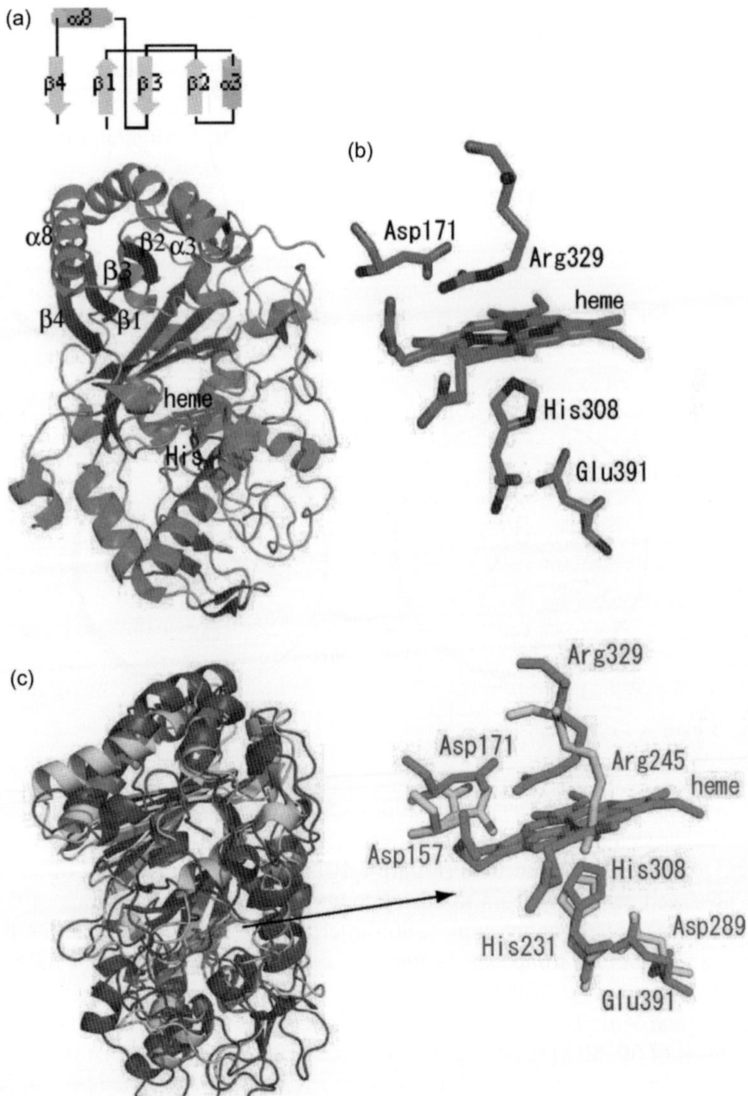

Figure 5.8. (a) Full structure of DyP. The α-helices and β-sheets are shown in light blue and violet, respectively. The proximal His and heme are shown in dark blue and green, respectively. (b) The heme and its surrounding residues important for the peroxidase activity of DyP. (c) Tertiary structural alignment of DyP and the hypothetical peroxidase, 2GVK. DyP and 2GVK are shown in blue and yellow, respectively. The heme and His308 of DyP are shown in light blue. Data on the heme position of 2GVK are not currently available through the PDB. The magnification of the heme and its critical residues is indicated by the arrow. Two possible phases of the Asp157 residue of 2GVK, which corresponds to Asp171 of DyP, are shown. (From Sugano et al. © 2007 The American Society for Biochemistry and Molecular Biology, with permission.)

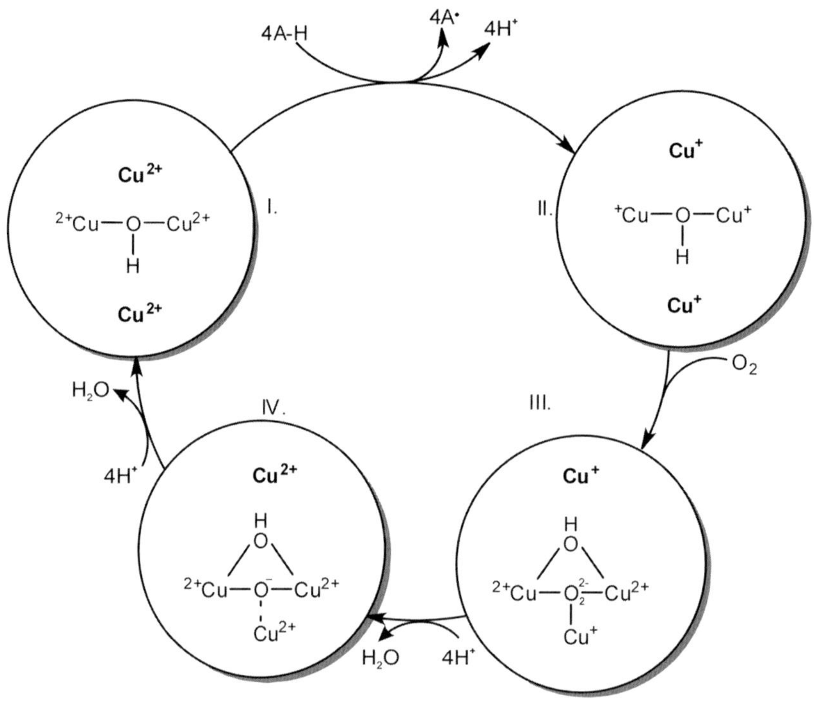

Figure 5.9. Catalytic cycle of laccase.

A phylogenetic analysis (Hoegger et al., 2006) separates the multicopper laccases into the true fungal "*sensu stricto*" laccases, fungal pigment MCOs, fungal ferroxidases, ascorbate oxidases, and plant and insect laccases (Figure 5.10). In this tree, a splitting of basidiomycete paralogs is observed, which appears to correlate with the different functions of the laccases in the respective species. Hoegger et al. (2006)interpreted this fact such that the ligninolytic laccases that are found in some basidiomycetes may be the consequence of independent duplication-divergence events during evolution.

Similar to LiP and MnP, the fungal laccases are also (mostly) extracellular glycoproteins with a molecular mass of 60–80 kDa and acidic pI values. The 3D structures of the laccases from several fungi (*Coprinopsis cinereus*, *T. versicolor*, *Pycnoporus cinnabarinus*, *Melanocarpus albomyces*, and *Rigidoporus lignosus*) are composed of three sequentially arranged β-barrel domains (Figure 5.11), which is generally typical for copper-containing proteins such as azurin and plastocyanin (Wong et al., 2009; Giardina et al., 2010).

Most of the fungal laccases contain four copper atoms located in two metal centers (T1, containing one Cu atom lying in a shallow depression on the surface of domain 3, and T2/T3, which form a trinuclear copper cluster at the interface between domains 1 and 3 with each domain providing ligand residues for the coordination of the copper atoms) and four substrate-binding loops. The copper atoms are coordinated by ten H and one C residues as ligands. In the resting enzyme, all four coppers are in the oxidation state Cu^{+2}(Malkin and Malmstrom, 1970; Dooley et al., 1979).

Laccases from different sources can exhibit a wide range of redox potentials. The T1 site in the fungal enzymes typically has a much higher redox potential (780–800 mV) than that in the

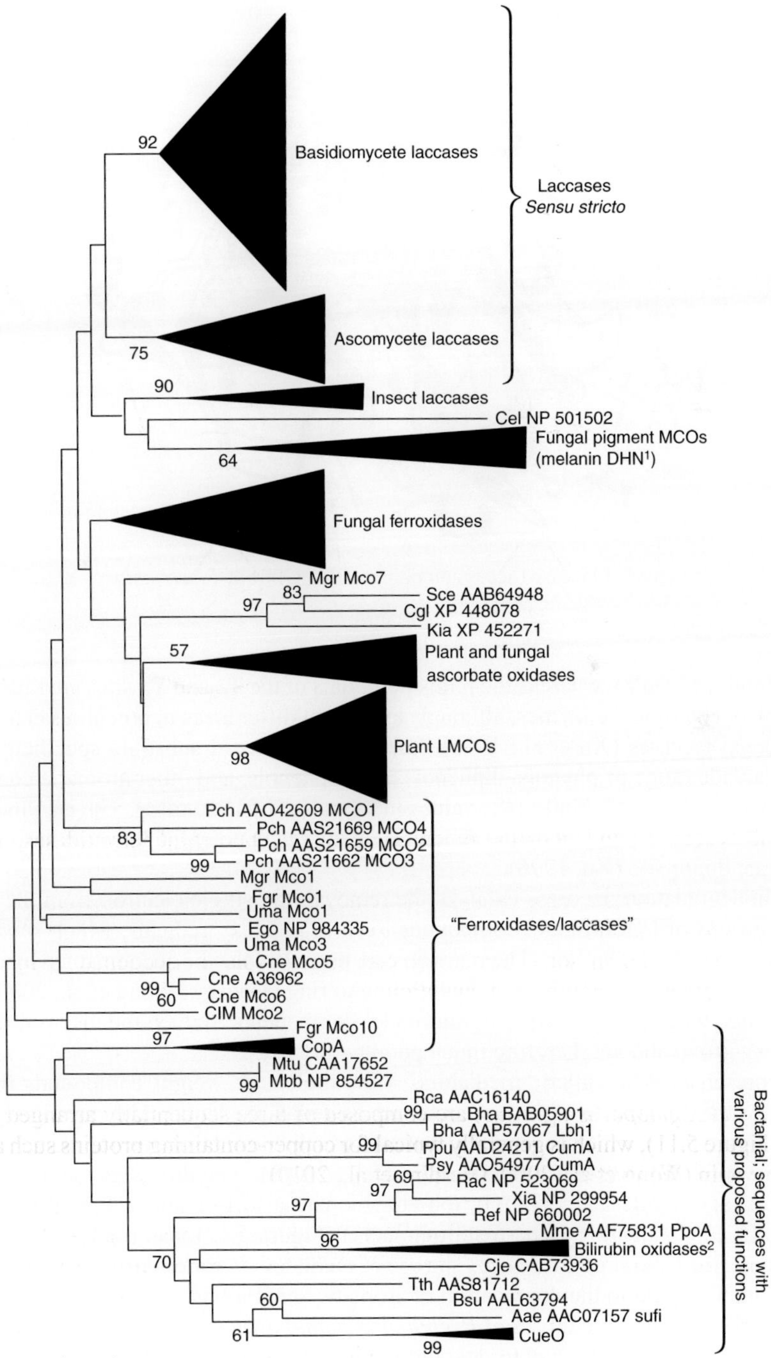

Figure 5.10. Neighbor joining tree of multicopper oxidase amino acid sequences. (From Hoegger et al. (2006), with permission from John Wiley & Sons, Inc.)

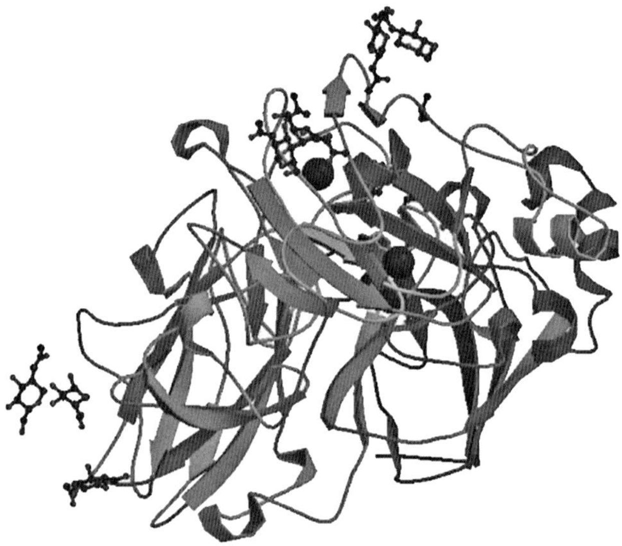

Figure 5.11. Structure of laccase (accession number 1GYC) from *T. versicolor.*

plant enzymes (420 mV), whereas the redox potentials of the T2 and T3 sites are rather similar for both plant and fungal enzymes, although significant differences in potentials can also exist among fungal laccases (Xu et al., 1996). Laccases have broad substrate specificity and are active on a wide range of phenols, diphenols, aminophenols, and other aromatic compounds, although with rather low affinity (K_m values in the range of 1–10 mM). The enzyme efficacy (k_{cat}/K_m) is directly dependent on the difference in the redox potential between the T1 site and the substrate donor site (Xu, 1996).

In lignin degradation, laccases catalyze the removal of a single electron from the phenolic hydroxyl groups of lignin model compounds to form phenoxy radicals, which subsequently polymerize via radical coupling. The reaction can, however, also be accompanied by demethylation, thus forming of a quinine and thus leading to ring cleavage (Wong et al., 2009). *o*- and *p*-substituents, when present in the lignin model components, reduce the electron density at the phenoxy group and are therefore much poorer substrates.

In the presence of so-called "mediators," (low molecular weight compounds that, once oxidized by the enzyme to stable radicals, act as redox mediators and permitting the enzyme to also oxidize other substances that are usually no substrates for it, for example, 2,2′-azino-bis-(3-ethylbenzthiazoline-6-sulfonate) (ABTS), 1-hydroxybenzotriazole (HBT), or 3-hydroxyanthranilic acid (HAA)), the enzyme can also β-1 attack lignin dimers. 2,2′-Azino-bis-(3-ethylbenzothiazoline-6-sulfonic acid) (Figure 5.12) was the first mediator that has been described and it is still in use. The use of mediators is particularly desired in the pulp and paper industry and in the degradation of aromatic xenobiotics.

HAA has been found in cultures of *Pycnoporus cinnabarinus* and could thus be the natural mediator that enables the fungus to degrade lignin only by the use of its laccases (Li et al., 2001). In the presence of such mediators, laccases can perform four types of attack on the lignin molecule: β-ether cleavage, Cα-Cβ cleavage, Cα-oxidation, and cleavage of the aromatic ring. The mechanism of this "mediation" is shown in Figure 5.13 in the case of ABTS: ABTS is

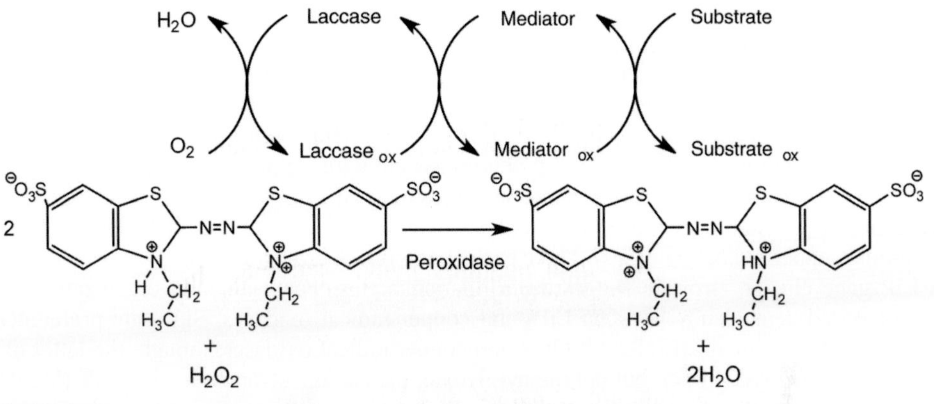

Figure 5.12. The laccase-mediator reaction, using ABTS as an example.

first oxidized to the radical cation ($ABTS^{+}$) and then to the di-cation ($ABTS^{2+}$), which is responsible for the oxidation of the nonphenolic substrate.

5.6 Enzymes Generating Hydrogen Peroxide

LiP and MnP require extracellular H_2O_2 for their *in vivo* catalytic activity. Several candidate enzymes for H_2O_2 have been discussed:

5.6.1 Glyoxal Oxidase

As discussed by Vanden Wymelenberg et al. (2006), among all candidates until now, GLX is the only hydrogen peroxide-producing oxidase that is actually secreted into the medium by white rot fungi. Also its upregulation under conditions of ligninolysis supports its role in this process. *P. chrysosporium* GLX is an acidic monomeric glycoprotein with a molecular mass of 57 kDa and requires one *g*-atom of copper for full activity. A free radical-coupled copper complex has been identified in the active site of GLX from *P. chrysosporium* (Whittaker et al.,

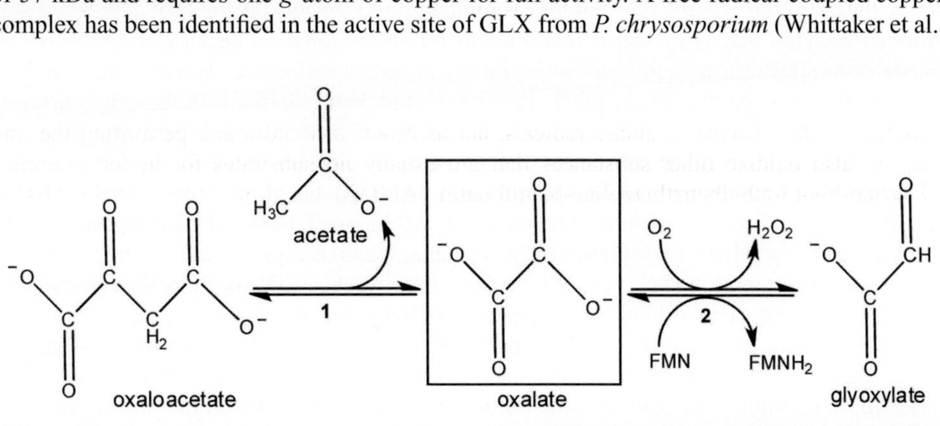

Figure 5.13. Oxalate formation in basidiomycetes: 1, oxaloacetate hydrolase (EC 3.7.1.1); 2, glyoxalate oxidase (EC 1.2.3.5).

1996). In analogy to galactose oxidase, the active site structure displays the coordination of the copper ion by two H residues (H496 and H581), a Y residue (Y495), and a C-Y-phenoxyl radical site in GLX. The latter is an unusual covalent modification forming a new, dimeric amino acid (cysteinyl-tyrosine) (Whittaker et al., 1996).

The temporal correlation of GLX, the substrates methylglyoxal and glyoxal, and peroxidase in liquid cultures of *P. chrysosporium* is consistent with a close physiological connection between these components. The genome sequence of *P. chrysosporium* reveals at least seven sequences predicted to encode copper radical oxidases (*glx*, encoding GLX and *cro1-cro6* encoding copper radical oxidases). Three of the latter (*cro3, cro4,* and *cro5)* are located within a LiP gene cluster, strongly suggesting a relationship between genomic organization and the proposed dependency between LiPs and copper radical oxidases. Substrate preferences suggest distinct biological roles for GLX and copper radical oxidases, though: the latter oxidizes glycolaldehyde dimer, but not methylglyoxal, the canonical substrate for GLX (Vanden Wymelenberg et al., 2006).

5.6.2 Other FAD-Dependent Oxidases

Other genes encoding FAD-containing oxidases in white rot fungi include the aryl alcohol oxidases (AAO) of *Pleurotus eryngii* (Gene bank accession numbers AF064069, AF143814) and a pyranose oxidase from *Coriolus versicolor* (D73369). In *P. chrysosporium*, at least four distinct AAO-like sequences, a pyranose 2-oxidase-like sequence, and a glucose oxidase-like sequence have been identified in the genome data. The pyranose 2-oxidase, which is a membrane-bound protein in ligninolytic fungi (Daniel et al., 1994), was found to be strongly upregulated when *P. chrysosporium* is cultivated on lignin, which underlines its potential role as a source of hydrogen peroxide (Manavalan et al., 2011). The precise roles and interactions of all these genes in lignin degradation remains yet to be determined, but when viewed together with the copper radical oxidase genes, it is clear that *P. chrysosporium* possesses an impressive array of genes encoding extracellular oxidative enzymes (Kersten and Cullen, 2007).

5.7 Cellobiose Dehydrogenase

Another enzyme that is of major relevance to lignin degradation because it connects cellulose decomposition to lignin decay is the extracellular cellobiose dehydrogenase (CDH, EC 1.1.99.18) (Zamocky et al., 2006; Peterbauer and Volc, 2010). It is the only currently known extracellular flavocytochrome and is formed by a number of wood-degrading fungi and also by various phytopathogenic fungi. Comprehensive sequence analysis of CDH-encoding genes and their translational products reveals significant sequence similarities along the entire sequences and also common domain architecture. All known CDHs fall into two related subgroups: class I members are represented by sequences from basidiomycetes, whereas class II comprises longer, more complex sequences from ascomycetes. CDH is typically a monomeric protein consisting of two domains (a larger one containing the dehydrogenase active center and the flavin, whereas the smaller domain binds heme), joined by a protease-sensitive linker region. The heme-binding domains contain unique sequence motifs that are conserved in all currently known flavocytochrome-binding proteins. Each cytochrome domain of CDH can bind a single heme b as prosthetic group. The dehydrogenase domain belongs to the so-called "glucose-methanol-choline oxidoreductase superfamily," a widespread evolutionary line of flavoproteins. It can be further divided into a flavin-binding subdomain and a substrate-binding subdomain. In addition, the class II (but not class I) proteins can possess a short cellulose-binding module belonging to the CBM1 family at their C-termini (Zamocky et al., 2006).

All CDHs oxidize cellobiose, cellodextrins, and lactose to the corresponding lactones and thereby transfer electrons to quinones and oxidized phenolic intermediates, which may be formed during lignin decomposition (Zamocky et al., 2006). Several hypothetical mechanisms of CDH involvement in the degradation of cellulose (and also hemicelluloses and lignin) have been proposed, such as the reduction of substrate inhibition by cellulolysis products, reduction of quinones to be used by ligninolytic enzymes or the support of a Mn-peroxidase reaction (Baldrian and Valášková, 2008). The hypothesis currently most accepted by the scientific community is that CDH degrades and modifies cellulose, hemicelluloses, and lignin by generating hydroxyl radicals in a Fenton-type reaction (Kremer and Wood, 1992; Mansfield et al., 1997). Depolymerization with CDH, cellobiose, Fe^{3+}, and H_2O_2 was demonstrated for carboxymethylated cellulose, water-soluble xylan, and radioactively labeled synthetic lignin; some depolymerization occurred even without added H_2O_2 due to its formation by the enzyme itself (Baldrian and Valaskova, 2008). The impact of this reaction, especially for brown rot fungi, has been described in Chapter 5.

5.8 Enzymes Essential for Oxalic Acid Formation

Evidence points to the possibility that decomposition of wood and non-wood lignocellulose is promoted by fungal secretion of oxalic acid (Baldrian and Valášková, 2008). White rot fungi usually produce less oxalate than brown rot fungi, which is interpreted by the fact that excess oxalate inhibits the activity of ligninolytic peroxidases (Baldrian and Valášková, 2008). However, because a stable concentration of oxalate is necessary for the functioning of the ligninolytic system of white rot fungi, the oxalate concentration is regulated by the production of oxalate decarboxylase (Kurek and Gaudard, 2000). As explained in Chapter 2, oxalate was also proposed to directly participate in cellulose hydrolysis in brown rot fungi by a Fenton-type mechanism, while H_2O_2 may also be produced during the synthesis of oxalate by glyoxalate oxidase (Shimada et al., 1997; but see later). Oxalate has also been proposed to function as an extracellular buffering agent, thus allowing the fungus to control the pH of its environment (Zapanta and Tien, 1997). It may also act as calcium sequester to increase the pore size of the plant cell wall and to facilitate the penetration of enzyme molecules. Oxidation of oxalic acid by Mn(III) produces a formate radical ($HCO_2.^-$) that reacts with dioxygen to form superoxide ($O_2.^-$) and, subsequently, H_2O_2 (Khindaria et al., 1994; Urzua et al., 1998). This process has thus also been implicated in contributing to the ability of the fungus to degrade lignin.

There are two routes for biosynthesis of oxalate: hydrolysis of oxaloacetate by oxaloacetate hydrolase (EC 3.7.1.1) or oxidation by an FAD-dependent glyoxalate oxidase (EC 1.2.3.5; Figure 5.13). Work in *Fomitopsis palustris* and *Sclerotium rolfsii* (Sakai et al., 2006; Schmid et al., 2010) provided strong support for the main operation of the former, which is also in good agreement with other findings that the most important pathway leading to oxalate formation in asco- and basidiomycetes is catalyzed via an oxaloacetate hydrolase and thus solely depends on oxaloacetate as precursor and not on glyoxylate. Irrespectively of which pathway operates, these enzymes are intracellular. There must therefore be an efficient mechanism for transport of oxalate out of the cell. It has been studied in *Fomitopsis palustris* and found to be ATP dependent and strongly inhibited by valinomycin and NH_{4+}, suggesting that it is a secondary transporter. It possesses six transmembrane domains but shows no similarity with known oxalate transporters from other organisms (Watanabe et al., 2010).

5.9 Glycopeptides

Low molecular mass proteinaceous compounds termed "glycopeptides" were first detected in brown rot fungi in 1980s and identified as compounds capable of catalyzing redox reactions between O_2 and electron donors to produce •OH, reduce Fe^{3+} to Fe^{2+}, and to strongly bind Fe^{2+} (for review, see Baldrian and Valášková, 2008). These compounds were initially described as 1.5–5 kDa peptides that contained iron and had 20–25% neutral carbohydrate bound (Hirano et al., 1995) from *Gloeophyllum trabeum* (Enoki et al., 1994) and *Irpex lacteus* (Tanaka et al., 1996). Later, larger glycopeptides have been described in white rot fungi *T. versicolor* and *P. chrysosporium,* where they were claimed to be the major source of hydroxyl radicals in wood-degrading cultures and to act synergistically with phenol oxidases in lignin degradation (Yamakawa et al., 2005).

Recently, the glycopeptide from *P. chrysosporium* was characterized in more detail (Tanaka et al., 2007). The preparation had a molecular mass of about 14 kDa and contained 25% neutral carbohydrate and 0.04% Fe. The glycoprotein contained 1-amino-1-deoxy-2-ketose (ketoamine) that is formed by the condensation of an amino acid side chain and a carbohydrate. Upon tautomerization, this structure can yield a 2,3-enediol that can reduce Fe^{3+} to Fe^{2+} and produce H_2O_2 from O_2 (see Chapter 2).

The size of glycopeptides, albeit small, still does not allow them to penetrate the intact wood cell wall and the reduction of their substrates thus probably occurs close to fungal hyphae although some diffusion into the cell wall was demonstrated (Hirano et al., 2000). To perform the complete catalytic cycle, oxidized saccharidic moieties of the glycopeptides have to be reduced again. This can potentially be performed by a cell wall-associated reductase that is most probably NADH-dependent (Enoki et al., 1994). However, the identity of the electron donor is not yet known.

Chapter 6

Catabolic Pathways of Soluble Degradation Products from Plant Biomass

The action of the cellulases and hemicellulases described in Chapters 3 and 4 result in the production of soluble oligo- and monosaccharides, which can be used by the respective fungi for generation of energy, growth, and development. The biochemical mechanisms and pathways by which these compounds are taken up into the cells and catabolized are described in this chapter.

6.1 Uptake of Mono- and Oligosaccharides

The uptake of oligo- and/or monosaccharides occurs by members of the sugar porter family, which is the largest family within the so-called major facilitator superfamily (MFS). The Transporter Classification system database (Saier et al., 2006; http://www.tcdb.org) details a comprehensive IUBMB-approved classification system for membrane transport proteins, analogous to the EC system for classification of enzymes, but it also incorporates both functional and phylogenetic information without which this classification would be impossible. Thereby, the transport systems are classified by the use of five numbers or letters—V.W.X.Y.Z: V (a number) corresponds to the transporter class (i.e., channel, carrier (porter), primary active transporter, or group translocator); W (a letter) corresponds to the transporter subclass that, in the case of primary active transporters, refers to the energy source used to drive transport; X (a number) corresponds to the transporter family (sometimes actually a superfamily); Y (a number) corresponds to the subfamily in which a transporter is found; and Z corresponds to the substrate or range of substrates transported.

Virtually all sugar transporters found in filamentous fungi are classified in this system and found to belong to the MSF family. They exhibit strong structural conservation, but share little sequence similarity (Leandro et al., 2009). They generally consist of a single integral membrane protein comprising two sets of six hydrophobic transmembrane-spanning α-helices

Fungi and Lignocellulosic Biomass, First Edition. Christian P. Kubicek.

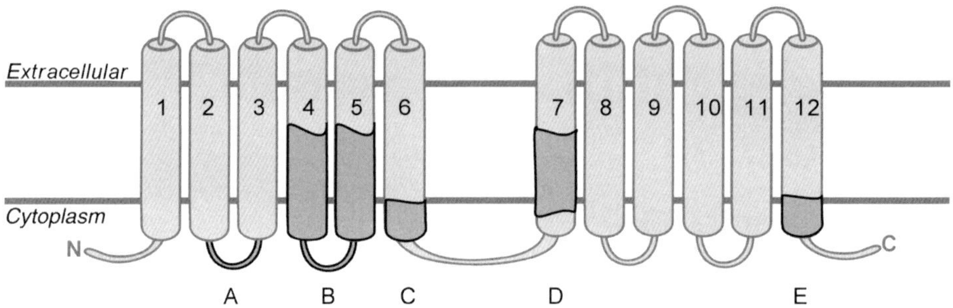

Figure 6.1. Topology of yeast sugar transporters. Generally, monosaccharide transporters in yeasts have 12 hydrophobic transmembrane domains (represented as cylinders in the cartoon). The position of the five conserved sequence motifs that have been recognized in sugar transporters is represented in the figure by the letters (A–E). Regions conserved within sugar transporters are marked in darker grey.

that are connected by a hydrophilic loop (Figure 6.1). They have likely evolved by gene duplication of a six-helix-containing bacterial ancestor (Katz and Kelly, 2010). Five conserved sequence motifs are present in these proteins, which aid in their identification (letters A–E in Figure 6.1): A, (R/K)XGR(R/K); B, (the so-called "diffused" motif) R-X_3-G-X_3-G-X_6-P-X-Y-X_2-**E**-X_6-R-G-X_6-Q-X_5-**G**; C, PESPRXL; D, [LI]-Q-X_2-Q-Q-X-[ST]-[GN]-X_3-Y-Y-E; and E, PETKGXXXE (Leandro et al., 2009). The conserved E and **G** in motif B are thought to have important catalytic and structural roles, respectively (Pao et al., 1998).

Most fungal sugar transporters operate either by facilitated diffusion, an energy-independent mechanism that catalyses the movement of a single solute molecule down its own concentration gradient across a membrane, or by proton symport, that is, the transport of a solute against its concentration gradient by coupling it to the movement of protons along their gradient (Abramson et al., 2004). On the basis of the number of MSFs detected in asco- and basidiomycetous genomes, multicellular fungi appear to have a large number of sugar transporters (although not all of the MFSs may necessarily transport sugars, but also, for example, polyols and other compounds). Unfortunately, only few of these transporters have been biochemically characterized, and most of these are hexose (or glucose) transporters.

In order to ensure an economic uptake of sugars from the medium, the transporters can have different affinities for the sugar to be transported, and they are usually distinguished as "low-affinity" (which have K_m values in the mM range) and "high-affinity" transporters (having K_m values in the μM range). The latter are frequently repressed by glucose. Their function is to transport the respective sugars under conditions when they are available only in low concentrations (e.g., also during the degradation of plant polysaccharides). The low affinity transporters in the presence of carbohydrate concentrations that accumulate in the environment of the fungus in this concentration (usually over 1 g/L).

Table 6.1 presents a summary of the sugar transporters that have been characterized from multicellular fungi so far. Because of the redundancy of transporters in their genomes, their function cannot be revealed by the use of knockout mutants, but only by expression in yeast, by overexpression in the original fungal strain, or by incorporating it into isolated membrane vesicles. Using such techniques, it was shown that some transporters are strictly specific for only glucose, whereas others (then more correctly called hexose transporters) also transport mannose and fructose. Galactose appears to require its own transporter, which has not yet been functionally identified for any multicellular fungus.

Table 6.1. Characterized fungal hexose transporters.

Organism	Name	Affinity for Glucose
Amanita muscaria	AmMST1	NI
Aspergillus nidulans	HktA	High
	MstE	Low
	SorA	High
Aspergillus niger	MstA	High
Neurospora crassa	RCO3	None[a]
	HGT1	High
Trichoderma harzianum	GTT1	High
Tuber borchii	TBHXT1	High
Uromyces fabae	Hxt1p	Medium

Data modified from Katz and Kelly (2010).
[a]RCO3 is supposed to bind glucose and act as a glucose sensor.

Some information is available on the transport of the pentoses xylose and arabinose, because they were specifically looked for to complement yeast strains for transport of pentoses during ethanol formation from hemicelluloses hydrolysates (Matsushika et al., 2009). Xylose permeases have thus been cloned from *Trichoderma reesei* (Saloheimo et al., 2007) and *Neurospora crassa* (Du et al., 2010) but none of them has been further characterized. To my knowledge, no transporters specific for L-arabinose have so far been characterized from any multicellular fungus. A similar situation occurs for the transport of the sugar acids entry of glucuronic and galacturonic acid into fungal cells.

Fungi also contain transporters for soluble di- and oligosaccharides. A β-glucoside transporter, that is induced during growth on cellulose, has been demonstrated in *T. reesei* (Kubicek et al., 1993), but its gene not yet been cloned. However, a gene encoding a cellooligosaccharide transporter has recently been cloned and characterized from *N. crassa* (NCU08114 and NCU00801), and orthologs of these genes have also been identified in other fungi (Galazka et al., 2010). The *T. reesei* ortholog is strongly induced under cellulase-inducing conditions (Ivanova et al., 2012). Transporters for β-galactosides and xylooligosaccharides are probably also present in at least some fungal species but so far none of them has been cloned or characterized.

6.2 Metabolism of D-Glucose and D-Mannose

D-Glucose is the major carbon source in the metabolism of most heterotrophic eukaryotes. Although most fungi will rarely encounter free glucose in their natural environment, it is the major component of cellulose and various glucans as I have outlined in the previous chapters. The canonical glycolytic pathway serves as the major catabolic route and also serves as an acceptor for other hexoses that arise as of breakdown products from extracellular polysaccharides (such as D-mannose or D-galactose). Annotations of the genomes from several asco- and basidiomycetes have demonstrated the presence of genes for all the enzymes required for glycolysis to occur. Interestingly, many fungi contain multiple genes for some of the steps, notably of aldolase, glyceraldehydes-3-phosphate dehydrogenase, triose-phosphate isomerase, and phosphoglycerate kinase. However, it is not known whether they all are indeed involved in glycolysis or may function in other pathways and whether they are consistently or differentially regulated.

Also, all the genes for the pentose phosphate pathway are present, and the occurrence of two or more genes for some steps is again common. The simultaneous operation of glycolysis and the pentose phosphate pathway during growth on glucose has been documented by ^{14}C-radiorespirometry and ^{13}C-NMR-based metabolic flux analysis, revealing that glycolysis accounts for 50–80% of glucose catabolism (Jouhten et al., 2009; Scandellari et al., 2009).

Pyruvate, the end product of glycolysis, can either enter the mitochondria or become subject to oxidative decarboxylation, leading to acetyl-CoA formation and followed by catabolism via the tricarboxylic acid (TCA) cycle and the respiratory chain. Alternatively, it may serve as an acceptor for the NADH-generated during glycolysis when the capacity of the TCA cycle and the respiratory chain is limited. In the yeast *Saccharomyces cerevisiae*, an excess of glucose represses the expression of some components of the respiratory chain even in the presence of oxygen, thus leading to aerobic ethanol formation (the so-called Crabtree effect; for review, see Nevoigt, 2008). However, such an effect is absent from multicellular fungi: several studies have investigated how filamentous fungi (*T. reesei, Aspergillus oryzae, N. crassa, and Fusarium spp.*) react to an excess of glucose or a shortage of oxygen, and while some indeed showed the formation of small amounts of ethanol, respiratory metabolism remained dominant in all of these cases.

Besides glycolysis and the pentose phosphate pathway, many fungi can use a further, oxidative route for glucose catabolism, which can be initiated either by the flavoprotein glucose oxidase (EC 1.1.3.4) or glucose-1-dehydrogenase (β-D-glucose:NAD(P)$^+$ 1-oxidoreductase). The former acts preferentially extracellular, whereas the latter is an intracellular enzyme. Further catabolism of the thus formed gluconate proceeds via gluconokinase, and the resulting 6-phosphogluconate can enter the pentose phosphate pathway. The physiological role of this pathway that bypasses glucose phosphorylation is not known. Glucose oxidase produces H_2O_2, and it could thus be a mechanism for antagonism of other organisms as has been shown for *Talaromyces emersonii* (Kim et al., 1990). Glucose dehydrogenase could become relevant when a large portion of the available glucose is in the β-D-anomeric configuration, that is, during catabolism of cellulose and other β-glucans. Somewhat surprising, glucose catabolism has not yet been studied under the conditions of plant biomass polymer degradation.

The catabolism of D-mannose is closely linked to glycolysis: mannose can be phosphorylated by the same hexokinase protein that phosphorylates glucose and fructose, and the resulting mannose-6-phosphate can be isomerized to fructose-6-phosphate by mannose-6-phosphate isomerase, whose gene is present in all fungal genomes released so far.

6.3 Catabolism of D-Galactose

The canonical catabolic pathway for D-galactose is the Leloir pathway, which involves the subsequent operation of galactokinase (EC 2.7.1.6), D-galactose-1-phosphate uridylyl transferase (EC 2.7.7.12), and UDP-galactose 4-epimerase (EC 5.1.3.2) to convert D-galactose to D-glucose 1-phosphate. This is then—by the action of phosphoglucomutase (EC 2.7.5.1)—transformed to D-glucose-6-phosphate, and thus glycolysis (Figure 6.2). Genes encoding the above-mentioned enzymes are present in all multicellular fungi, but in contrast to yeast (Slot and Rokas, 2010), none of them are clustered in the genome.

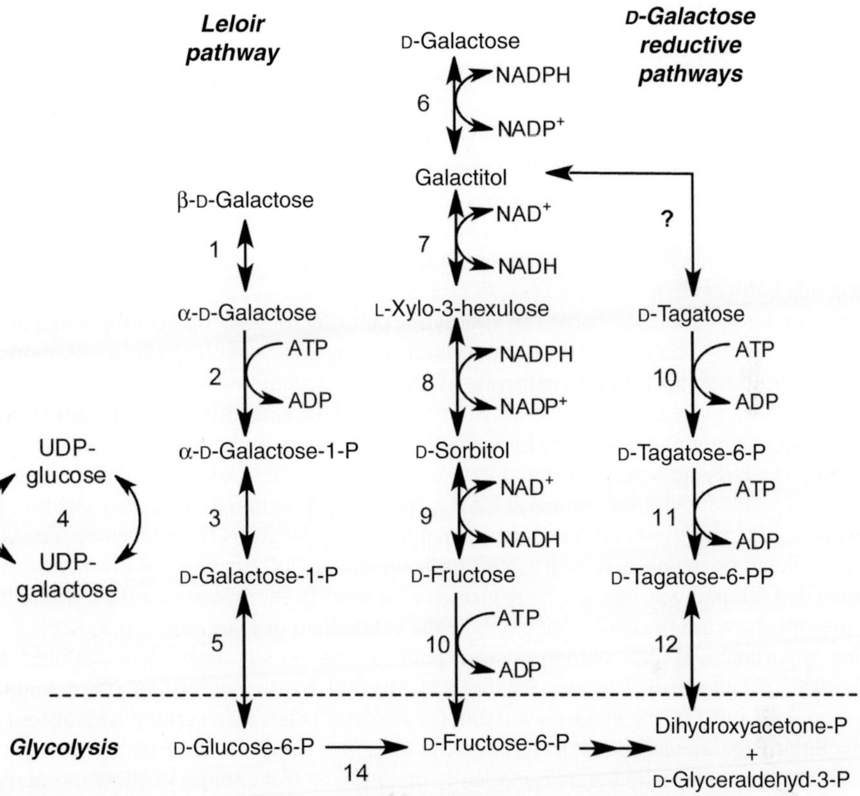

Figure 6.2. Pathways for D-galactose catabolism in fungi. Involved enzymes are indicated by numbers: 1, aldose epimerase; 2, galactokinase; 3, galactose-1-phosphate-uridyltransferase; 4, UDP-galactose-4-epimerase; 5, phosphoglucomutase; 6, aldose reductase; 7, L-arabinitol dehydrogenase; 8, L-xylulose reductase; 9, D-sorbitol/xylitol dehydrogenase; 10, hexokinase; 11, phosphofructokinase; 12, D-tagatose-1,6-bisphosphate aldolase; and 13, phosphoglucoisomerase.

A significant difference between the Leloir pathway in multicellular fungi and yeast, yet, resides in the structure of UDP-galactose-4-epimerase: the respective enzyme from yeasts consists of two domains, of which the C-terminal one encodes a mutarotase, which is lacking from multicellular fungi. Interestingly, two putative UDP-glucose-4-epimerases (GenBank accession numbers T40321 and CAC21414, respectively) are present in *Schizosaccharomyces pombe*, of which one resembles the longer "yeast" version of the protein, whereas the other is similar to the short version of the multicellular fungi. Parsimony analysis of the aa-sequences from position 1–32,000 reveals that the shorter gene from *S. pombe* clusters in a basal position to both its own longer gene and the genes from the other yeasts and filamentous fungi, indicating that the long version has been developed from the shorter one during evolution of the yeasts by incorporating an ancestral mutarotase gene (Seiboth et al., 2002). This hypothesis is supported by the occurrence of respective mutarotase (= aldose-1-epimerase) genes with high similarity to the C-terminal domain in yeast Gal10p in the genomes of multicellular fungi. For example, *T. reesei* contains three putative aldose 1-epimerase genes (*aep1-3*), of which two encode intracellular proteins (AEP1-2) and one a putative extracellular protein (AEP3) (Fekete et al., 2008).

What is the necessity for these mutarotase genes? The key to understanding this phenomenon is based on the stereospecificity of the D-galactopyranose cleaved by β-galactosidases from plant polysaccharides: D-galactose arising by such hydrolysis is in the β-anomeric form. Unfortunately, the first enzyme of the Leloir pathway—galactokinase—is absolutely specific for α-D-galactose and will not phosphorylate the β-D-anomer. While the β-anomer will be converted to the α-anomer by chemical mutarotation, this is a slow process and would be rate limiting to the Leloir pathway. The presence of aldose-1-epimerases is thus necessary to bypass this bottleneck.

However, fungi have also evolved an alternative pathway for galactose catabolism that works with both α- and β-D-galactose: the oxidoreductive galactose catabolic pathway (Seiboth et al., 2004, 2007a; Figure 6.4). In this pathway, whose existence and operation has been proven in *T. reesei* and *A. nidulans*, D-galactose is first reduced to galactitol by an NADPH-specific aldose reductase, and then oxidized by an NAD-dependent polyol dehydrogenase to L-xylo-3-hexulose. The enzyme acting on this uncommon sugar has been identified in *T. reesei* as the NADPH-dependent xylulose reductase LXR4. It converts L-xylo-3-hexulose to sorbitol, which finally is oxidized by another NAD-dependent polyol dehydrogenase to D-fructose. The latter is phosphorylated by hexokinase to fructose-6-phosphate, an intermediate of glycolysis. It must be noted that this pathway has also been termed a "stowaway shunt" because the oxidoreductive enzymes involved are the same that catalyze the catabolism of L-arabinose (see Section 6.4.).

The importance of this oxidoreductive pathway has nicely been demonstrated in the catabolism of lactose in *T. reesei* (Seiboth et al., 2007b). In contrast to other fungi such as *A. nidulans*, most of the lactose available for *T. reesei* is not taken up but hydrolyzed by an extracellular β-galactosidase BGA1 (Seiboth et al., 2005). The arising β-D-galactose is apparently neither chemically nor enzymatically mutarotated fast enough to allow its catabolism via the Leloir pathway because mutants in the aldose reductase XYL1 that catalyzes the first step in the oxidoreductive galactose catabolic pathway display a strongly reduced growth on galactose. In contrast, mutants in the first enzyme of the Leloir pathway (galactokinase GAL1) show almost normal growth on lactose (Seiboth et al., 2007b).

In addition, genes encoding galactose oxidase (EC 1.1.3.9) and galactolactonase (EC 3.1.1.25) have been detected in several fungal genomes, particularly *Aspergillus* species, and could point to the presence of an oxidative galactose catabolic pathway as well (Flipphi et al., 2009; Figure 6.2). A galactonate dehydratase or D-galactonate hydrolyase (2-dehydro-3-deoxy-D-galactonate-forming, EC 4.2.1.6) are also present in the *Aspergillus* genomes, but the kinase that phosphorylates this latter compound for further metabolism (EC 2.7.1.58) has not been identified. The relevance of this pathway for the catabolism of D-galactose is thus unclear. Some of the genes mentioned above are also involved in the degradation of pectin-derived D-galacturonate (Kourelathi et al., 2006; see later). A non-phosphorylated pathway for catabolism of D-galactonate has been proposed for *A. terreus* by Elshafei et al. (1995), based on enzyme activities in cell-free extracts, but its function *in vivo* has not yet been proven.

Flipphi et al. (2008) have recently hypothesized on a yet further pathway for D-galactose utilization, based on the identification of a Class-II D-tagatose bisphosphate aldolase (EC 4.1.2.40) in the *Aspergillus* genomes. These enzymes have a wider substrate range and bear similarity to two *Escherichia coli* proteins involved in the degradation of galactitol and *N*-acetylgalactosamine, and could therefore split L-xylo-3-hexulose into two molecules of glyceraldehydes-3-phosphate. Whether or not this pathway indeed is functional and relevant for galactose catabolism has to proven first.

6.4 Catabolism of Pentoses

The catabolism of pentoses by fungi is unique in the kingdom of life. In prokaryotes, the predominant pathway is based on isomerase steps by which D-xylose is converted to D-xylulose, and L-arabinose to L-ribulose (Figure 6.3). D-xylulose 5-phosphate, an intermediate of the canonical pentose phosphate pathway, can therefore be formed by xylulokinase (in the case of D-xylose) and a sequence of L-ribulokinase and L-ribulose-5-phosphate 4-epimerase (in the case of L-arabinose; Mishra and Singh, 1993). Some bacteria have also developed alternative oxidative pathways, which result in either the TCA cycle metabolite α-ketoglutarate or pyruvate and glycolaldehyde, but which are not relevant here.

Fungi, in contrast, have elaborated an oxidoreductive pathway, which ultimately also produces D-xylulose 5-phosphate (Figure 6.4; Seiboth and Metz, 2011). This pathway is characterized by alternating oxidoreductive reactions, such as (for D-xylose) xylose to xylitol and xylitol to xylulose, and (for L-arabinose) L-arabinose to L-arabinitol, L-arabinitol to L-xylulose, L-xylulose to xylitol, and xylitol to D-xylulose. Thus, the last enzymatic step in the xylose and arabinose pathway is the same and the two pathways are therefore linked. In the final step, D-xylulose is phosphorylated to D-xylulose 5-phosphate.

A noteworthy property of these two reaction chains is that (in the direction of catabolism) the reductive step is always catalyzed by an NADPH-dependent enzyme, whereas the oxidative step is due to an enzyme acting with NAD+. Thus, although the overall redox balance of these two pathways is neutral, it results in an imbalance of the redox cofactors and thus the rate of catabolism is slow when compared to hexoses like D-glucose, and the accumulation and excretion of polyol intermediates is a commonly observed phenomenon. Genes encoding the enzymes for these pathways are present in most fungal genomes, and in some cases, an

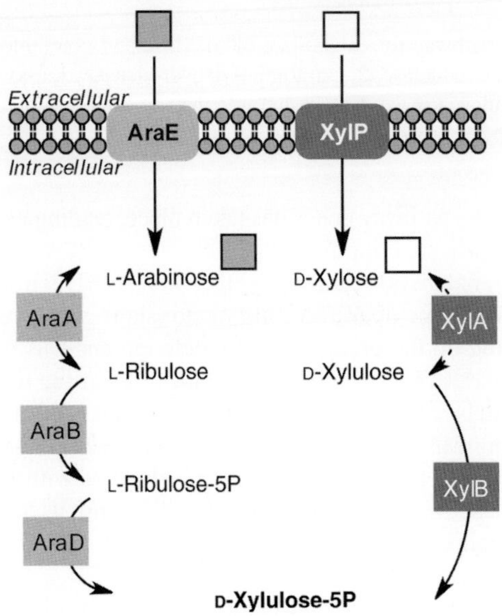

Figure 6.3. Bacterial pathway for catabolism of D-xylose and L-arabinose. The respective permeases and proteins are given by their three letter code: AraE, arabinose transporter; XylP, xylose transporter; AraA, L-arabinose isomerase; AraB, L-ribulose kinase; AraC, L-ribulose-5-phosphate isomerase; XylA, D-xylose isomerase; and XylB, D-xylulose kinase.

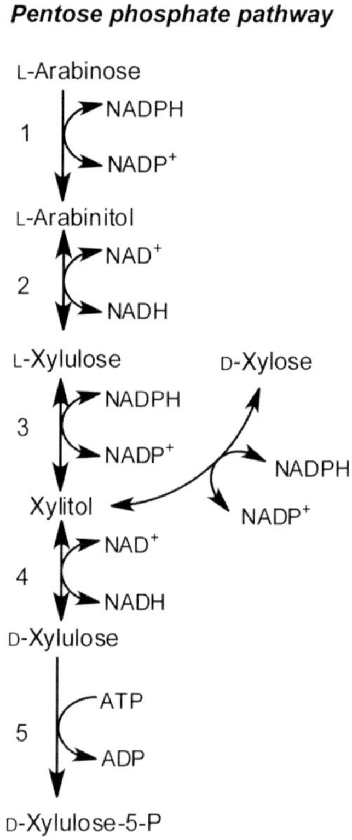

Figure 6.4. The fungal pathway for catabolism of D-xylose and L-arabinose. Enzymes involved are given by numbers: 1, Aldose reductase; 2, L-arabinitol dehydrogenase; 3, L-xylulose reductase; 4, xylitol dehydrogenase; and 5, xylulose kinase. Note reaction 1 may be catalyzed by different enzymes in some fungi (e.g., *Aspergillus* spp.).

amplification of some of the pathway genes has taken place, leading to a number of paralogous genes (e.g., in the Aspergilli; cf. Flipphi et al., 2008).

The nature of the enzyme(s) catalyzing the first NADPH-dependent reduction of L-arabinose and D-xylose—a member of the aldo-keto (short chain) reductase/dehydrogenase family—differs depending on the organism: gene deletion analysis of *T. reesei xyl1* (encoding the D-xylose reductase XYL1, which accounts for most of the reductase activity on both L-arabinose and D-xylose) is responsible for both D-xylose and L-arabinose catabolism because its absence results in significantly impaired growth on D-xylose and L-arabinose (Seiboth et al., 2007a; Akel et al., 2009). In *A. niger*, however, a reductase with preferred specificity for L-arabinose has been described (de Groot et al., 2005) and its gene (*larA*) been cloned (Mojzita et al., 2010). Strains bearing a *larA* gene knockout show significantly reduced growth on L-arabinose but not D-xylose.

The polyol dehydrogenases catalyzing the oxidation of arabinitol and xylitol have been identified in *T. reesei* and *A. niger* and can compensate for each other: L-arabinitol 4-dehydrogenase was first cloned in *T. reesei,* and deletion of its gene leads to a loss of growth on L-arabinose and L-arabinitol. Genetic evidence further showed that the *T. reesei* LAD1 can partially compensate for the loss of the xylitol dehydrogenase step in an *xdh1*-negative strain during growth on D-xylose and xylitol (Seiboth and Metz, 2011). A xylitol dehydrogenase was first characterized in *T. reesei*: it is induced by D-xylose and L-arabinitol, but not xylitol or L-arabinose. Deletion of both *lad1* and *xdh1* in *T. reesei* fully impairs growth on L-arabinose or D-xylose and its corresponding polyols (Seiboth and Metz, 2011).

The identification of the enzyme that catalyzes the second reducing step, that is, L-xylulose reductase (LXR) proved to be difficult because of the presence of many potential candidates in the fungal genomes: the first gene encoding a fungal enzyme with LXR activity (LXR1) was identified in *T. reesei* by Richard et al. (2002), but subsequently turned out to be a D-mannitol dehydrogenase involved in developmental processes (Metz et al., 2009). Only recently, an NADPH-dependent LXR was identified from *T. reesei*, whose deletion results in reduced growth on L-arabinose and L-arabinitol (B. Metz, C.P. Kubicek and B. Seiboth, unpublished data).

Phosphorylation of D-xylulose is catalyzed by xylulokinase, which has been investigated in *A. niger* and shown to be essential for L-arabinose and D-xylose catabolism.

Interestingly, genome annotation showed the presence of a gene encoding a putative phosphoketolase (EC 4.1.2.9). This enzyme could cleave xylulose-5-phosphate to glyceraldehyde-3-phosphate and acetyl phosphate, and thus bypass the pentose phosphate pathway (Flipphi et al., 2008). However, proof for the actual operation of this shortcut is still lacking.

Berghäll et al. (2007) identified an NADP-dependent D-xylose dehydrogenase in *T. reesei*, which could be part of an oxidative catabolic pathway for D-xylose. However, the role of this enzyme is still unclear because deletion of the *xyl1* and *lad1*/*xdh1* genes have already cause an almost complete decrease in growth on D-xylose.

6.5 Catabolism of Hexuronic Acids

D-galacturonic acid is the major component of the plant cell wall polysaccharide pectin. Its catabolism occurs via a nonphosphorylating pathway in fungi, which has been studied in detail in *T. reesei, A. niger,* and *Botrytis cinerea* (Richard and Hilditch, 2009). The pathway is sum-

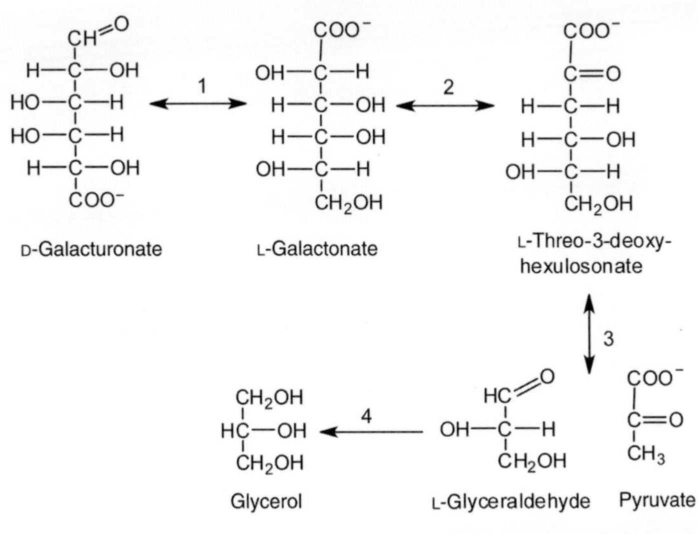

Figure 6.5. Fungal pathway for D-galacturonate catabolism. Enzymes involved are given by numbers: 1, D-galacturonate reductase; 2, D-galactonate dehydratase; 3, L-threo-deoxy-hexulosonate aldolase; and 4, glyceraldehyde reductase.

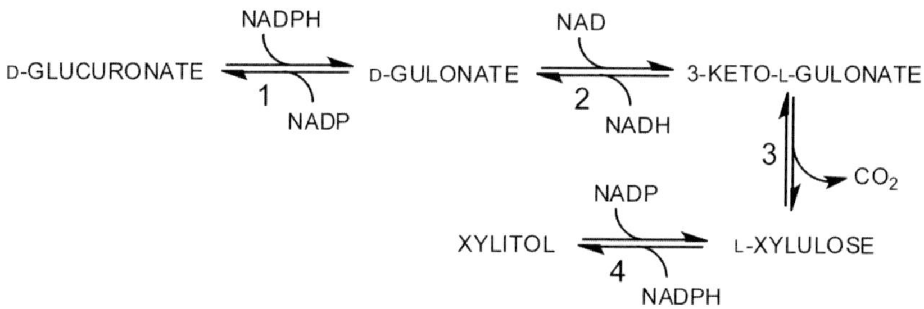

Figure 6.6. Hypothetical pathway for D-glucuronate catabolism in fungi, based on findings in mammals. The respective enzymes are given by numbers: 1, D-glucuronate reductase; 2, L-gulonate-3-oxidoreductase; 3, 3-keto-1-gulonate carboxylase; and 4, xylitol dehydrogenase.

marized in Figure 6.5. In the first step, D-galacturonic acid is first converted to L-galactonate by a galacturonate reductase. Two nonorthologous isoenzymes exist in the good pectin degraders *B. cinerea* and *A. niger*, whereas *T. reesei,* which degrades pectin rather poorly, contains only a single enzyme. The second enzyme of the pathway is L-galactonate dehydratase, which converts L-galactonate to 2-keto-3-deoxy-L-galactonate. 2-Keto-3-deoxy-L-galactonate aldolase finally splits 2-keto-3-deoxy-L-galactonate into pyruvate and L-glyceraldehyde (Hilditch et al., 2007).

In contrast, the fungal pathway for catabolism of glucuronate, arising from xylan degradation, has not received any attention so far. The catabolic pathway in mammals proceeds via D-glucuronate reductase, an aldehyde reductase of the aldo-keto reductase superfamily, L-gulonate NAD(+)-3-oxidoreductase, L-iditol-NAD(+)-5-oxidoreductase, and xylulose reductase to D-xylulose (Figure 6.6; Goode et al., 1996). Thus, this pathway leads into the pentose catabolic pathway that has been described in Section 6.4. Owing to the fact that xylose and arabinose induce the α-glucuronidase from *A. niger* and *T. reesei* (de Vries et al., 1998; Kubicek and Penttilä, 1998), it would be tempting to test for the presence of this pathway in filamentous fungi.

Chapter 7

Regulation of Formation of Plant Biomass-Degrading Enzymes in Fungi

The transcription, translation, and secretion of the array of extracellular enzymes needed for the efficient degradation of the plant biomass polymers is a process that requires considerable energy. Consequently, fungi have developed sophisticated mechanisms that ensure that these enzymes are formed only under conditions where these plant polymers are available as an energy and carbon source. Clearly, this necessitates the presence of efficient systems to sense the type (or composition) of the plant biomass in order to ensure that all the enzymes that are needed for the synergistic degradation are indeed formed. Since different fungi have become specialized for different niches comprising differently composed plant polymers, it is not surprising that this cross-talk in the induction of expression of the genes encoding different classes of enzymes can be different in different fungi, for example, white rot and soft rot fungi, but also, for example, *Aspergillus niger* and *Trichoderma reesei*. In addition, most of the respective polysaccharide catabolic genes are repressed by the presence of easily metabolizable carbon sources that arise from the degradation of these plant polymers (notably glucose, but also galactose or mannose). This mechanism ensures that the rate of hydrolysis of the polymers does not occur in excess of the rate of the uptake and intracellular metabolism of the mono- or oligomers, as this would attract competing organisms that are not able to attack the polymers themselves.

In this chapter, I will try to outline the major strategies that have evolved in different fungi to achieve the goal of economic plant polymer degradation.

7.1 The Cellulase Inducer Enigma

Because of the reasons outlined earlier, cellulases and hemicellulases are not formed in the absence of plant biomass polymers as a substrate and are thus adaptive enzymes. However, since these polymers are insoluble and cannot be taken up by the fungi, what is signaling their presence to the fungi? This problem was studied mainly with respect to cellulase formation

Fungi and Lignocellulosic Biomass, First Edition. Christian P. Kubicek.

in *Trichoderma* and several lines of explanations have been offered, most of which emphasize the formation of a low molecular weight and soluble-inducing molecules from celluloses and hemicelluloses: a popular model postulates that a low level of cellulase activity is formed constitutively and can start the degradation of cellulose whenever it becomes available, thereby releasing small amounts of oligosaccharides that then induce further cellulase biosynthesis. Evidence for this theory was provided by the findings that the addition of anti-cellulase antibodies to *T. reesei* blocked the induction of cellulase formation by cellulose (El-Gogary et al., 1989). In support of this, Carle-Urioste et al. (1997) demonstrated the presence of transcripts of *cel7A* and *cel7B* in uninduced *T. reesei* cells, and further 1,100-fold induction in the presence of cellulose. Expression of a hygromycin B phosphotransferase gene under the *cel7A* promoter conferred hygromycin B resistance to *T. reesei* in the absence of cellulose. Foreman et al. (2003) identified several further genes that displayed a low basal expression level in the absence of cellulose, which could play a role in primary inducer formation for cellulase expression. One of them—the cellulase CEL5B—is particularly interesting as it is expressed under all conditions tested and contains the consensus sequence for membrane-anchoring via a glycosylphosphatidylinositol residue. Its constitutive presence and possible location close to the cell membrane make this enzyme an interesting candidate for inducer formation, which has however not yet been tested by reverse genetics.

Another hypothesis also supported by experimental evidence emphasizes the fact that conidia of *Hypocrea jecorina* contain an array of plant polymer-degrading enzymes (Table 7.1). Their removal by nonionic detergents impairs germination of the conidia on cellulose (Kubicek et al., 1988). Metz et al. (2011) showed that they are transcribed during the early phase of conidiation in the absence of any exogenous inducer. A search of the conidiation-specific transcriptome of *Neurospora crassa* or *A. fumigatus* did not detect any upregulation of genes encoding cellulases or hemicellulases, suggesting that this mechanism is specific for *T. reesei* and may provide a selective advantage to this fungus to use a lignocelluloses-rich habitat upon eventual arrival of its spores.

A third line of explanation of how insoluble polysaccharides trigger the induction of the respective CAZom for their hydrolysis is based on the detection of cellulase transcripts 20–30 hours after glucose exhaustion (Ilmén et al., 1997). This phenomenon could be misunderstood as a relief from carbon catabolite repression (CCR; see later) but is in fact not related to it. Also, an induction of cellulases by carbon starvation in general can be ruled out as the incubation of *T. reesei* in media lacking any carbon source does not lead to cellulase gene transcription. Thus, this finding still lacks a conclusive interpretation, but it is possible that cellulase-inducing components are derived from the cell wall polymers of the fungal cell wall that is partially recycled under starvation conditions. Alternatively, *T. reesei* is known to start forming chlamydospores rapidly after exhaustion of a fast metabolizable carbon source, and thus the observed cellulase transcripts could be due to those formed by the chlamydospores.

While these three explanatory models are in essence not really different and rather may reflect basically similar mechanisms operating under different physiological conditions (Kubicek et al., 2009), they all imply that an inducer is formed. Since the product of the synergistic action of cellulases is cellobiose and (in the case of high β-glucosidase activities) glucose, it would be logic to assume that one of these two sugars (or a further conversion product) would be the inducer of cellulases. Glucose can be ruled out as an inducer for two reasons: first, its accumulation causes CCR (see later), and second, glucose is a molecule that is also formed from other polysaccharides such as starch and therefore not a specific signal for the presence of plant cell wall components.

Table 7.1. CAZy transcripts found in conidia of *T. reesei*.

	Cellulases	Old Name
120312	GH5 endo-β-1,4-glucanase Cel5A	EGL2
72567	GH6 cellobiohydrolase CEL6A	CBH2
123989	GH7 cellobiohydrolase Cel7A	CBH1
122081	GH7 endo-β-1,4-glucanase CEL7B	EGL1
123232	GH12 endo-β-1,4-glucanase Cel12A	EGL3
49976	GH45 endo-β-1,4-glucanase Cel45	EGL5
Auxiliary proteins		
73638	CBM1 cellulose-binding domain protein	CIP1
123940	CBM1 cellulose-binding domain protein	CIP2
123992	CBM1 expansin domain protein swollenin	SWO1
73643	GH61 carbohydrate monooxygenase CEL61A	EGL4
120961	GH61 carbohydrate monooxygenase CEL61B	
β-glucosidases		
120749	GH1 β-glucosidase CEL1A	BGL2
22197	GH1 β-glucosidase CEL1B	
108671	GH3 β-glucosidase	
104797	GH3 β-glucosidase	
76672	GH3 β-glucosidase CEL3A	BGL1
82227	GH3 β-glucosidase CEL3C	
46816	GH3 β-glucosidase CEL3D	
Xylanases		
120229	GH10 endo-β-1,4-xylanase XYN10	XYN3
123818	GH11 endo-β-1,4-xylanase XYN11B	XYN2
112392	GH11 endo-β-1,4-xylanase XYN11C	XYN5
111849	GH30 endo-β-1,4-xylanase XYN30	XYN4
49081	GH74 xyloglucanase	
121127	GH3 β-xylosidase	BXL1
3739	GH43 β-xylosidase	
Xylan esterases		
65215	CE4 polysaccharide deacetylase	
73632	CE5 acetyl xylan esterase	AXE1
54219	CE5 acetyl xylan esterase	
Arabinases		
123283	GH54 α-arabinofuranosidase	ABF1
76210	GH62, α-L-arabinofuranosidase	ABF2
α-Galactosidases		
55999	GH27 α-galactosidase	
75015	GH27 α-galactosidase	
72704	GH27 α-galactosidase	AGL3
65986	GH27 α-galactosidase	
59391	GH27 α-galactosidase	
72632	GH27-α-galactosidase	AGL1
Endo-β-galactanases		
110894	GH5 β-1,4-endogalactanase	
Pectinases		
112140	GH28 exo-polygalacturonase	PGX1
122780	GH28 exo-rhamnogalacturonase	RGX1
Accessory enzymes		
59689	GH2 Predicted β-mannosidase	
69245	GH2 β-mannosidase	
5836	GH2 β-mannosidase	
56996	GH5 mannan endo-1,4-β-mannosidase	

(*continued*)

Table 7.1. (*Continued*)

	Cellulases	Old Name
5807	GH95 α-fucosidase	
73179	GH95 α-fucosidase	
72488	GH95 α-fucosidase	
72526	GH67 α-glucuronidase	AGU1

From Metz et al. (2011).

Growth in the presence of cellobiose has indeed been shown to induce cellulase expression in many fungal species, including *T. reesei*, *Volvariella volvacea*, *Penicillium janthinellum*, and *A. nidulans*, but not so in others (Aro et al., 2005). A reason for this could be that cellobiose could be hydrolyzed to glucose by the extracellular β-glucosidase, which would stop further cellulase formation either by inhibiting cellobiose transport (Kubicek et al., 1993) or by repressing cellulase gene transcription (see later). Inhibition of transport may be the more relevant mechanism because it has been shown that altering the relative ratio of the uptake and hydrolysis rate (e.g., by the addition of β-glucosidase inhibitors, pH lower than 3 that leads to low activities of β-glucosidase or the use of β-glucosidase knockout strains) increases cellulase induction by cellobiose (Sternberg and Mandels, 1979; Fritscher et al., 1990; Fowler and Brown, 1992). Since growth of the fungus on cellulose releases cellobiose in concentrations that favor the uptake rather than hydrolysis (the K_m of β-glucosidase for cellobiose is 20–40-fold higher than that of the β-glucoside permease; Kubicek et al., 1993), cellobiose could well be the *in vivo* inducer.

Probably, the most well-studied inducer of cellulase formation in *T. reesei* is sophorose (a glucosyl-β-1,2-D-glucose disaccharide), because it is the strongest known soluble inducer, and its presence in the culture medium during growth on cellulose has been demonstrated (Mandels et al., 1962). Moreover, its formation by β-glucosidase via transglycosylation has been demonstrated (Vaheri et al., 1979). The major extracellular β-glucosidase CEL3A has been investigated in this direction: disruption of the *cel3A* gene resulted in a delay in induction of the other cellulase genes by cellulose, but not by sophorose, and a *cel3a*-multicopy strain formed higher amounts of cellulases than the parent strain under nonsaturating concentrations of sophorose (see Kubicek et al., 2009). However, the intracellular β-glucosidase CEL1A has also been shown to produce sophorose and cellobiose from glucose and could therefore be involved in inducer formation as well. Sophorose is also an effective cellulase inducer in *A. terreus* and *P. purpurogenum* but not in *P. janthinellum*, *P. chrysosporium*, *A. nidulans*, and *A. niger* (Aro et al., 2005). A potentially important property of sophorose is that it is much less efficiently hydrolyzed by β-glucosidase, while still efficiently transported by the *T. reesei* cellobiose permease. Thus, sophorose could be a gratuity inducer that mimics in the laboratory experiments the role of cellobiose in nature.

In *P. chrysosporium*, on the other hand, cellulase gene transcription has been reported to be induced by cellotetraose and less efficient by cellotriose, whereas cellobiose was not effective (Suzuki et al., 2010). This could be another variant of the above-described β-glucosidase-dependent scenario of inducer uptake because the affinity of β-glucosidases decreases with increasing cellooligodextrin chain length.

Another disaccharide, lactose (D-galactosyl-1,4-β-D-glucoside), has also shown to be a potent inducer of cellulases and—because of its cheapness, as a by-product of cheese manufacture—has become the only currently available soluble inducer for commercial cellulase production (Seiboth et al., 2007b). Although cellulase induction by lactose was mainly known for *T. reesei*, cellulase formation by several other fungi (e.g., *Acremonium cellulolyticus* and *V. volvacea*; Fang et al., 2008; Zhang et al., 2011) is also induced by lactose. The mechanism by which this compound induces cellulase gene transcription has been investigated in some detail: evidence points toward a major role of the β-D-galactose anomer that is released from lactose by β-galactosidase BGA1 in cellulase induction by lactose. Yet, despite the fact that β-D-galactose generated from lactose must be catabolized via the reductive galactose catabolizing pathway (see Chapter 6), disruption of the *gal1* gene still results in a strong decrease of cellulase formation on lactose (Seiboth et al., 2004). Thereby, not the GAL1 protein but only the galactokinase activity itself is important for cellulase induction by lactose (Hartl et al., 2007). Inactivation of the subsequent step, in which GAL7 transforms D-galactose-1-phosphate to UDP-galactose, has no effect on cellulase induction except that the cellulase transcripts display a longer half-life. A knockout in the aldose reductase gene *xyl1* also results in a decrease in cellulase gene expression on lactose, although not as severe as a knockout in *gal1* (for review, see Seiboth et al., 2007b). A consensus explanation for these findings would be that the inducer formed during lactose catabolism is an oligosaccharide composed of metabolites both from the Leloir and the alternate pathway. In fact, a metabolomic analysis of intracellular oligosaccharides formed in *H. jecorina* QM 9414 (parent strain) and the various pathway deletion strains (manuscript in preparation) detected several galacto-oligosaccharides, whose intracellular concentrations changed in a consistent way with cellulase formation (C.P. Kubicek, L. Karaffa and P.J. Punt, unpublished data). However, further investigations are needed to prove that any of these functions is an inducer of cellulase gene expression by lactose. The physiological relevance of induction by lactose may be that β-galactosidase has been shown to also cleave other β-D-galactosides in plant polymers, and thus the actual components that trigger the induction of cellulases *in vivo* may be β-galactooligosaccharides.

There are also reports that oxidized cellulose degradation products such as cellobiono-1,5-lactone stimulate cellulase formation (Iyayi et al., 1989). Yet, the physiological relevance of this observation was unclear. However, more recently, Langston et al. (2011) showed that a combination of the action of the GH61 cellulase-enhancing protein and cellobiose dehydrogenase led to the formation of oxidized breakdown products from cellulose. GH61 and CDH proteins are coexpressed and secreted by some fungi such as *Thielavia terrestris* in response to environmental cellulose and could thus act as cellulase inducers in these organisms.

7.2 Inducers for Hemicellulases

In contrast to cellulose, the hemicelluloses are composed of monomers that usually do not occur in the fungal cell, and therefore these monomers could well serve as the inducers of hemicellulases. The fact that some of them (xylose, mannose, and galactose) can exert CCR does not interfere with this speculation because they may not accumulate in the medium at concentrations high enough to cause this effect. In fact, the genes encoding the xylanolytic enzymes in *Aspergillus* species are induced by low concentrations of D-xylose, and this sugar further also induces expression of various genes encoding the side chain-cleaving enzymes α-glucuronidase (*aguA*), acetylxylan esterase (*axeA*), and feruloyl esterase (*faeA*) (Aro et al., 2005). In *A. niger* and *T. reesei*, high xylose concentrations can act as a repressing carbon source (Aro et al., 2005; Mach-Aigner et al., 2010b). Also disaccharides (xylobiose, sophorose, glucose-β-1,2-D-xylose, lactose) have been reported to induce xylanases in *Aspergillus* and/or *Trichoderma*. However, their effect may be due—and in analogy to the situation with cellobiose/sophorose as explained earlier—to their different rates of cleavage by β-xylosidase. A good example for this are the *T. reesei* xylanases: in a wild-type strain, xylose induces only *xyn1*, whereas *xyn2* expression is partially constitutive and inducible by xylobiose and sophorose. However, this pattern is reversed in a mutant strain blocked in the intracellular catabolism of xylose, thus implying that the respective rate of xylose catabolism is responsible for observed effect. In fact, Herold et al. (2012) demonstrated that all five xylanases of *T. reesei* are induced by D-xylose in a *T. reesei* strain defective in the first step of xylose catabolism, thus demonstrating that D-xylose is likely the natural inducer of the *T. reesei* xylanolytic enzyme system.

Likewise, arabinose and L-arabitol have been shown to induce expression of the genes encoding enzymes involved in the degradation of arabinans in *A. niger*, and there is indirect evidence that L-arabinitol could be the true inducer for them (Aro et al., 2005). Interestingly, induction of the L-arabinofuranosidase genes in L-arabinose pathway mutants of *T. reesei* by L-arabinose and L-arabinitol shows that it is strongly stimulated in the absence of L-arabinitol dehydrogenase and blocked in both cases in the absence of the aldose reductase XYL1 that also acts on L-arabinose (Akel et al., 2009). Thus, the presence of the XYL1 protein is required for either L-arabinose or L-arabinitol to act as an inducer.

Less detailed information is available for enzymes acting on polymers composed of other monosaccharides. For example, D-galactose has been shown to induce the β-galactosidase BGA1 and all of its α-galactosidases in *T. reesei* (C.P. Kubicek, unpublished data). In *A. niger*, galactose and mannose are involved in the induction of α-galactosidases and β-mannosidases (Aro et al., 2005). D-galacturonic acid has been shown to induce expression of nearly all the genes encoding for enzymes involved in the degradation of pectin and its side chains in *A. niger* (Aro et al., 2005).

There is also evidence that single mono- and disaccharides can induce almost the whole spectrum of cellulases and hemicellulases. An example for this is lactose: transcriptome analysis of *T. reesei* growing on lactose as a sole carbon source revealed the expression of not only cellulases but also xylanases, arabinofuranosidases, galactosidases, mannanases, and α-fucosidases (Table 7.2). In this case, it seems that β-D-galactose signals the presence of appropriate plant polymers to the fungus, thus suggesting that polysaccharides with this sugar component may be particularly rich in the habitat of *T. reesei*.

7.3 Transcriptional Regulation of Cellulase and Hemicellulase Gene Expression

The obligatory presence of an inducer for cellulase and hemicellulase gene expression to occur implies a tight co-regulation of the respective promoters. In fact, most of the cellulase genes in *T. reesei* and other fungi are regulated in a consistent way, although the relative ratio of their expression is somewhat different in higher producer mutants probably due to promoter titration effects (Kubicek et al., 2009). The identification of genes involved in transcriptional regulation of cellulase and hemicellulase gene expression has been a major effort in the past 10 years, and major breakthrough have been achieved for *T. reesei* and *A. niger*/*A. nidulans*, where several positive as well as negative-acting genes were identified and described in detail.

Table 7.2. Cellulase and hemicellulase genes expressed during growth of *T. reesei* on lactose.

	Protein ID	Upregulated[a]
GH1 β-glucosidase CEL1A	120749	21.2
GH1 β-glucosidase CEL1B	22197	4.1
GH10 xylanase xyn3	120229	8.4
GH11 xylanase XYN2	123818	28.6
GH12 endoglucanase Cel12A	123232	72.6
GH16 endo-β-1.3/1.4-glucanase (GPI anchor)	55886	4.6
GH16 endo-β-1.3-glucanase	69751	5.1
GH2 β-mannosidase	69245	101.2
GH2 β-mannosidase	5836	21.3
GH2 β-mannosidase	57857	6.0
GH27 α-galactosidase	65986	35.5
GH27 α-galactosidase	59391	23.3
GH27 α-galactosidase	55999	5.7
GH27 α-galactosidase AGL3	72704	7.1
GH27 α-galactosidase AGL1	72632	76.4
GH28 exo-rhamnogalacturonase RGX1	122780	125.2
GH3 β-glucosidase	104797	16.9
GH3 β-glucosidase CEL3C	82227	8.1
GH3 β-glucosidase CEL3D	46816	5.3
GH3 β-xylosidase BLX1	121127	10.7
GH30 endo-β-1.4-xylanase XYN4	111849	16.9
GH30 glucuronyl-xylanase	69276	4.7
GH36 α-galactosidase	124016	12.7
GH43 β-xylosidase	3739	233.3
GH45 endoglucanase Cel45A	49976	12.9
GH5 endoglucanase Cel5A	120312	163.5
GH5 glycosyl hydrolase	81087	62.9
GH54 α-arabinofuranosidase ABF1	123283	4.2
GH6 cellobiohydrolase 2 CEL6A	72567	273.1
GH61 carbohydrate monooxygenase CEL61A	73643	21.0
GH61 carbohydrate monooxygenase CEL61B	120961	62.7
GH62 α-L-arabinofuranosidase ABF2	76210	10.8
GH67 α-glucuronidase AGU1	72526	275.1
GH7 cellobiohydrolase 1 Cel7A	123989	246.5
GH74 xyloglucanase	49081	5.4
GH76 α-1.6-mannanase	122495	19.4
GH79 methyl-α-glucuronidase	106575	5.0
GH95 α-L-fucosidase	58802	15.4
GH95 α-L-fucosidase	72488	5.0
GH95 α-L-fucosidase	5807	34.0

Data from Ivanova et al. (2012).
[a]Fold over a control on glucose.

7.3.1 Activators of Cellulase and Hemicellulase Gene Expression

The general main activator of cellulase and hemicellulase gene expression has been cloned and characterized from *Aspergillus* (here termed XlnR) and *T. reesei* (XYR1). Its deletion virtually eliminates their induction by all known inducers (van Peij et al., 1998; Stricker et al., 2006). The XlnR gene was originally cloned by complementation of an *A. niger* mutant lacking xylanolytic activity (van Peij et al., 1998). It encodes a binuclear zinc finger protein, which is a class of transcription factors specific for fungi. In addition, it contains a fungal-

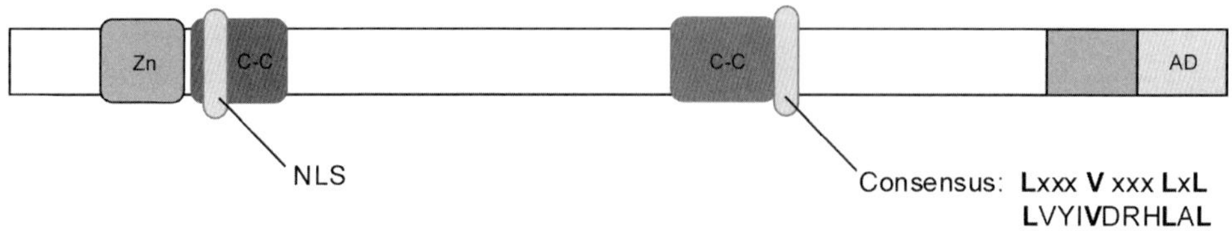

Figure 7.1. Structural organization of the XlnR/XYR1 protein. Zn, Zn2Cys6-zinc finger; NLS, putative nuclear localization sequence; CC, coiled-coil domain; AD, putative activation domain; the region preceding it (darker grey) is also conserved but the function is unknown. An L-rich region putatively binding the karyopherin KarA, which is conserved in all XLR1/XlnR othologs, is indicated by its consensus.

specific transcription factor domain (PF04082), which is typically found in many transcription factors of PF00172 that are involved in metabolic processes, although its function is unknown (Stricker et al., 2006). Using green fluorescent protein fusions of mutated XlnR proteins, Hasper et al. (2004) further showed that a putative coiled-coil domain is important for the XlnR function, as the disruption of the respective α-helix structure by a L650P mutation leads to a loss of function and permanent cytoplasmic localization of XlnR and, as a result, to a loss of the transcription of the genes under control by XlnR. Hence, this structure appears to be involved in the nuclear localization of the protein (Hasper et al., 2004). Finally, a deletion of 78 C-terminal amino acids leads to an increased expression of the XlnR target genes, also under D-glucose repression conditions (Hasper et al., 2004), suggesting this region is a modulation domain. Single amino acid mutations in this region lead to a loss of function of XlnR. A model of the structure of XlnR/XYR1 is given in Figure 7.1. Interestingly, the XYR1 protein of *T. reesei* seems to differ in this respect: genome sequence analysis of a *T. reesei* mutant (QM 9136) that is unable to form any cellulases or hemicellulases contains a *xyr1* gene with an amino acid deletion that truncates the XYR1 protein just before the putative activation domain at the C-terminus (Figure 7.1). Retransformation of the mutant with the intact *xyr1* gene fully restores cellulase and hemicellulase production (Hartl et al., 2012).

As a transcriptional activator, XlnR/XYR1 must be able to receive the respective signals from the inducer. In *Aspergillus*, Noguchi et al. (2011) showed that in the absence of D-xylose, XlnR was present as a mixture of various phosphorylated forms and that the presence of D-xylose triggered additional phosphorylation of the protein. D-Xylose-triggered phosphorylation was a rapid process that occurred within 5 minutes prior to the accumulation of xylanase mRNA, and removal of D-xylose caused slow dephosphorylation of XlnR. Unfortunately, it is not known where these phosphorylation sites are located and whether they also affect the nuclear localization of XlnR.

Binding of XlnR/XYR1 to the target promoters has been mainly studied with XYR1, and it is currently believed that a $GGC(T/A)_4$ consensus is functional (Furukawa et al., 2008).

Noguchi et al. (2009), using strains of *A. oryzae* either deleted in or overexpressing *xlnR*, identified 75 genes whose expression (in DNA microarray analysis) was dependent on XlnR function, of which 32 genes were predicted to encode a glycoside hydrolase. Sixty-one of the 75 genes possessed putative XlnR-binding sites in their promoter regions. The other 43 genes encoded enzymes for the conversion of D-xylose to xylulose-5-phosphate, arabinose catabolism, the pentose phosphate pathway, and for sugar transport. Microarray analysis of an *xlnR* wild type and *xlnR* knockout strain of *A. niger* produced largely consistent results (Trindade, van der Veen, and de Graaff, unpublished data; cited by Stricker et al., 2006).

In *T. reesei*, XYR1 has also been shown to be the major regulator of their induction by lactose and is also required for the induction of the aldose reductase XYL1 and the β-galactosidase BGA1 (Seiboth et al., 2007a; Stricker et al., 2007). In addition, the expression of the arabinanolytic genes *abf2* and *bxl1* require XYR1 (Akel et al., 2009). The authors also showed that the induction mediated by XYR1 requires XYL1 in a yet unknown way.

While XYR1 is also responsible for the induction of genes encoding arabinan-degrading enzymes, a second transcriptional activator (AraR) is responsible for the induction of arabinan-degrading enzymes and also for catabolizing the resulting L-arabinose in the *Aspergilli* (Battaglia et al., 2011). AraR has high similarity to XlnR and seems to be the result of a gene duplication, which occurred specifically in the Eurotiales and is, therefore, not present in *T. reesei*.

In contrast, *Trichoderma* spp. have an additional transcriptional activator (ACE2) that is absent from most other fungi. It also belongs to the class of zinc binuclear cluster proteins found exclusively in fungi. During growth on cellulose, deletion of the *ace2* gene led to lowered induction kinetics of cellulase mRNAs and 30–70% reduced cellulase activity, whereas induction by sophorose was not affected by *ace2* deletion (Saloheimo et al., 2001). The DNA-binding domain of ACE2 was shown to bind *in vitro* to the 5′-GGCTAATAA site present in the *cbh1* promoter. Therefore, both XYR1 and ACE2 are apparently able to bind to the same motif. Stricker et al. (2008), using the *xyn2* xylanase gene expression as a model, suggested that ACE2 acts in a dual role, on the one hand as an antagonist of early induction and on the other hand as an enhancer of a continuous extension of expression of *xyn2* and concluded that this dual role implies the action of posttranslational modification of Ace2. This was further supported by the finding that ACE2 needs to be phosphorylated to be able to bind DNA *in vitro*. They also found that ACE2 binds only as a dimer. Since *Trichoderma* XYR1 is responsible for the induction of most of its glycosyl hydrolases, it is possible that ACE2 has evolved as a fine-tuning component that binds to XYR1 and enables its response to the different inducing signals.

Endo et al. (2008) identified a region in the *A. nidulans* endoglucanase gene *eglA* spanning from −153 to −138 (CCGTACCTTTTTAGGA), which was essential for induction by cellobiose and cellulose. Since this motif does not correspond to any already known fungal upstream-activating sequence, it is likely that they have identified a novel *cis*-element responsible for XlnR-independent cellulase induction.

Promoter mutation and *in vivo* footprinting analysis of the *T. reesei cel6A* cellulase gene further revealed that induction by sophorose is partially dependent on a CCAAT box located adjacently to the XYR1/ACE2-binding site and is bound by the ubiquitously occurring HAP2/3/5 protein complex (Zeilinger et al., 1998). Like their other eukaryotic orthologs, the *T. reesei* HAP2, HAP3, and HAP5 contain a histone fold motif, suggesting that NF-Y might be involved in the organization of the chromatin structure, eventually by association with acetyltransferases. Consequently, the HAP2/3/5 complex may function in opening the chromatin structure necessary for full transcriptional activation (Kato, 2005). In fact, Zeilinger et al. (2004) detected a nucleosome-free region around the XYR1/ACE2/HAP2/3/5-binding area in the *cel6a* promoter, which is flanked by strictly positioned nucleosomes. Induction by sophorose results in a loss of positioning of nucleosomes −1 and −2 downstream of the binding area, thus making the TATA box accessible, and this positioning is shifted by a mutation in the CCAAT box.

7.3.2 Specific Repressors of Cellulase and Hemicellulase Gene Expression

Repressor proteins acting specifically on the expression of cellulase or hemicellulase genes have so far only been reported for *T. reesei*, that is, ACEI and XPP1 (Saloheimo et al., 2000; Aro et al., 2003; Mach-Aigner et al., 2011).

ACE1 was originally cloned in the yeast one-hybrid system as an activator of *cel7a* gene transcription. However, its deletion in *T. reesei* led to a 2–30-fold increase in cellulase transcription under inducing conditions (Aro et al., 2003). Aro et al. (2005) explain this discrepancy by the fact that ACEI expressed from the yeast library plasmid was truncated and lacked 242 amino acids from the N-terminus. Since this N-terminus is conserved between the putative ACEI proteins that are present in other fungi, it may be of function functional importance for ACE1 and its absence convert it into an inducer. Genetic evidence for this is still lacking, however. The importance of ACE1 to cellulase production by *T. reesei* is also emphasized by the finding that the hyperproducing mutant *T. reesei* CL847 exhibits strongly reduced *ace1* expression (Portnoy et al., 2010). ACE1 contains three Cys_2His_2-type zinc fingers, and it was shown to bind *in vitro* to eight sites in the *cbh1* promoter, all of which contain the core 5′AGGCA followed by a sequence rich in A and T (Aro et al., 2005). Its deletion causes an increase in the expression of all the main cellulase and hemicellulase genes in sophorose- and cellulose-induced cultures, and also of the expression of *xyr1* during growth on D-xylose (Mach-Aigner et al., 2008).

ACE1 has been identified as an ortholog of the *A. nidulans stzA* gene that encodes a stress response factor (Chilton et al., 2008), and the promoters of the *A. nidulans* and *A. fumigatus cpcA* (cross pathway control regulator of amino acid biosynthesis) genes have seven, spatially conserved ACE1/StzA binding sites, suggesting a link between intracellular amino acid availability and cellulase gene expression. While this still needs to be studied in detail, Gremel et al. (2007) reported an enhancement of cellulase gene expression by the addition of methionine, thus illustrating a cross talk between an amino acid and cellulase formation.

In vitro and *in vivo* analysis of the *xyn2* promoter suggested the binding of still another putative repressor protein (XPP1) to an AGAA-box under glucose-repressing conditions (Würnleitner et al., 2002). The protein binding to the motif and its gene have recently been identified by Mach-Aigner et al. (2010b) using a combination of electrophoretic mobility shift assays and Nano-HPLC/tandem MS-coupled protein detection. It encodes a protein (Trire2:122879) of 505 amino acids with a predicted molecular mass of 55 kDa that appears to be conserved among Sordariomycetes. XPP1 is a member of the E-box helix-loop-helix (HLH) domain superfamily, which typically would bind to a hexameric palindrome. It is therefore possible that the AGAA element is only part of the actual binding site. A biochemical analysis of this possibility and the effect of a *xpp1* gene knockout on xylanase gene expression are still required to fully understand the role of this protein in *xyn2* (and eventually also other) gene transcription. *Xpp1* expression is upregulated under repressive conditions, which is in agreement with its functions as a repressor of hydrolase transcription as demonstrated by promoter deletion experiments (Würnleitner et al., 2002).

7.3.3 Wide Domain Regulators of Cellulase and Hemicellulase Gene Transcription

As mentioned earlier, expression of almost all of the cellulase and hemicellulase genes that have been studied in *T. reesei* and other filamentous fungi does not occur during growth on glucose. This has been shown to be due to both inducer exclusion (i.e., inhibition of inducer uptake by D-glucose; Saier and Crasnier, 1996) and glucose repression. The latter specifies a mechanism by which the transcription of genes encoding enzymes for the catabolism of carbon sources that yield less energy, allowing only slow growth or need complicated cascades for making them available to the cell, is inhibited by D-glucose or other monosaccharides whose catabolism provides a high yield of ATP at a faster rate and is more generally called CCR.

Consequently, one of the earliest attempts for engineering cellulase production was removal of CCR (see also Chapter 9). Classical mutagenesis combined with selection for 2-desoxyglucose resistance (an agent believed primarily to enrich carbon catabolite-resistant mutants; see Chapter 9; Zimmermann and Scheel, 1977) has led to increased cellulase producers such as *T. reesei* (RUT C30, RL-P37 and CL847), thus supporting the possible importance of CCR in cellulase formation (see also Chapter 9; Kubicek et al., 2009).

In the ascomycetes, the key player in this glucose repression is the Cys_2His_2-type transcription factor CRE1/CreA (Dowzer and Kelly, 1991; Ilmen et al., 1991). The molecular mechanisms involved in CRE1/CreA-mediated regulation has been studied in most detail in *A. nidulans* (Felenbok et al., 2001). Numerous cellulase, hemicellulase, and pectinase genes have been shown to be regulated by CreA/CRE1 proteins in *T. reesei* and *Aspergillus* species by using *creA/cre1* mutant strains in the studies (Aro et al., 2005). In general, mutations of the *creA/cre1* gene lead to (partial) derepression of enzyme gene expression on glucose. Interestingly, one of the cellulase-overproducing strains isolated by selection for 2-deoxyglucose resistance (*T. reesei* RUT C30) was indeed shown to contain a loss of a 2478-base pair fragment, which starts downstream of the region encoding the CRE1 zinc finger and reaches into the 3′-noncoding region (Ilmén et al., 1996).

A. nidulans CreA was shown to bind to a 5′-SYGGRG motif (Kulmburg et al., 1993). *In vivo* function of the CRE1-binding sites that were demonstrated in the *T. reesei cbh1* and *xyn1* promoters were proven by showing that mutations in these binding sequences resulted in a constitutive expression level of the respective genes in the presence of D-glucose (Aro et al., 2005; Kubicek et al., 2009). Functional CREI/CreA-binding sites have been shown to consist of two closely spaced 5′-SYGGRG motifs, and it has been suggested that direct CREI/CreA repression would occur only through such double binding sites.

Phosphorylation of a serine residue in a conserved short stretch of *H. jecorina* CREI has been demonstrated to regulate its binding to its target sequence (Cziferszky et al., 2002). This phosphorylation target site would be typical for that of a casein kinase 2, which has been reported to play a role in the regulation of a large number of transcription factors in other eukaryotes (Kubicek et al., 2009). Also, the genomes of *T. reesei, A. nidulans*, and other ascomycetes contain casein kinase orthologs. However, genetic proof for their involvement in the modulation of CreA/CRE1 action is not yet available. In *Sclerotinia sclerotiorum*, the CreA/CRE1 ortholog CreI was demonstrated to be localized in the nucleus under conditions of CCR and transported into the cytoplasm upon carbon catabolite derepression (Vautard-Mey et al., 1999). Regulation of transcription factor activity by nuclear translocation is a frequently observed phenomenon. However, the mechanism for this observation remained unexplained. Frequently, such a nuclear traffic is initiated by altering the phosphorylation status of the respective transcription factor. However, mutation of the serine residue in *S. sclerotiorum* CreI, whose corresponding serine in *T. reesei* CRE1 has been shown to be phosphorylated, did not affect the nuclear localization of CreI (Vautard-Mey and Fèvre, 2000).

Interestingly, knockout strains in *creA/cre1* often still do not produce significant amounts of cellulases and hemicellulases. In *A. niger*, expression of the arabinofuranosidase encoding genes *abfA* and *abfB* and the endo-arabinase gene *abnA* is only up to ten times higher in the *creA* mutants than in the wild type when the fungus is grown in inducing conditions (i.e., arabinose or arabitol as carbon sources; Ruijte et al., 1997; Gielkens et al., 1999b). Similarly, the expression of the *A. nidulans* xylanase genes *xlnA* and *xlnB* are induced to a higher level in a *creA*^{d} 30 mutant than in the wild-type strain only when xylanase expression is induced upon transfer to xylose-containing medium (Orejas et al., 1999, 2001; Flipphi et al., 2003). Expression levels of the hemicellulase genes *xlnB*, *xnlD*, *faeA*, and *aguA* of *A. niger* also are all higher in the *creA* mutant than in the wild-type strain during growth in inducing conditions on various xylose concentrations (de Vries et al., 1999). Only Orejas et al. (2001) have shown that the *A. nidulans xlnB* gene promoter is repressed by glucose and that its expression is higher in the *creA*^{d}30 mutant than in the wild type.

The way in which the presence of glucose triggers CCR is still only poorly understood in filamentous fungi. In *Saccharomyces cerevisiae*, the D-glucose and D-fructose kinases are involved in D-glucose and carbon catabolite sensing: of them, the hexokinase Hxk2p is responsible for glucose repression (Entian and Fröhlich, 1984). The mechanism by which Hxk2p contributes to glucose repression has not yet been fully elucidated, but its catalytic activity seems to be dispensable and thus signal transmission may rather depend on substrate binding-induced conformational changes in the Hxk2p protein or a direct regulatory role of the Hxk2p in the nucleus (for discussion see Moreno and Herrero, 2002). However, in filamentous fungi, the enzymatic activity appears to be relevant for CCR: for example, *A. nidulans* and *T. reesei* contain only a single glucokinase and a single hexokinase. Only a knockout in both of them renders the respective mutants carbon catabolite derepressed (Flipphi et al., 2003). Strauss et al. (1999) have shown data that suggested that *A. nidulans* requires a certain intracellular concentration of glucose-6-phosphate to stimulate CreA-mediated CCR. Interestingly, Limón et al. (2011) showed that a deletion of the *pgi1* (phosphoglucose isomerase-encoding) gene in *T. reesei* led to a higher total cellulase activity under carbon catabolite repressing conditions. This may be interpreted such that this knockout prevents the fructose-6-phosphate that has been formed by the pentose phosphate pathway from glucose to contribute to the repressing glucose-6-phosphate pool.

In *A. nidulans*, three further genes (*creB, creC*, and *creD*) have been isolated, which also participate in CCR. Orthologs of these genes can be found in all Pezizomycetes, for which genome sequences are available. Interestingly, all three of them are related to protein deubiquitinylation. Deubiquitinating enzymes are cysteine proteases that target proteins for degradation by the proteasome. CreB encodes such a deubiquitinating enzyme and forms a complex with the WD40-repeat protein CreC, which protects CreB against proteolysis in the absence of CCR. An *A. nidulans* mutant in *creB* and disruption of *creB* homolog *cre2* in *H. jecorina* led to carbon catabolite derepression of cellulases (reviewed by Kubicek et al., 2009). CreD is a protein with arrestin domains and PY motifs, which is highly similar to the *S. cerevisiae* proteins Rod1p and Rog3p, which interact with the ubiquitin ligase Rsp5p, and mutations in *creD* suppress the phenotypic effects of mutations in *creC* and *creB* (Boase and Kelly, 2004). Ubiquitinylation of the activation domains in several transcription factors has been shown to regulate gene transcription (Salghetti et al., 2001). It will be interesting to directly test whether CCR in filamentous fungi also involves this mechanism. Figure 7.2 summarizes the current knowledge about CCR in fungi.

The pH of the external environment with which the fungus is in contact can affect and, in extreme cases, inactivate the enzymes that are secreted into this medium. Therefore, fungi have developed a regulatory system that senses the external pH and correspondingly regulates the synthesis of enzymes with regards to their stability under these conditions. This regulatory system has been studied in detail in *A. nidulans* (Peñalva and Arst, 2004; Peñalva et al., 2008; Figure 7.3). The zinc finger transcription factor PacC is the main player in this regulatory

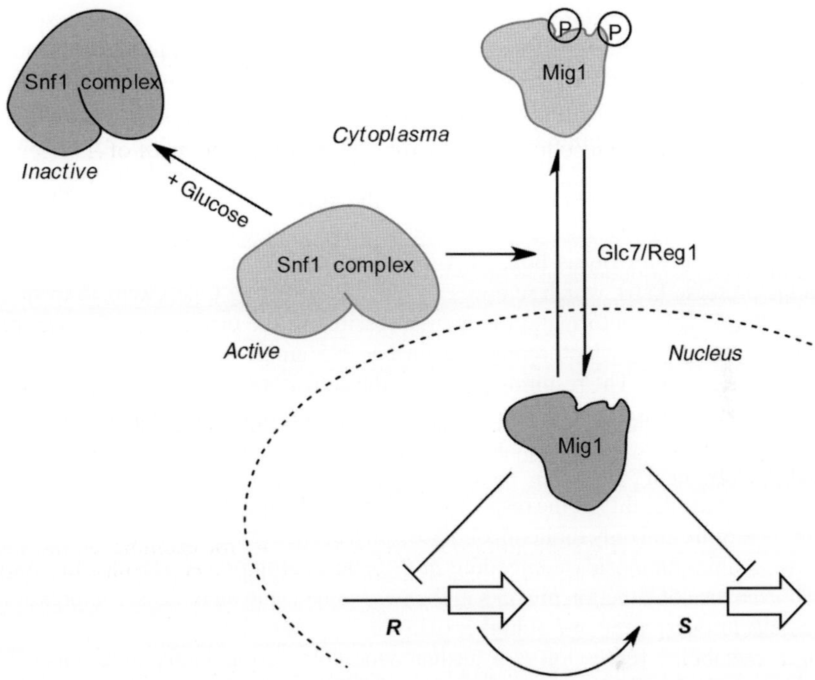

Figure 7.2. Basic features of carbon catabolite repression in yeast: R, regulatory genes; S, structural genes.

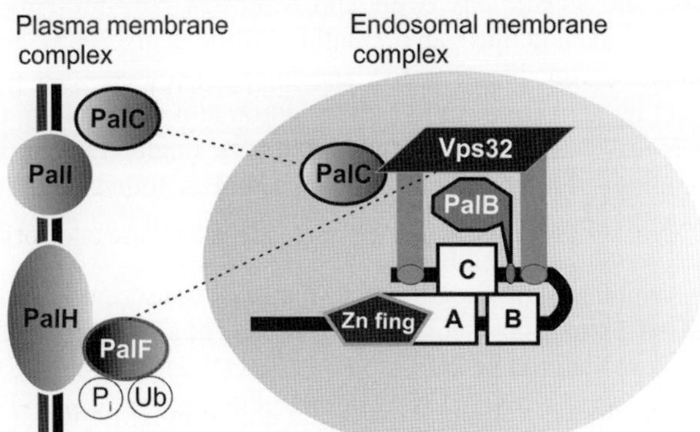

Figure 7.3. Generalized and abbreviated scheme of pH regulation in *A. nidulans*. Under alkaline conditions, the respective pH signal is transmitted from the plasma membrane by endocytosis of PalH and/or PalI/F and PalC to the endosomal membrane complex on the surface of the endosome. Here, PacC is in a processing protease-inaccessible conformation. Upon binding of PalC, PacC—bound by PalA through YPXL/I motifs (grey bars)—undergoes pH-dependent proteolysis within the signaling proteolysis box (grey ellipsoid) by the cysteine protease PalB. This conformation is then accessible to the processing protease, which probably is a part of the proteasome, and removes further 245 amino acid residues from PacC in a pH-independent manner consisting only of the Zn-finger and domain A (not shown). This form is then the activator of alkaline-expressed genes and repressor of acid-expressed genes. (From Peñalva et al., 2008.)

system by acting as an activator for alkaline-expressed genes and by preventing expression of acid-expressed genes. It contains three Cys_2His_2 zinc fingers that bind to the hexanucleotide sequence 5′GCCARG (Peñalva et al., 2008). Under acidic conditions, PacC is held in a processing protease-inaccessible closed conformation. Under alkaline conditions, the alkaline ambient pH signal is transmitted from the plasma membrane complex by a process that probably involves endocytosis of the proteins PalH and/or PalIF and the participation of PalC to the endosomal membrane complex on the surface of the endosome. Here, PacC, bound by PalA through YPXL/I motifs, undergoes pH-dependent proteolysis, which occurs within the signaling proteolysis box. This signaling proteolysis is very likely mediated by the cysteine protease PalB, which removes the negatively acting C-terminus to form an open conformed PacC. This conformation is then accessible to the processing protease, probably a part of the proteasome, which removes a further 245 amino acid residues from PacC in a pH-independent manner. The resulting protein is the activator of alkaline-expressed genes and repressor of acid-expressed genes. Under acidic conditions, trace amounts of PacC exist in the open conformation accessible to the processing protease, bypassing the requirement for the pH signal (Peñalva et al., 2008).

Because of the availability of the respective mutants, pH regulation via *pacC* has been shown mainly in Aspergilli, and only for hemicellulases such as the *xlnA* and *xlnB* encoding xylanases, and the α-L-arabinofuranosidase-encoding *abfB* gene of *A nidulans*, Besides the Aspergilli, *S. sclerotiorum* loss-of-function mutants in the *pac1* gene led to an increased expression of the endopolygalacturonase gene *pg1* at higher pH (see Aro et al., 2005).

Nitrogen metabolite repression is a further wide-domain regulatory mechanism in fungi that enables preferential utilization of easily assimilated nitrogen sources (such as ammonium) prior to the use of secondary nitrogen sources such as nitrate, nitrite, and purines. In filamentous ascomycetes, regulation of nitrogen catabolic genes is mediated by GATA-type transcription factors. The *A. nidulans* AreA and *N. crassa* Nit2 proteins are the positively acting GATA factors involved in nitrogen regulation (Wong et al., 2008), and loss-of-function *areA* and *nit-2* mutants are unable to utilize any nitrogen source other than glutamine and ammonium. They act by activating transcription, chromatin remodeling, and by interacting with the repressor NmrA/Nmr1 (Scazzocchio, 2000). A single AreA ortholog with significant sequence homology throughout the entire protein is found in all other filamentous ascomycetes. Cross-complementation experiments have shown that many of these orthologs are functionally conserved. During nitrogen sufficiency, AreA activation is prevented by the co-repressor NmrA, which is in turn regulated by the bZIP transcription factor MeaB by binding to 5′-TTGCACCAT-3′ that is present in the promoters of all ascomycete NmrA homologs. Expression of *meaB* is modulated by the nitrogen status (Wong et al., 2008).

In addition to the conserved GATA DNA-binding domain, the AreA/NIT-2 orthologs contain two highly conserved blocks of sequences, RMENLTWRMM and WEWLTMSL, at the N- and C-termini, respectively. The role of the N-terminal conserved domain is still unknown, but the C-terminal domain has been shown to be important for modulation of AreA/Nit-2 by the NmrA/Nmr1 protein.

As far as cellulase or hemicellulase gene expression is concerned, data on a potential regulation by the nitrogen source are only available for AreA: in *A. nidulans* and in the presence of ammonium, formation of endoglucanase activity was increased in an *areA*-overexpressing mutant and decreased in an *areA* loss-of-function mutant, suggesting positive regulation by AreA (Lockington et al., 2002). While preliminary, these data may—besides the above-explained role of ACE1/ StzA—hint toward a cross-talk between cellulase induction and nitrogen utilization in fungi, which needs to be still illustrated by detailed experiments.

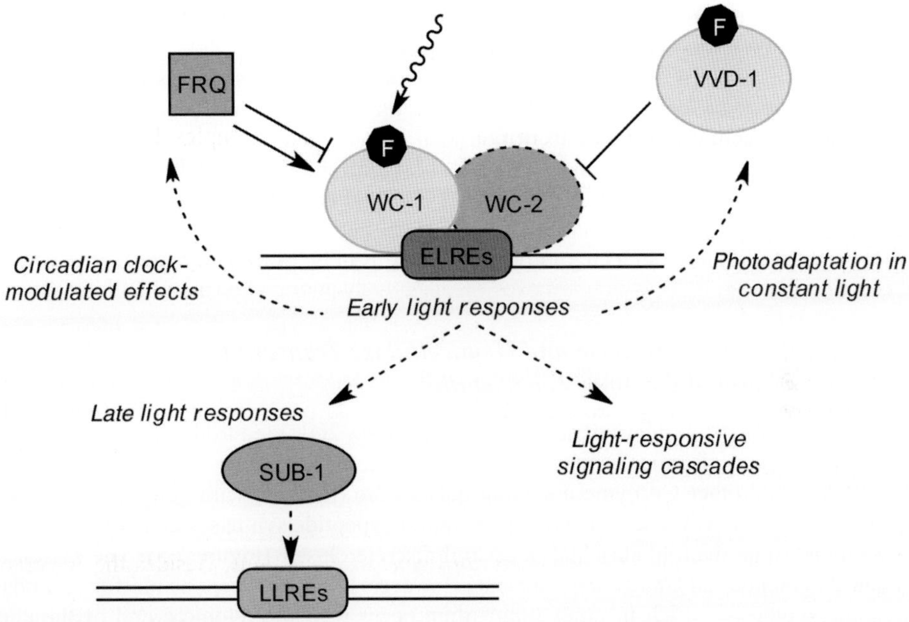

Figure 7.4. Mechanism by which *N. crassa* gene expression is regulated by light (spiral arrow). In the presence of light, the White-Collar Complex (WCC, consisting of WC-1 and WC-2), binds to its target sequence (early light-responsive elements; ELREs) in the respective promoters including FRG, the transcription factor SUB-1, and VVD ("vivid"). FRQ ("frequency") modulates the WCC activity in a circadian manner to determine the amplitude of the light responses. VVD protein highly expressed in light represses this activation by interaction with the WCC. SUB-1 activates the expression of late-light responsive genes by binding to another conserved nucleotide sequence in their promoters (late light response elements; LLRE's) on their promoters. F indicates a flavin moiety on the respective proteins. Arrows indicate activation, whereas arrows ending with a vertical bar indicate repression. (Adapted from Chen et al. (2010a), where more details can also be found.)

Light is another environmental parameter that influences the behavior and lifestyle of all organisms. In filamentous fungi, it has been shown to modulate and trigger the expression of a large number of enzymes and physiological traits (Corrochano, 2007). The molecular genetic basis of reception and reaction to light has been most extensively studied in *N. crassa* (Chen et al., 2010a). The key players are the products of the *white collar* (*wc-1* and *wc-2*) genes that form the transcriptionally and photoactive white collar complex (WCC; Figure 7.4). WC-1, which is mainly localized in the nucleus, is both a FAD-binding photoreceptor and a GATA zinc finger transcription factor. Together with WC-2, another GATA zinc finger transcription factor, it forms the WCC by association through their PER-ARNT-SIM (PAS) domains. After exposure to light, the WCC binds to the promoter of early light-responsive targets at defined consensus sequences and activates transcription. VIVID (VVD) is an additional blue-light photoreceptor that functions downstream of the WCC to negatively regulate and thus fine tune the light responses initiated by the WCC. VVD is a small 21-kDa FAD-binding protein consisting of a LOV domain. Once activated by blue light, the formation of a covalent cysteinyl-flavin adduct within the LOV domain induces an N-terminal conformational change, which is believed to be required for its action as a repressor. Chen et al. (2010b) have recently demonstrated a direct molecular connection between VVD and WCC.

Schmoll et al. recently identified that cellulase formation by *T. reesei* is inhibited by UV light and that the *Trichoderma* orthologs of the *N. crassa* photoreceptor proteins participate in the regulation of cellulase gene transcription in *T. reesei* (reviewed in Schmoll et al., 2010). They showed that a deletion mutant in the *T. reesei vvd* ortholog *env1* leads to higher cellulase activity, whereas knockout mutants in *blr1* and *blr2* had no significant effect. However, the relevance of these findings to induction of cellulase or hemicellulase genes by *T. reesei* under fermentation conditions or in its natural habitat remains to be proven.

7.3.4 Regulation of Cellulase and Hemicellulase Transcription at the Level of Genome Accessability

The sequencing of the *T. reesei* genome (Martinez et al., 2008) led to the interesting finding that—in contrast to all other ascomycetes whose genome had been sequenced—cellulase, hemicellulase and other CAZyme-encoding genes were (together with genes encoding secondary metabolic enzymes such as non-ribosomal polypeptide synthases and polyketide synthases) found to be fivefold enriched in several discrete cluster (Figure 7.5). The reason for this and its potential advantage was unclear. However, the genes responsible for secondary metabolite synthesis, which in other fungi often occur near the telomere end of the chromosomes (Keller and Hohn, 1997), have been demonstrated to be regulated at the level of histones by the putative protein methyltransferase LaeA that is believed to reverse the repression of gene expression at the level of heterochromatin structure (for review, see Strauss and Reyes-Dominguez, 2011; Figure 7.6). Because of the co-clustering of cellulase, hemicellulases, other CAZymes, and secondary metabolite synthesis genes in the *T. reesei* genome, we hypothesized that cellulase expression may be regulated by a *T. reesei* LaeA ortholog. To test this hypothesis, Seiboth et al. (2012) have produced *lae1*-knockout and *lae1*-overexpressing mutants of *T. reesei*. Production of cellulases and other CAZymes was indeed reduced almost to zero in the knockouts and dramatically increased in the overexpressors, showing that LAE1 is essential for cellulase and hemicellulase formation in *T. reesei*.

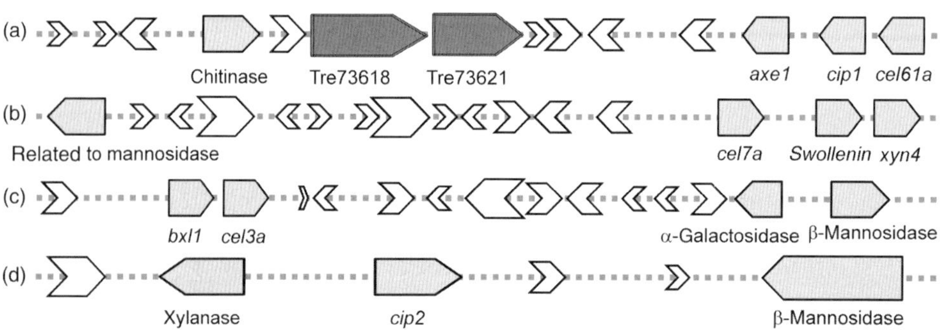

Figure 7.5. Examples for the clustered organization of CAZys in the genome of *T. reesei*. Grey arrows indicate CAZyme genes; the two genes indicated by Tre73618 and Tre73621 specify two polyketide synthases. Other genes, not encoding CAZymes, are given by empty arrows. (Adapted from Martinez et al., 2008.)

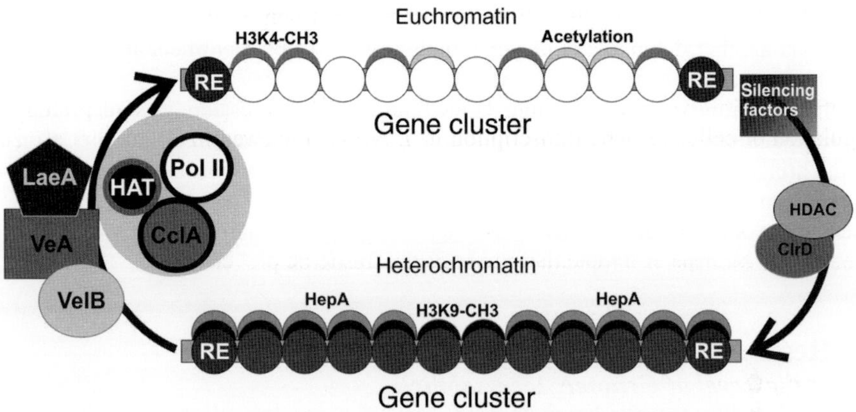

Figure 7.6. Regulation of expression of clustered genes in fungi at the histone and chromatin level. RE, repetitive elements. (Adapted from Palmer and Keller, 2010.)

7.3.5 Signal Transduction to Cellulase and Hemicellulase Gene Expression

While our knowledge about the transcriptional regulators involved in cellulase and hemicellulase gene expression is already considerably advanced, there is still a significant gap in understanding how the presumptive cellulase and hemicellulase inducers activate these transcription factors, for example, how they cause the reported phosphorylation of XlnR/XYR1 and ACE2. The earliest, yet preliminary, report on this topic was the demonstration that exogenous addition of dibutyryl-cyclic AMP or of inhibitors of phosphodiesterase that enhance the intracellular cyclic AMP level leads to an increase of endoglucanase induction by *T. reesei* (Sestak and Farkas, 1993). Cyclic AMP is a classical messenger in signal transduction and formed by the action of adenylate cyclase. Adenylate cyclase, in turn, is activated by the G-alpha subunit of the G-protein complex that is associated with the cytoplasmic C-terminus of seven-transmembrane helix receptors that receive the external signal to be transmitted (Lafon et al., 2006). Consistent with a role of G-proteins in signal transduction between plant biomass and fungi, Wang and Nuss (1995) showed that the G-alpha protein CPG-1 is necessary for the induction of *cel7A* gene expression by cellulose in *Cryphonectria parasitica*.

In *T. reesei*, the G-alpha proteins GNA1 and GNA3 are only partially involved in cellulase gene expression: while loss of function strains and strains bearing constitutively activated *gna1* and *gna3* alleles showed a decreased and increased *cel7A* and *cel6A* transcription under some conditions, respectively, cellulase gene expression was still dependent on the presence of an inducer, thus implying that the signaling molecule that binds to the receptor and activates GNA1 and GNA3 cannot be the cellulase inducer (Schmoll et al., 2009; Seibel et al., 2009). The main target of cAMP in this signaling cascade is cAMP-dependent protein kinase A (PKA), which—after binding of cAMP to the regulatory subunit that inhibits the activity of PKA—dissociates from it and phosphorylates its target proteins. Consistent with the data of Sestak and Farkas (1993), Schuster et al. (2012) have shown that the adenylate cyclase of *T. reesei* is required for cellulase induction, because disruption of the gene encoding the catalytic subunit strongly impairs cellulase gene transcription. Interestingly, however, PKA1 appears to influence cellulase gene transcription negatively because a knockout in the *pka1* gene actually increases *cel7A* and *cel6A* expression. Thus, the positive effect of cyclic AMP must act via another downstream player, which remains to be identified, and PKA1 may be involved via yet another mechanism.

Another signaling pathway leading to cellulase gene expression was identified in the maize leaf pathogen *Cochliobolus heterostrophus* (Lev and Horwitz, 2003). Chk1, a mitogen-activated protein kinase that functions in virulence, mating, and sporulation, was found to be required for induction of two cellulolytic enzymes, that is, CEL7A and CEL7B.

The paucity of data available clearly shows that signal transduction of the presence of plant lignocellulosic biomass is a topic that needs further research.

7.4 Regulation of Ligninase Gene Expression

As reviewed earlier, the regulation of expression of genes for cellulose and hemicellulose degradation is already quite known in some detail, and while there are still gaps in the in-depth understanding, the inducers, their metabolism, and regulatory genes are already known in detail. This contrasts strikingly with the situation concerning the lignin-degrading enzymes. One of the main reasons for this is certainly the fact that basidiomycetes (the only fungi capable of producing these enzymes; see Chapter 5) are more difficult to treat by gene manipulatory techniques, and no regulatory mutants that would facilitate the isolation of the respective genes are known. Most of the little information that is available today is derived from gene expression and promoter fragmentation analysis.

It is well known that high-level production of lignin-degrading enzymes requires the presence of carbon and energy sources under conditions that limit growth, suggesting that ligninase-producing cultures must be in the stage of secondary metabolism, that is, derepressed of carbon, nitrogen, or sulfur repression (reviewed by Singh and Chen, 2008). However, none of these wide domain circuits has a yet been studied in white rot fungi, nor have the respective genes been identified. In general, regulation of the genes encoding ligninolytic enzymes has been studied mainly in *Phanerochaete chrysosporium*. The respective data confirmed that depletion of nutrient nitrogen, carbon, or sulfur is obligatory for expression and that these conditions trigger the expression *via* a stress response. The presence of putative metal response elements (MREs) has been reported in the manganese peroxidase (*mnp*) promoters (Gold and Alic, 1993; Asada et al., 1995). The expression of *mnp* is also transiently induced by a heat shock in nitrogen-limited cultures, and sequences matching to the consensus heat shock element (CN_2GAAN_2 $TTCN_2G$) are also found in the promoters of *mnp* genes. Furthermore, addition of H_2O_2 and other chemical stress compounds into the cultures also induce *mnp* expression in *P. chrysosporium*. Belinky et al. (2003) showed that the induction of an oxidative stress response enhances the formation of lignin and manganese peroxidases in *P. chyrsosporium*. The *lip* genes of *P. chrysosporium* are also regulated in a similar way by various factors, for example, as a response to carbon and nitrogen starvation (Aro et al., 2005).

Furthermore, manganese peroxidase production is strictly dependent on the presence of sufficient Mn^{2+} in culture (Singh and Chen, 2008). A 33 bp-long putative manganese-responsive element has been found in the *P. chrysosporium mnp1*, *mnp2*, and *mnp3* promoter regions (Ma et al., 2004). Moreover, in *C. subvermispora*, this metal causes a differential regulation of *mnp* genes (Manubens et al., 2003). Growing evidence suggests that copper may also affect the production of MnP. For instance, copper has been shown to increase MnP activity in *Trametes trogii* and *Stereum hirsutum, and* in *C. subvermispora*, *mnp1* and *mnp2*-mRNA levels increase with copper (Manubens et al., 2003).

In addition, conserved *cis*-acting promoter elements, including inverted CCAAT boxes, MREs, heat shock elements, and a binding site for activator protein 2 (AP-2; GCGTTGGG) have been identified in the promoter regions of *P. chrysosporium* manganese peroxidase genes (Aro et al., 2005). However, the involvement of these sequences in the regulation of *mnp* gene expression has not been demonstrated. Dhawale (1993) hypothesized that AP-2 may be important for the regulation of LiP- and MnP-encoding gene expression under nitrogen limitation. However, Ma et al. (2004) showed that at least the two AP-2 sites that are located within the manganese-responsive element in the *mnp1* promoter of *P. chrysosporium* are not involved in the nitrogen-dependent regulatory response.

Knowledge of the mechanisms by which nitrogen limitation and manganese ions signal their presence to lignin and manganese peroxidase gene expression is even less scarcely known than for cellulases and hemicellulases. There are some hints, though, for a role of cyclic AMP and calmodulin: Sakamoto et al. (2010) showed that the addition of the calmodulin inhibitor W-7 repressed the transcription of LiP and MnP-encoding genes. In addition, atropine, an inhibitor of cyclic AMP formation, also repressed LiP and MnP activities. The authors proposed that calmodulin may stimulate adenylate cyclase and so transfer the signal toward *lip* and *mnn* gene expression.

In contrast to the lignin and manganese peroxidase genes, expression of laccase-encoding genes is not dependent on nutrient depletion. Laccases are expressed constitutively in many basidiomycetous fungi, and this—albeit rather low—expression level can be further enhanced by inducers. Increase in laccase transcription in response to aromatic compounds such as 2,5-xylidine has been reported for the laccase genes from, for example, *T. versicolor*, *Trametes villosa*, and *A. bisporus* (Aro et al., 2005). It has been proposed that the laccases may actually initiate the attack on the lignin and that the respective degradation products released from lignin would then act as inducers for further laccase and ligninase gene expression (Scheel et al., 2000).

Being copper-containing enzymes, copper ions also have a significant activating effect on laccase gene expression. The mechanism how copper activates transcription of laccase genes has been investigated in some detail in *C. subvermispora.* Several fungal laccase genes contain a putative ACE1 element (5′-HTHNNGCTGD-3′) to which the transcriptional activator ACE1 (activation of *cup1* expression protein; not to be mixed with the *T. reesei* ACE1 repressor of cellulase formation), which was first identified in *S. cerevisiae* as a regulator of expression of metallothionein-encoding genes, would bind (Alvarez et al., 2009). As for basidiomycetes, orthologs of ACE1 have been found in *P. chrysosporium* (Polanco et al., 2006) and *C. subvermispora* (Alvarez et al., 2009).

Chapter 8
The Fungal Secretory Pathways and Their Relation to Lignocellulose Degradation

8.1 The Fungal Secretory Pathway

Protein secretion is of enormous importance to the degradation of lignocellulosic biomass because of the insolubility of the substrate. In order to attack the substrates, the enzymes must be transported out of the cell. To this end, the enzymes must not only be formed and cargoed along the hyphae to the cellular border and secreted out of the cell but also become posttranslationally modified to be recognized by several of the molecular machineries involved. In addition, extracellular enzymes must be protected against the adverse conditions that prevail exterior of the cell, such as very low protein concentration, adverse pH, and proteolytic attack (Figure 8.1). Understanding of the ultrastructure, biochemistry, and genetics of fungal protein secretion is, therefore, not only important to understand how the extracellular enzymes are formed but also pivotal for the development of strategies how to improve the fungal strains that secrete these enzymes at an industrial level for plant biomass hydrolysis. Most of our detailed information on the fungal secretory pathway comes from studies on the yeast *Saccharomyces cerevisiae*, pioneered by Randy Schekman and his *sec* mutants (Schekman, 2010), but knowledge based on selected topics with some filamentous fungi (mostly *Trichoderma* and *Aspergillus*) has considerably advanced over the last years.

8.1.1 It Starts at the Endoplasmic Reticulum

In order to target potential secretory proteins to the cellular secretory machinery, these proteins are already distinguished from the other proteins during their translation: mRNAs encoding secreted proteins (which from the perspective of the cell are also proteins that are integral components of the membrane) are targeted to the endoplasmic reticulum (ER) to serve as a template for nascent polypeptide synthesis. This can involve two independent pathways: cotranslational translocation and posttranslational translocation, which, in *Saccharomyces cerevisiae,* are both equally important. Yet, the cotranslational pathway was discovered much earlier and is therefore much better known. Key to its operation is the presence of a

Fungi and Lignocellulosic Biomass, First Edition. Christian P. Kubicek.

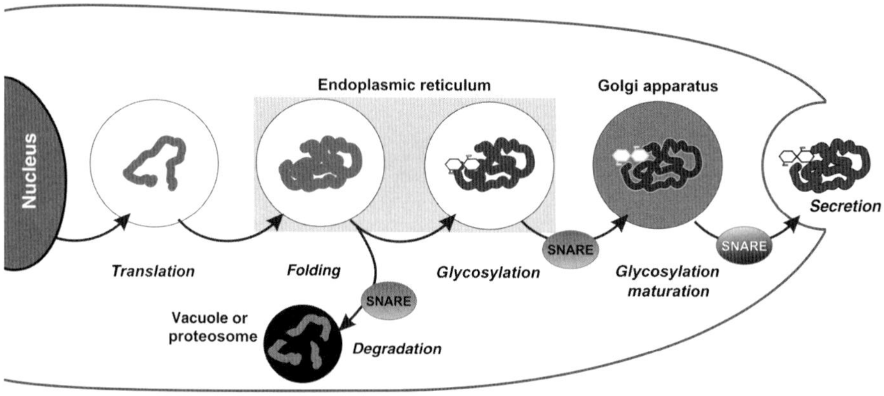

Figure 8.1. Schematic model of the subsequent steps involved in protein secretion in filamentous fungi.

so-called "signal peptide" (SP) at the N-terminus of a protein destined for secretion. While SPs generally do not share sequence homology, they do have three conserved domains (Paetzel et al., 2002): a positively charged amino-terminal domain, a central hydrophobic domain, and a carboxy-terminal hydrophilic domain containing the SPase processing determinants. These sequences are similar to bacterial SPs. Also, like their bacterial counterparts, the hydrophobic character is important for function as a translocation signal. The most important key feature is a minimum hydrophobicity. ER SPs have a higher content of leucine residues relative to bacterial SPs (Figure 8.2). Interestingly, the cotranslational pathway has been shown to require protein precursors with more hydrophobic SPs.

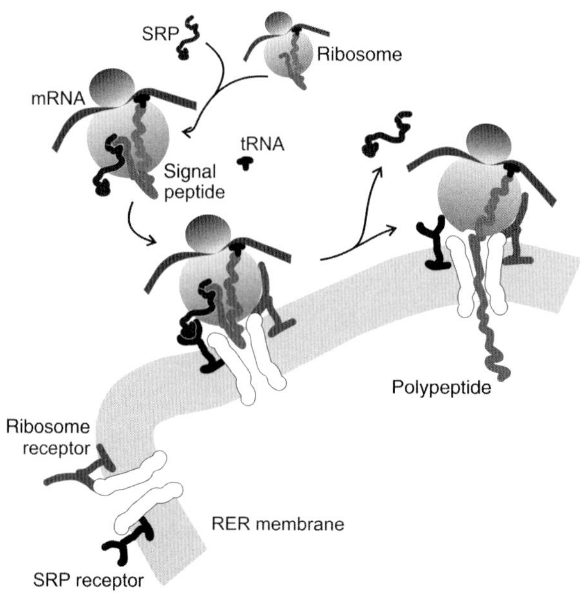

Figure 8.2. Schematic overview of cotranslational targeting of proteins destined for secretion or membrane insertion (SRP cycle). SRP, signal recognition particle. Details are described in the text.

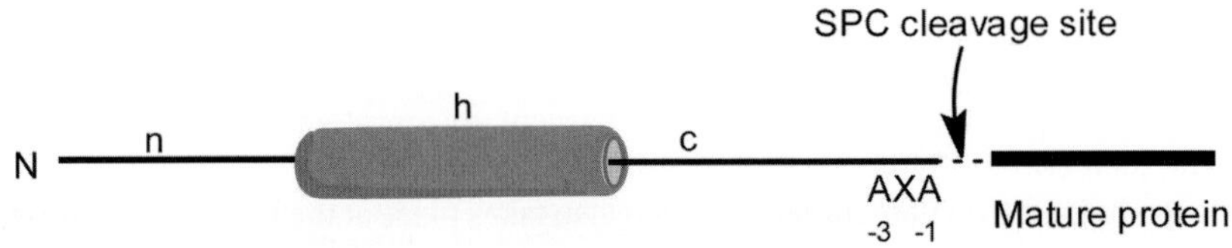

Figure 8.3. Structure of the prepropeptide of a fungal secretory protein: n, N-terminal positive charged region; h, hydrophobic helical region; c, C-terminal region containing the target sequence for cleavage by the signal peptide processing protease (AXA, where X can be any amino acid).

When the mRNA is translated on ribosomes and the nascent protein chain exhibits such an SP, it is recognized by the signal recognition particle (SRP; Walter and Johnson, 1994; Figure 8.3) and peptide elongation is retarded. This complex is then targeted to the ER membrane where it binds to the signal recognition particle receptor (SR). GTP has to be bound to both SRP as well as SR to achieve this. Once recognition by SR has taken place, the ribosome–mRNA–nascent chain complex binds to the to the Sec61 translocon complex, which functions as a translocation channel for the newly synthesized polypeptides to enter the ER lumen. Energy for this translocation comes from hydrolysis of the GTP bound to SRP and SR, and subsequent dissociation of the SRP–SR complex (Figure 8.3). When the N-terminus of the protein precursor has reached the lumen of the ER, the SP is removed by a signal peptidase (SPase). The eukaryotic signal peptidase is an integral membrane protein complex that contains four subunits: Sec11, Spc3p, Spc2p, and Spc1p.

After the termination and translocation of polypeptide synthesis, both the mRNA and the ribosomal subunits recycle to the cytosol for further rounds of protein synthesis (via the SRP pathway) or for degradation (Gerst, 2008). For more details about the structure and function of the SRP, see Wild et al. (2004).

In the other pathway (posttranslational transport), already fully synthesized precursor polypeptides associate with cytosolic molecular chaperones of the Hsp70 and Hsp40 heat shock protein family, which function in keeping the precursor polypeptides in an appropriate folding to remain soluble. A heterotrimeric complex of three membrane proteins (Sec62p, Sec71p, and Sec72p) then serves as a SP receptor to enable the passage through the Sec61p-translocon complex (Zimmermann et al., 2011).

In both pathways, about 70 amino acid residues of the nascent polypeptide chain remain still buried in the ribosomal tunnel and the polypeptide conducting channel of the Sec61 complex after termination of translation. Complete translocation of this carboxy-terminal peptide into the ER lumen therefore requires helper proteins. This function is fulfilled by the ER-chaperon BiP that is recruited to the ribosomal tunnel and the polypeptide-conducting channel within the heterotrimeric Sec61 complex. BiP is an ATP-regulated Hsp70 chaperone, which, together with cochaperones, mediates folding of the protein in an ATP-dependent cyclic process. The functional cycle of Hsp70s is determined by conformational transitions that are required for substrate binding and release (Marcinowski et al., 2011). Further proteins required for proper folding are calnexin, peptidyl-prolyl-isomerases, and the protein disulfide isomerases (PDIs) that catalyze the formation of disulfide bonds. The latter are essential because the vectorial nature of translocation process causes formation of several disulfide bonds that are not present in the final native protein. Thus, many initial disulfides are subsequently isomerized by the PDIs. To this end, the Cys-X-X-Cys motif in the thioredoxin domain of PDIs accepts oxidizing equivalents, primarily from the proteins Ero1 and Erv2 (but additional mechanisms have recently been uncovered; cf. Fass, 2010). These oxidizing equivalents are then transferred to client proteins via mixed disulfide intermediates. Then the correctly folded protein is released to transport vesicles.

Calnexin plays a special role in the folding of nascent glycoproteins by binding to them and by mediating their correct folding and processing of the *N*-glycans.

An early modification on nascent proteins that also takes place at the ER involves the transfer of a core oligosaccharide ($GlcNAC_2Man_9Glc_3$) to Asn-X-Ser/Thr. *N*-linked glycosylation improves protein solubility, decreases aggregation, provides a binding site for calnexin, and facilitates interaction with PDI. Glycosylation of secretory proteins is described in detail in Section 8.2.

8.1.2 Quality Control in the ER

The ER also has a central role for quality control of proteins; misfolded proteins are recognized and delivered to proteasomes (Hampton, 2002) or to vacuoles (Coughlan et al., 2004) for degradation. In this regard, the ER seems to be the major rate-limiting step in protein secretion and to be one of the bottlenecks in production of heterologous proteins by fungi.

Two mechanism monitor the proper folding of ER proteins: one is the UDP-glucose:glycoprotein glucosyltransferase (UGT), a central player of glycoprotein quality control in the ER (Kleizen and Braakman, 2004). After addition of the core glycan (GlcNac2-Man9Glc3) to specific asparagine residues of the nascent polypeptide, as mentioned above, the three terminal glucose residues have to be removed in order to allow the protein to exit the ER (see Section 8.3.1). In case the resulting protein has not folded properly, UGT will add a single glucose residue to the deglucosylated core glycan, which labels it for reassociation with calnexin to retry folding. Upon persistent misfolding, the *N*-glycosylated polypeptides will dissociate from calnexin and enter the second mechanism of quality control exerted by the BiP chaperone complex (Molinari et al., 2005). This complex preferentially binds to hydrophobic patches of the misfolded protein. Prolonged binding to either calnexin or the BiP complex is indicative of intrinsic problems with this protein and thus targets the polypeptides to ER-associated protein degradation (ERAD, see below). An accumulation of unfolded proteins in the ER activates the unfolded protein response (UPR) pathway (Figure 8.4.). The UPR allows the ER to trigger a comprehensive gene expression program that adjusts the protein folding capacity of the cell according to need (Patil and Walter, 2001). In mammals, the UPR genes are regulated by IRE1α and ATF6 (an ER stress-regulated transmembrane transcription factor), which activate transcription factor XBP1 to induce transcription of the UPR genes (Lee et al., 2002). In addition, unfolded proteins in the ER activate the ER-resident protein kinase (PERK) that phosphorylates the α-subunit of the translation initiation factor 2 (eIF2α) (Harding et al., 1999). This phosphorylation leads to attenuation of general translation initiation, but also to a concomitant activation of translation of selected proteins, including ATF4 (Harding et al., 2000), a homolog of the regulator of general amino acid control Gcn4p.

In *S. cerevisiae,* this response is mediated by a transmembrane kinase/endonuclease Ire1p that is required for generation of an actively translated form of the transcript encoding Hac1p, a bZIP transcription factor essential for the induction of the UPR target genes (Figure 8.5). This active form of Ire1p arises by the excision of a translation-inhibiting large intron from the *HAC1* mRNA. Removal of the *HAC1* intron and subsequent ligation of the two liberated exons by a tRNA ligase produces a spliced mRNA that is efficiently translated (Patil and Walter, 2001). Hac1p then binds to the UPR element (a palindromic CANCNTG motif, separated by a single nt; "UPRE," Ogawa and Mori, 2004) in the promoters of UPR target genes. The

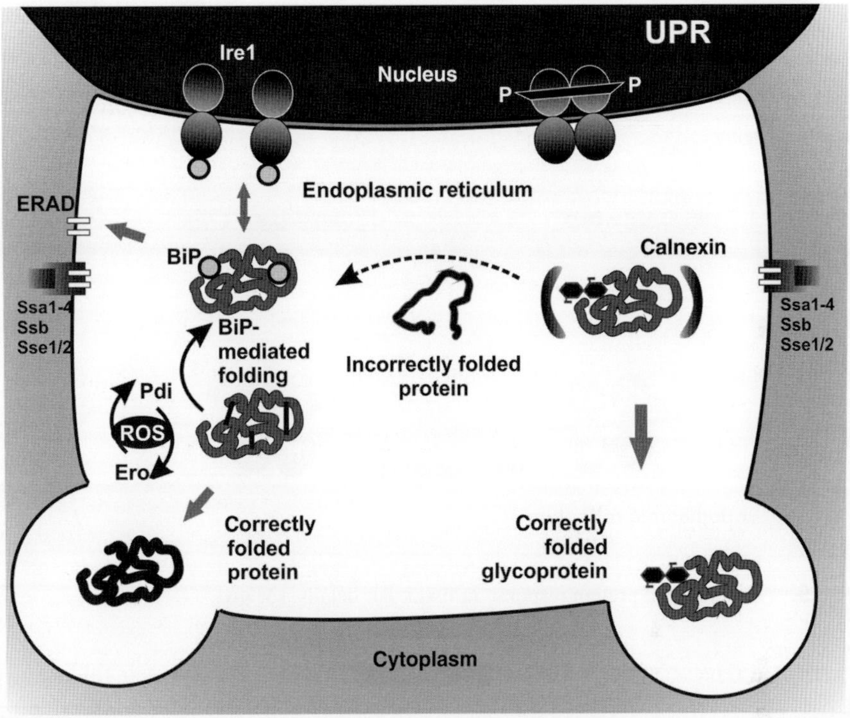

Figure 8.4. Summary of the events accompanying protein folding, quality control, and unfolded protein response in lower eukaryotic cells, based on data from yeast. For details, see text. Ssa1-4, Ssb, Sse1/2: chaperons that support solubility and prevent aggregation of the polypeptide chains; ROS, reactive oxygen species.

orthologs of *IRE1* and *HAC1* have been cloned and functionally characterized from *T. reesei* (Valkonen et al., 2004).

One class of genes targeted by UPR is that involved in ERAD. ERAD is a process by which misfolded ER proteins are detected by ER-resident factors and prevented from progressing along the secretory pathway. They are directed to the translocon for retrotranslocation into the cytosol, where they undergo ubiquitin- and proteasome-dependent degradation (for reviews, see Plemper and Wolf, 1999). Whole-genome expression profiling by DNA microarrays has shown that more than 350 out of its total 6,300 genes; i.e., >5%) are regulated by the UPR. They comprised genes encoding proteins implicated in protein translocation, lipid metabolism, glycosylation, ERAD, ER to Golgi traffic, and protein targeting to the vacuole and to the cell surface (Travers et al., 2000). Interestingly, UPR is also activated by a nitrogen-rich environment, possibly because rapid translation in a nitrogen-rich environment results in accumulation of unfolded proteins (Schröder et al., 2000). The transcription factor of general amino acid control Gcn4p plays an essential role in the UPR, and both Gcn4p and its activator Gcn2p are required for induction of the majority of UPR target genes during ER stress (Patil et al., 2004). Regulation of Gcn4p levels in response to changing physiological conditions may function as an additional means to modulate the UPR.

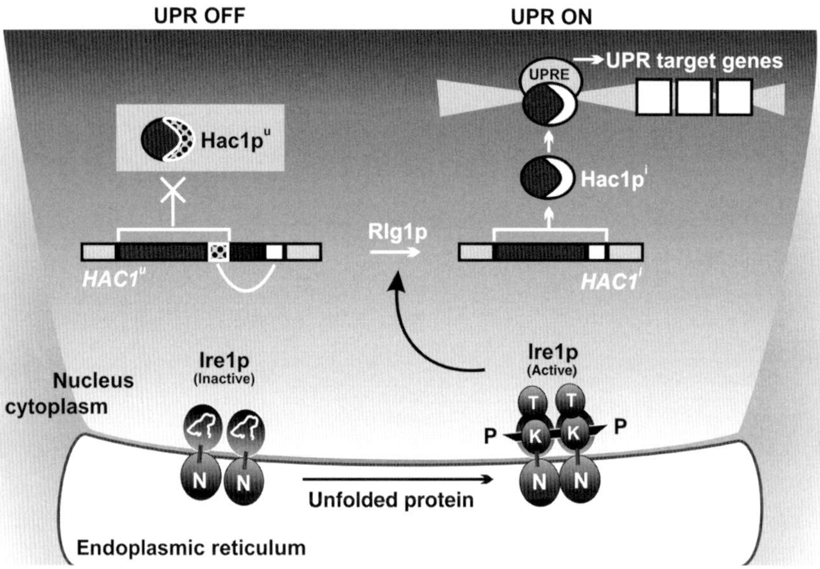

Figure 8.5. The unfolded protein response in fungi. For details, see text.

Treatment of *T. reesei* cultures with chemical agents known to inhibit protein folding revealed yet another further UPR response: repression under secretion stress. It is reflected in a rapid decline in the transcription of genes encoding secreted proteins (such as the cellobiohydrolase CEL7A), but not intracellular proteins, and this decline was shown to be dependent on specific consensus sequences in the promoter of *cel7A* (Pakula et al., 2003). A similar repression mechanism has also been demonstrated in *Aspergillus niger*. It is currently not known if (and if so how) the UPR pathway is involved in this mechanism, and it has been suggested that the ER stress response signaling could be more diverse in filamentous fungi and involve additional not yet discovered components (Aro et al., 2005).

8.1.3 Golgi Apparatus

After being synthesized and folded, secretory proteins exit the ER and are delivered to the Golgi apparatus for further modification and sorting for the final destination (i.e., plasma membrane, export, or vacuole).

The Golgi apparatus in filamentous fungi was recognized as punctate structures that—like the ER—are typically found more abundantly at the apical regions of hyphae and are significantly different from the Golgi apparatus originally detected in plants (Cole et al., 2000).

To travel to the Golgi, proteins synthesized in the ER must be packed into vesicles (Figure 8.6). The coat protein complex II (COPII) is responsible for the biogenesis of these ER-derived vesicles (Kirchhausen, 2007). It consists of five subunits: the small GTPase Sar1 (secretion-associated, ras-related), a small GTPase that inserts its N-terminal amphipathic helix into the cytoplasmic side of the ER membrane bilayer when bound to GTP; the "inner coat" complex, comprising a heterodimeric tight complex of Sec23 and Sec24; and the Sec13–Sec31 "outer coat" heterotetrameric complex. For assembly, the Sec23–Sec24 complex is recruited

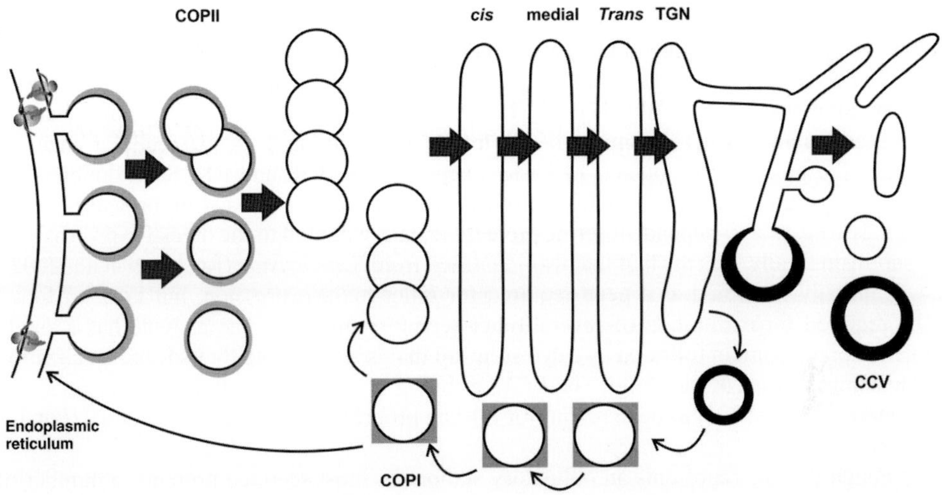

Figure 8.6. Model for protein transport from the endoplasmic reticulum across the Golgi apparatus. TGN, trans Golgi network; CCV, clathrin-coated vesicle.

by Sar1 and binds directly to the ER membrane as an "inner coat," which then recruits the Sec13–Sec31 "outer coat." Sec24p is primarily responsible to recruit cargo proteins. Many of them have specific export-signal sequences in their cytoplasmic domains to mark them for COPII transport, such as di-hydrophobic, di-acidic, C-terminal hydrophobic, and aromatic motifs (Jensen and Schekman, 2011). However, not all cargo proteins contain a signal for binding to Sec24. Those without such a signal may interact with a transport adaptor, be included in the COPII vesicle, and enter COPII vesicles by simple diffusion through an indirect interaction. Once the COPII complex, it deforms the membrane enough to bud a vesicle off (Fromme et al., 2008; Jensen and Schekman, 2011).

The *cis*-Golgi cisternae are believed to be formed directly *de novo* from ER-derived, cargo-containing vesicles, a process called "cisternal maturation model." This is facilitated by an appropriate spatial organization: in the yeast *Pichia pastoris*, the Golgi stacks are formed just in front of the ER exit sites, and a spatial correlation between the ER exit sites and the cis-Golgi has been observed in *S. cerevisiae* (Czlapinski and Bertozzi, 2006). However, before fusion with the *cis*-Golgi, the COPII vesicles must first disassemble their coat. This is aided by Sec23, which has a GTPase-activating activity toward Sar1, and this activity is strongly increased by binding of the outer coat. Upon reaching the Golgi, hydrolysis of Sar1-bound GTP to GDP therefore results in depolymerization of the COPII coat.

The mechanism by which proteins then pass through the Golgi apparatus is still not completely understood, and this is particularly true for multicellular fungi. Having traversed the Golgi, the cargo molecules are sorted at the TGN (*trans*-Golgi network) to various cellular destinations. A detailed description of the processing and sorting at the TGN in higher eukaryotes is available (De Matteis and Luini, 2008; Wilson et al., 2011). While the yeast Golgi apparatus does not feature defined layers of cisternae typical for the Golgi of animal or plant cells, it can be separated into three compartments (early, mid, and late), which can be distinguished by specific enzymes: α-1,6-mannosyltransferase (early), α-1,3-mannosyltransferase (mid), and Kex2p protease (late) (Nothwehr and Stevens, 1994). Proteins that are not destined to rest in the Golgi apparatus are sorted for cargo to two different destinations: the plasma membrane or the vacuole. The Golgi is also the place for final modification of glycosylated proteins (see below) and protease cleavage of proteins to their mature form (Rockwell et al., 2002).

As for the latter, many fungal secretory proteins contain a second cleavage site for a protease after the SP cleavage site (Calmels et al., 1991). A comparison of the dipeptide distributions of 33 secreted and 34 cytoplasmic proteins from fungal producers of extracellular enzymes indicated the presence of doublets of basic dipeptides KR, RR, and RK. Such doublets are typical recognition sites for the KEX2 protease: this proprotein processing protease (kexin, EC 3.4.21.61), a Ca^{2+}-dependent serine protease distantly related to the digestive proteases of the subtilisin family, was the first that was identified from *S. cerevisiae* (Rockwell et al., 2002). It was initially identified as a gene required for *k*iller toxin *ex*pression, but later on found to be required for maturation of several other secretory proteins. The enzyme has a single transmembrane domain followed by a cytosolic tail that is responsible for its Kex2 localization to late compartments of the Golgi. The *KEX2* ortholog of *A. niger* (*kexB*) has been cloned and characterized and shown to be essential for dibasic proprotein cleavage of secretory proteins (Punt et al., 2003).

Although the GC represents an obligatory station for most secreted proteins, a number of proteins reach the cell surface by bypassing the Golgi in what is known as unconventional trafficking, a process that appears to occur in all organisms (Nickel and Rabouille, 2009; Wilson et al., 2011).

8.1.4 The Plasma Membrane

Secreted and plasma membrane proteins leave the Golgi by exocytosis, which involves delivery of Golgi apparatus-derived secretory vesicles to the apical plasma membrane via an apical vesicle cluster within a multicomponent complex called the Spitzenkörper (Virag and Harris, 2006). These secretory vesicles contain membrane proteins and lipids, enzymes involved in cell wall synthesis, and possibly cell wall precursors. Docking of exocytic vesicles to the plasma membrane involves a protein complex named “exocyst” (Lipschütz and Mostov, 2002; Munson and Novick, 2006). In *S. cerevisiae*, the exocyst comprises eight proteins: Sec3p, Sec5p, Sec6p, Sec8p, Sec10p, Sec15p, Exo70p, and Exo84p, which are all hydrophilic and interact with each other to form a 19.5S complex, which was shown to bind to regions of active cell surface expansion (Figure 8.7).

Several small GTPases regulate the attachment of the exocyst to the correct place: Sec4, the first identified member of the Rab (regulators of vesicular traffic events) family of GTPases, was also the first GTPase for which a function in regulating the exocyst was shown (Lazar et al., 1997). Thereby, Sec15p (see above) interacts with Sec4-GTP, located on the surface of the vesicular membrane, which triggers the docking and fusion of secretory vesicles with specific domains of the plasma membrane.

Membrane-associated proteins called SNAREs (soluble *N*-ethylmaleimide-sensitive factor attachment protein receptors) facilitate the subsequent fusion between vesicles and target membranes. SNAREs are divided into two classes according to their localization: vesicle (v)-SNAREs and target (t)-SNAREs (Malsam and Söllner, 2011). Each SNARE protein contains an α-helical structure that is complimentary to a similar α-helix in other SNAREs and thus brings the vesicle and target membranes into close proximity and so enables membrane fusion. SNARE complex formation in the yeast secretory pathway has been shown to begin with the formation of binary complexes between two t-SNAREs, Ssop1/2p, and Sec9. This complex then binds the v-SNARE Sncp. *S. cerevisiae* has two functionally redundant plasma

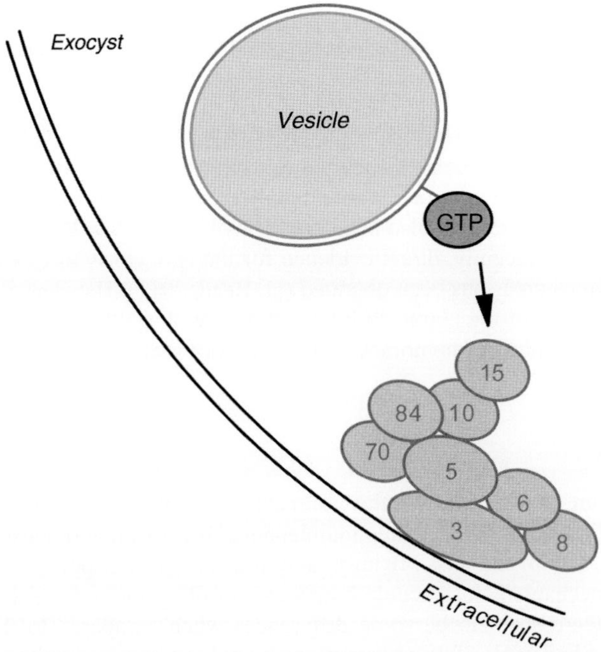

Figure 8.7. Model of docking of vesicles via Sec4-GTP (indicated by "GTP") to the exocyst complex. The eight subunits of the complex and their interactions are indicated by the respective "Sec" numbers, that is, Sec84 is "84" ands so on. For more details see Lipschütz and Mostov (2002).

membrane t-SNAREs (Sso1p and Sso2p; Aalto et al., 1993), and two syntaxin-like t-SNAREs are also present in several filamentous fungi, including *N. crassa*, *A. oryzae*, *A. niger*, and *T. reesei.* Some (e.g., *A. nidulans* and *A. fumigatus*) have only one protein though (Shoji et al., 2008). In *T. reesei*, SSO1 interacts with the v-SNARE Snc1p at subapical/basal regions of the hyphae, while SSO2 interacts with SNC1 at apical regions (Valkonen et al., 2007). The authors hypothesized that the two SSO proteins are involved in distinct secretion pathways in filamentous fungi, which would be further supported by the findings that Sec4 of *A. niger* is not an essential protein (Punt et al., 2001). Shoji et al. (2008) concluded that "if the existence of several secretion pathways is really the case, it may suggest an explanation for the high secretion ability of filamentous fungi, and it might be even possible to elevate production of homologous and heterologous proteins without affecting hyphal tip growth."

Furthermore, *A. oryzae* seems to have an additional v-SNARE for secretion. A putative vacuolar v-SNARE, AoNyv1p, localizes to the plasma membrane in addition to vacuoles, suggesting a role for it as an exocytic v-SNARE (Kuratsu et al., 2007). Although it is generally believed that exocytosis of extracellular enzymes (i.e., in contrast to "constitutive" secretion at growth cones or buds) into the external medium occurs mainly from hyphal tips, a few studies have actually localized the subcellular sites of secretion. However, immunogold and green fluorescent protein labeling of glucoamylase in *A. niger* supported the hypothesis that the secretion of this enzyme primarily takes place from hyphal tips. Using a fusion protein of BipAp with enhanced green fluorescent protein (BipA-EGFP) in *A. oryzae* revealed that ER-resident BipA-EGFP has a tip-high gradient, suggesting that requirement of molecular chaperones is more severe at apical regions (Maruyama et al., 2006). Similar distribution gradients were found for ER-resident SNARE proteins (Kuratsu et al., 2007). These data strengthen the view that ER-resident protein folding, modification, and transport to Golgi apparatus are processes that are more active at apical regions. However, the ER in *A. oryzae* has been shown to be often located at the cell periphery of subapical regions and near septa as well (Maruyama et al., 2006; Kuratsu et al., 2007), suggesting that the ER has roles in these regions in addition to hyphal tips. It has been long established that proteins are not only secreted at the periphery of the mycelium but also in the colony centre. It is also been established that enzymes are commonly secreted from hyphal tips. Very recently, direct evidence for the exocytosis of extracellular enzymes from subapical hyphal areas has been obtained (cf. Read, 2011). Hayakawa et al. (2011) have shown that the a fusion of the extracellular α-amylase with GFP rapidly accumulates in the periplasm between the plasma membrane and cell wall of septa of *A. oryzae*.

8.1.5 Vacuoles

Vacuolar proteins are also identified at the Golgi apparatus and delivered to vacuoles via endosomes. Functions of the vacuole include maintenance of cytosolic homeostasis, degradation, storage (Klionsky, 1997), and possibly intra- and intercellular transport (Darrah et al., 2006).

Although the mechanisms of vacuole biogenesis in filamentous fungi have only scarcely been elucidated, there seem to be some differences when compared to those operating in budding yeast (cf. Shoji et al., 2008): filamentous fungi have only one protein that corresponds to both the late endosomal t-SNARE Pep12p (Becherer et al., 1996) and vacuolar t-SNARE Vam3p of *S. cerevisiae*, as shown by the fact that the respective Pep12p ortholog of *A. oryzae* can be found both in the vacuolar membrane and putative late endosomes (Kuratsu et al., 2007). Similar findings have also been obtained for the *A. oryzae* ortholog of *S. cerevisiae* Vam7p, another vacuolar t-SNARE (Ungermann and Wickner, 1998), which localizes both to the vacuoles and the late endosome (Kuratsu et al., 2007).

In yeast, vacuolar proteins can reach their target organelle by two defined pathways: the carboxypeptidase Y (CPY) pathway, which reaches the vacuole via a prevacuolar compartment, and the alkaline phosphatase pathway, which leads directly from the Golgi to the vacuole (Conibear and Stevens, 1998). For the former, a sorting signal is needed, and mutational analysis defined the residues "QRPL" as crucial elements of this sorting signal. It is only needed for soluble proteins and not for integral membrane proteins because the latter are transported to the vacuole as a default (Roberts et al., 1992). In yeast, the protein Vps10 recognizes the QRPL target signal of pro-CPY and—together with clathrin—forms vesicles that head for the vacuole. Its knockout leads to a mislocation of vacuolar proteins to the plasma membrane (Conibear and Stevens, 1998).

In filamentous fungi, vacuoles have been observed to contain enzymes that were destined for secretion into the medium (e.g., ligninases of *Phanerochaete chrysosporium*, Kuan and Tien, 1989; and xylanases of *T. reesei*, Kurzatkowski et al., 1996). These findings have recently been substantiated by the findings that the genome of the hypercellulolytic *T. reesei* strain RUT C30 contains mutations in three genes encoding proteins necessary for vacuolar protein sorting (VPS1, VSP13, and 16-kDa proteolipid subunit of the vacuolar ATPase, Le Crom et al., 2009), suggesting that mutations in genes of vacuolar function may increase cellulase production by this fungus. In *A. oryzae*, disruption of the gene encoding the vacuolar protein sorting receptor VPS10 resulted in missorting and secretion of the vacuolar carboxypeptidase CpyA, but also increased the secretion of heterologous proteins, suggesting that such proteins may be (partially) degraded in the vacuole (Yoon et al., 2010). In summary, there are indications available that the vacuole could act as a dead end street for part of the population of proteins destined for secretion in fungi. However, solid proof for this requires further detailed studies.

8.2 Protein Glycosylation

Most of the proteins that are destined for secretion by fungi are glycosylated, which improves solubility of the protein, and decreases the tendency for aggregation. Two types of protein glycosylation can be distinguished: *N*-glycosylation, in which an oligosaccharide chain is linked via an *N*-acetylglucosamine to a protein asparagines residue (that occurs in the consensus sequence N-X-S/T (wherein X can be any amino acid but a proline), and *O*-glycosylation, in which short glycan chains are attached directly to a serine or threonine residue. As described above, attachment of *N*-glycans already starts in the ER by attaching a presynthesized glycan chain ($NAcGlc_2Man_9Glc_3$) to the amino group of asparagine. *O*-glycosylation also starts in the ER and continues in the Golgi.

8.2.1 N-Glycosylation

The transfer of a $NAcGlc_2Man_9Glc_3$ starts during translocation of the nascent polypeptide into the ER. This core oligosaccharide is presynthesized in the cytoplasm and then transferred into the ER (Helenius and Aebi, 2004), thereby the first cytoplasmic step—the synthesis of GDP-mannose from GTP and mannose-1-phosphate—is catalyzed by GDP:α-D-mannose-1-phosphate guanyltransferase (EC 2.7.7.13; MPG1). Subsequently, dolichyl-phosphate mannose synthase (DPM1, EC 2.4.1.83) transfers the mannose residue from GDP-mannose to dolichol phosphate, thus forming dolicholphosphate mannose (Dol-P-Man). Dolichol is an isoprenoid derivative, characterized by a saturated isoprene moiety at the α-position and 17–21 isoprene units, and functions to associate the nascent oligosaccharide to the ER (Schenk et al., 2001). DPM1 from *T. reesei* has been studied in detail (Kruszewska et al., 2000) and shown to belong to the "human" DPM1 class. Members of this class, unlike the DPM1 proteins from yeast, lack the hydrophobic C-terminus that is characteristic and necessary for insertion into the ER membrane, and also do not contain any other transmembrane domain (Maeda and Kinoshita, 2008). Since Dol-P-Man is synthesized at the cytoplasmic face of the ER membrane, it requires a protein that translocates it into the ER lumen where it is further used. To this end, the "mammalian" DPM synthase complex consists of three proteins (DPM1, DPM2, and DPM3; Figure 8.8), of which DPM3 interacts with the DPM1 C-terminus and tethers it to the ER membrane. DPM2 then interacts with and stabilizes or increases the expression of DPM3.

The transfer of the first *N*-acetyl-glucosamine unit (NAcGlc) occurs via NAcGlc-1-P (derived from UDP-NAcGlc) onto Dol-P to form GlcNAc-PP-dolichol is then catalyzed by UDP *N*-acetylglucosamine:dolichol phosphate *N*-acetylglucosaminyl phosphoryl transferase (EC 2.7.8.15).

The enzymes involved in the addition of the first two NAcGlc residues, the five mannose residues, and the three terminal glucose residues bind to the cytosolic face of the ER membrane are conserved across the filamentous fungi. The membrane protein RFT1, a flippase, then translocates the oligosaccharides across the membrane into the ER lumen (Deshpande et al., 2008). There the attachment of the glycan to the nascent protein is catalyzed by the SST3

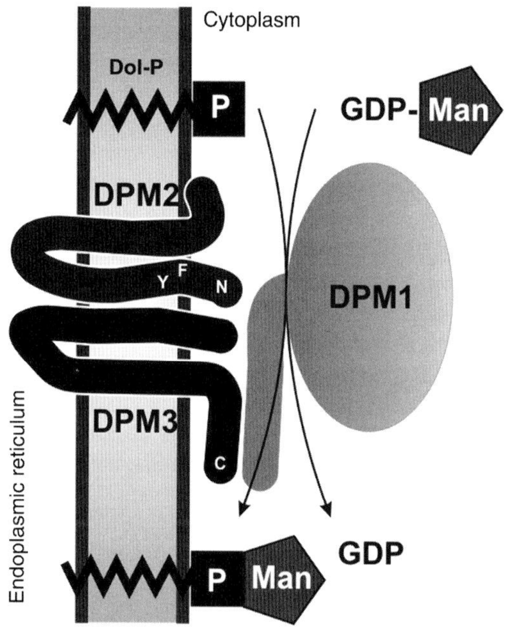

Figure 8.8. The dolichol-phosphate-mannose synthase complex.

subunit(s) of oligosaccharyltransferase, a heterogeneous membrane complex that associates with the translocon, scans for *N*-glycosylation consensus sites and attaches the preassembled high-mannose oligosaccharide ($Glc_3Man_9GlcNAc_2$) from the dolichol-linked pyrophosphate donor to the side chain of asparagines in the N-X-S/T motif of the still translated protein.

Deshpande et al. (2008) performed a detailed comparative genomics analysis of the oligosaccharyl transferase enzyme complex: in *S. cerevisiae,* five genes in this complex (*OST1*, *OST2*, *STT3*, *WBP1*, and *SWP1*) are essential. Out of these, *OST1*α, *STT3,* and *WBP1ß* are highly conserved among filamentous fungi. The genes *OST3* and *OST6*, which are essential for recruiting the fully active complex, are much less conserved, however, although the functional catalytic domains are present. Orthologs for *OST2*, *SWP1*, and *OST4* were not found in filamentous fungi.

After attachment of $Glc_3Man_9GlcNAc_2$ to the polypeptide, the three terminal glucose residues are removed. The respective glucosidases, glucosidase I and glucosidase II, are conserved between fungi and other eukaryotes. The resulting $NAcGlc_2Man_9$ structure can then be further trimmed by the action of an α-1,2-mannosidases, which remove four α-1,2-linked mannose residues, resulting in the formation of $GlcNAc_2Man_5$ (Akao et al., 2006). However, expression of this enzyme is low and thus only a minor part of the *N*-glycans in filamentous fungi actually contain the $Man_5GlcNAc_2$ structure.

After transfer of the glycoprotein to the Golgi, a single mannose residue is removed from the $GlcNAc_2Man_9$ by the action of a second α-1,2-mannosidase (Yoshida et al., 2000). This oligosaccharide can then act as a substrate for Golgi-located α-1,6-mannosyltransferases. Deshpande et al. (2008) identified five orthologs for the complex mannosyltransferases Mnn9, Van1, Mnn10, Mnn11, and Anp1 in the genomes of filamentous fungi *in silico,* suggesting that the respective *N*-glycosylated proteins will bear high-level mannosylation.

Research on *A. fumigatus* has further led to the identification of *N*-linked galactomannan structures in its glycoproteins, and the respective galactofuranose residues are antigenic. They are typically found at the nonreducing end of *N*-glycans and have been hypothesized to act like a stop signal for further mannose addition (Morelle et al., 2005). Galactofuranose arises from UDP-galactose by the enzyme UDP-galactopyranose mutase, and the gene encoding this enzyme has been identified in *A. fumigates* (Oppenheimer et al., 2010).

8.2.2 O-Glycan Synthesis

O-glycosylation also occurs frequent in fungal extracellular proteins, especially in the hinge regions that connect catalytic and substrate binding domains in cellulases (cf. Chapter 3). They are typically linear and one to five residue long mannooligosaccharides, but branched *O*-glycans with mannose residues linked by α-1–6, α-1–2, and α-1–3 linkages have also been found. In addition, glucose, galactopyranose, and galactofuranose residues were found in *A. niger* glucoamylase (Pazur et al., 1980; Neustroev et al., 1993). In *T. reesei* CEL7A, sulfate or phosphate esters have been found to be attached to these short *O*-glycan chains (Harrison et al., 1998; Hui et al., 2001).

O-glycosylation starts in the ER. So far, a consensus sequence for attachment has not been identified; rather, proline rich, serine- and threonine-containing regions seem to serve as a recipient. Genes encoding the protein-*O*-mannosyltransferase (PMT, EC 2.4.1.109) enzyme have been isolated from *T. reesei* (*pmt1*), *A. nidulans (AnPmtA)*, and *A. awamori (AnPmtA)* (for references, see Deshpande et al., 2008). In contrast, none of the other enzymes of the *O*-glycosylation pathway has yet been characterized from filamentous fungi, but three putative α-1,2-mannosyltransferase-encoding genes (*mnt1*, *ktr1*, *ktr3*) have been identified in their genomes based on similarity with the known proteins from *S. cerevisiae* (Deshpande et al., 2008).

8.3 Strategies for Improvement of the Fungal Secretory Pathway

While the fungal lifestyle is essentially dependent on the continuous secretion of proteins for making nutrients available to it and also other purposes, they usually do well with a low level of protein secretion to which the energy demanding machinery of protein secretion and modification is adapted. However, these levels rarely satisfy the needs and expectations of industrial processes, which today manage to secrete final extracellular protein concentrations of more than 120 g/L (see Chapter 9). Consequently, the strains in use today are the result of a long-term breeding program that presumably led to a number of changes in the protein secretory and glycosylating pathway.

The high capacity for protein formation and secretion of fungi like *T. reesei*, *A. niger*, or *A. oryzae* has challenged their use as heterologous hosts for the production of proteins from other organisms ranging from basidiomycetes to man (for review, see Gasser et al., 2008; Shoji et al., 2008; Lubertozzi and Keasling, 2009). A comparison of the quality of heterologous protein secretion by filamentous fungi with other organisms is given in Table 8.1. However, the yields of heterologous secreted proteins by filamentous fungi remained so far much below the expectations. In order to identify the bottlenecks, the factors that limit and restrict the expression and secretion of heterologous proteins in fungi has been investigated in some detail.

Table 8.1. Factors influencing the choice of the most appropriate host for heterologous production of mammalian proteins.

					N-Glycosylation		*O*-Glycosylation
	Media Costs	Expression Level	Secretion	Protein Refolding Required	Mammalian Type	Sialic Acid Present	Non-human Sugars
Escherichia coli and other bacteria	Low to medium	High	Periplasmic space or inclusion bodies	Mostly	No	No	No
Filamentous fungi	Low to medium	High	Medium	Sometimes	Yes	No	Yes
Pichia pastoris	Low to medium	High	Medium	Sometimes	No (high mannose)	No	Yes
Insect cells	High	High	Medium	No	Complex	No	Yes
Mammalian cells	High	High	Medium	No	Complex	Yes	Yes
Plant cell cultures	Medium to high	Low	Medium	No	Complex	No	Yes

8.3.1 Folding and UPR

As explained in Section 8.1, the early folding steps in the ER have been shown to be rate limiting. Fusing heterologous proteins to homologous "cargo proteins" has been shown to increase the yield considerably, indicating a potential problem at the folding step (Archer, 2000). First attention was consequently paid to the chaperone BiP. Indeed, overexpressing this protein in *S. cerevisiae* led to promising result: the secretion of human erythropoietin was increased fivefold and that of bovine prochymosin 26-fold (cf. Gasser et al., 2008). However, other studies did not succeed in any increase in protein secretion and in some cases even a reduction in the final yield was observed. Similar observations were made with filamentous fungi: although thaumatin secretion by *A. oryzae* was increased, as expected, secretion of cutinase in *A. awamori* remained unaffected. It is likely that the presence of increased levels of BiP results in prolonged binding to the target proteins and that this results rather in protein degradation than a channeling into the secretory pathway (Gasser et al., 2008).

Similar controversial results were obtained using calnexin overproducing strains: Conesa et al. (2002) compared the impact of overexpression of calnexin and BiP on the secretion of *Phanerochaete chrysosporium* glycoprotein manganese peroxidase (MnP) in *A. niger*. While BiP overproduction severely reduced the secretion of MnP, overexpression of calnexin resulted in a four- to fivefold increase in the extracellular MnP levels. Positive results with calnexin overexpression were also reported for protein secretion in *Hansenula polymorpha*, even with nonglycosylated proteins, whereas other attempts in *S. pombe* and *S. cerevisiae* remained unsuccessful (cf. Gasser et al., 2008).

The ER PDI is responsible for the correct formation of disulfide bonds during oxidative folding and for the isomerization of incorrectly folded disulfides (*vide supra*). To this end, it receives disulfides—generated by the flavoenzyme Ero1, which itself is oxidized by molecular oxygen—and transfers them to substrate proteins by a dithiol-disulfide exchange reaction. Apart from this foldase activity, PDI also acts as a chaperone. Interestingly, overexpression of *pdiA* up to a certain threshold stimulated thaumatin production in *A. niger*, but decreased it when expressed above this threshold level. Similar findings have also been made with *bipA* (Moralejo et al., 2001). Thus, although the results with PDI and with Ero1 overexpression generally yielded more positive and consistent results, it appears that the effect of co-expression of chaperone and foldase genes strongly depends on the properties of the target protein and also seems to require a carefully fine-tuned overexpression to actually improve foreign protein overproduction.

So far, the best results were obtained by overexpressing the transcriptional activator HAC1 that regulates UPR (Table 8.2).

8.3.2 Engineering of Protein Glycosylation

Another point of investigation has been the tailoring of *N*-glycan chains: the biosynthetic glycosylation pathways in fungi diverge from that in mammals after the step where the glycoprotein has left the ER and is shuttled through the Golgi apparatus. Although filamentous fungi, unlike yeasts, do not produce the overglycosylated high-mannose-type *N*-glycans that contain up to 100 mannose sugars, their *N*-glycan chains nevertheless differ in some of the linkages and residues from those present in mammalian *N*-glycosylated proteins such as *N*-acetylglucosamine, galactose, fucose, and sialic acid (Roth et al., 2010). Altering the glycosylation pattern of secreted glycoproteins is of particular importance when human proteins are produced in heterologous systems, as the precise structure of the carbohydrate antenna

Table 8.2. Heterologous protein production in fungal strains overexpressing UPR pathway genes.

Protein	Organism	Manipulated Gene	Effect
Plant thaumatin	*Aspergillus awamori*	OE-*bipA*	2.5-fold increase
Phanerochaete chrysosporium MnP	*Aspergillus niger*	OE-*bipA*	Decrease
		OE-calnexin	4–5-fold increase
Plant thaumatin	*A. niger*	OE-*pdiA*	Expression-dependent increase/ decrease
T. versicolor laccase	*A. awamori*	OE-*hacA*	7-fold increase
Bovine prochymosin	*A. awamori*	OE-*hacA*	2.8-fold increase

Data compiled from Gasser et al. (2008), for original references see there.
OE, overexpression; MnP, manganese peroxidase.

may influence the *in vivo* function of the protein and also be required by legislative acts. Attempts to "humanize" heterologous proteins were therefore made (Kalsner et al., 1995; Maras et al., 1999; Kainz et al., 2008). The strategy in this direction makes use of the complementation of the fungal hosts with enzymes acting on the *N*-glycosylation structures and attaching further carbohydrate residues. Glycoengineering in the yeast *P. pastoris* and the expression of therapeutic glycoproteins with complex "humanized" *N*-glycosylation structures have shown significant progress (Wildt and Gerngross, 2005). As for filamentous fungi, human *N*-acetylglucosaminyl transferase I was expressed in *T. reesei* and shown to lead to the incorporation of *N*- acetylglucosamine into the α-1,3-linked mannose of the core oligosaccharide $NAcGlc_2Man_5$. *A. niger* was engineered by introducing the α-1,2-mannosidase of *T. reesei* fused to the HDEL domain, which resulted in a strain modifying its glycoproteins mainly with $Man_{5-6}GlcNAc_2$ glycan structures (Vervecken et al., cited as unpublished data by Deshpande et al., 2008). Kainz et al. (2008) functionally expressed heterologous chimeric fusion proteins containing different localization peptides and catalytic domains in *Aspergillus* spp. They first overexpressed a fusion protein composed of the *P. pastoris* Sec12p leader and the *Caenorhabditis elegans* α-1,2-mannosidase catalytic domain. This strain produced increased amounts of $NAcGlc_2Man_5$-type *N*-glycans. Subsequent introduction of a human NAcGlc transferase I yielded a strain producing a fraction of the $NAcGlc_2Man_5NAcGlcN$-glycan. This strategy allowed the isolation of a strain with a functional α-1,2-mannosidase producing increased amounts of *N*-glycans of the $NAcGlc_2Man_5$ type. This strain was then further engineered by the introduction of a functional *N*-acetylglucosaminyl transferase I, which led to strains producing $NAcGlc_2Man_5NAcGlc$ *N*-glycans. Additionally, they deleted the genes encoding the early step of the fungal glycosylation pathway and obtained strains producing $NAcGlc_2Man_3$ *N*-glycans.

Chapter 9

Production of Cellulases and Hemicellulases by Fungi

As I have outlined in Chapter 2, there are numerous organisms whose successful development in their natural habitat depends on the use of plant biomass polymers as carbon sources and therefore could be potential sources of cellulolytic enzymes. However, all the currently available commercial enzyme preparations used in lignocelluloses hydrolysis are exclusively derived from fungi. This is mainly due to the fact that fungi produce an enzyme mixture that is characterized by both high productivity and also the highest catalytic efficiency compared with other organisms. A further advantage is that fungi secrete their enzymes as free enzymes into the medium (see Chapter 8), which facilitates their recovery after production. This is also the main advantage over anaerobic bacteria, which express very active carbohydrate-degrading enzymes on cellulosomes (see Chapter 2) that remain physically attached to the bacterial cell wall.

9.1 Fungal Producer Strains

There is a plethora of literature about the ability of many, mostly imperfect, fungi to produce cellulases, too large to be cited here in detail, but most of these studies have not gone beyond laboratory investigations. Thus, most of the commercial enzyme products available today for lignocelluloses hydrolysis are produced by submerged fermentation with a limited number of only a few fungi, among which *Trichoderma reesei* dominates (Tolan and Foody, 1999).

9.1.1 Trichoderma Reesei

T. reesei was originally isolated as "*T. viride*" during World War II at Guadalcanal on the Solomon Islands, now Papua New Guinea, because it deteriorated tents, patron belts, and clothing, which were all made from cotton, of the US Army (Reese, 1976). Its identification as *T. viride* was not a mistake but due to the fact that fungal taxonomy before 1969 believed

Fungi and Lignocellulosic Biomass, First Edition. Christian P. Kubicek.

Trichoderma to consist only of a single species (*T. viride*; Bisby, 1939). Since biotechnologists have usually little interest in taxonomy, the fungus isolated at the Solomon Islands was published as "*T. viride*" until the late seventies, which led to the general claim—even maintained in text books—that *T. viride* would be a good cellulase producer. Unfortunately, "true" *T. viride* is a rare species and in fact a poor cellulase producer (Jaklitsch et al., 2006).

The US Army consequently set up several efforts to combat this unexpected enemy and started a long-term basic research program within the Quartermaster (QM) Corps at Natick, MA, lead by Elwyn T. Reese, to identify the responsible organism, its mechanism of action, and how it could be controlled. Apart from identifying the organism, Reese et al. pioneered by finding that it degrades cotton by secreting extracellular enzymes that depolymerize cellulose. During the "oil shock" in the early seventies, the research interest in cellulase increased dramatically because of the idea that cellulase can be used to produce sugar from cellulose, which could be fermented to ethanol by yeast as an alternative fuel. In a path making study, Reese and his long-time collaborator Mary Mandels screened more than 14,000 fungal isolates for cellulase formation and found that the original "*T. viride*" isolate QM 6a was indeed the best producer (reviewed by Reese, 1976). Consequently, the Natick team started a mutagenesis program for strain improvement and setting up fermentation conditions for production of cellulases by the fungus, which within 10 years led to the production of cellulases at >10 g/L, a milestone at that time (Reese, 1976). The isolate QM 6a was subsequently used by many research groups and industries in North America, Europe, and Asia for their own strain breeding programs. An excellent review about Mary Mandels and the work on cellulase improvement at Natick has recently been published in a special volume of Biotechnology for Biofuels, and the reader is referred to this article for more information (Bayer, 2009).

In the meantime, taxonomy of *Trichoderma* had progressed: Rifai (1969) recognized that the single-species genus indeed consisted of nine-species aggregates. However, none of them matched the phenotypical characters of "*T. viride*" QM 6a, and Simmons (1977) therefore described it as a new taxon and called it, in honor of E.T. Reese's work, "*T. reesei.*" Unfortunately, this situation became rapidly further confused by Bissett (1984): in the endeavor of refining Rifai's work on *Trichoderma* taxonomy and dissect his "species aggregates" into true species, he concluded that *T. reesei* is indistinguishable from *T. longibrachiatum* and therefore considered the name *T. reesei* illegitimate. This led to the situation that *T. reesei* was subsequently called *T. longibrachiatum* in several papers in the eighties, and cellulase preparations from some companies are still sold under this (wrong) name. In fact, while *T. longibrachiatum* is indeed not a bad cellulase producer (Kubicek et al., 1996), it has become notorious as a facultative human pathogen of immunocompromised animals and humans (Druzhinina et al., 2011), and its use as a commercial enzyme producer would thus bear considerable risk.

A further—and so far last—change in the name of "isolate QM6a" occurred as a consequence of the work by G. J. Samuels and colleagues (Kuhls et al., 1996), who identified that *T. reesei* was in fact the asexual form ("anamorph") of the well-known and pantropically distributed saprophytic and sexually reproducing ascomycete *Hypocrea jecorina* (Berkeley and Broome, 1875). The rules of botanical nomenclature—to which fungi were subjected that time—dictated that the name of the sexual form (if known) prevails over that of the asexual form and thus should be used. Consequently, most researchers therefore then used the name *H. jecorina* instead of *T. reesei*. However, at the time of this writing, there are strong attempts to change this rule and it is likely that the name *T. reesei* will—as therefore already practiced in this book—be again used.

Nevertheless, the leading position of *T. reesei* in cellulase production has recently been questioned by several findings. The most often raised argument is based on the findings that the genome sequence of *T. reesei* contains only a poor repertoire of genes for cellulases and hemicellulases (i.e., eight endoglucanases and two cellobiohydrolases; Martinez et al., 2008), and other fungal species in which many more genes were found (see Chapters 3 and 4) would thus produce a cellulase preparation of higher activity. While this argument ignores the fact that the activity of the cellulase preparation is not the only criterion for a good producer but growth rate and final concentration of secreted protein are also of high (maybe even higher) importance, it is also weakened by recent findings (described in detail in Chapter 10) that a limited number of 11 enzymes, when optimally selected (and which all are formed by *T. reesei*), can hydrolyze cellulose as effectively as the currently used commercial preparations.

T. reesei mutants are the producers of cellulase preparations from Genencor-Danisco A/S-DuPont (www.genencor.com; Accelerase®), Iogen Corp. (www.iogen.ca), Novozymes (www.novozymes.com; Cellic CTec2), Roal Oy (www.roal.fi), and Meiji Seika Kaisha Ltd. (www.meiji-seika-pharma.co.jp/english/corporate/index.html; Meicelase CEP).

Nevertheless, some other fungi have at least good perspectives as industrial cellulase producers (some actually being already produced), which shall be described below.

9.1.2 Other Fungal Producer Strains

Aspergillus niger and other Aspergilli are industrial enzyme producers (such as of glucose oxidase, pectinases, and hemicellulases; Polizeli et al., 2005; Wong et al., 2008; Lara-Márquez et al., 2011), but they have never been considered as producers of cellulases, probably owing to the fact that their cellulases are not strongly expressed.

Humicola insolens is used by Novo Nordisk as a producer of lipases and cellulases (Godfrey and West, 1996; Schülein, 1997). However, while these cellulase preparations are used for various applications such as for textile finishing, detergent, and papermaking, they are to the best of my knowledge not used for biomass conversion.

Penicillium spp. have recently been emphasized as serious alternatives to *T. reesei* (Skomarovsky et al., 2005; reviewed by Gusakov, 2011). Table 9.1 shows a comparison of the biomass hydrolyzing potential of several cellulase preparations from *Penicillium* spp. with commercial *T. reesei* enzymes. The main features that are used to explain the reported superiority of the *Penicillium* cellulase preparations are a high β-glucosidase activity (as in *Aspergillus* spp.), a higher specific activity of cellobiohydrolases (mainly Cel7A), and a reduced affinity for lignin and decreased susceptibility to inhibition by lignin-derived compounds (Gusakov, 2011). The latter property is an important feature for economic biomass hydrolysis technologies (see Chapter 10). In addition, the *Penicillium occitanis* Cel7A was shown to be much less inhibited by the natural product cellobiose than *T. reesei* CEL7A (K_i = 2 mM versus 0.02 mM), which was interpreted by poor hydrogen bonding and a more open configuration of the active site in Cel7A (Bhiri et al., 2010). Yet, protein production by these *Penicillium* spp. is low (0.3–2.2 g/L), although 47 g/L have been documented for *P. verruculosum* in fed-batch fermentations (reviewed by Gusakov, 2011), which makes their application still inferior to T. reesei.

Acremonium cellulolyticus is used as an industrial cellulase producer by Meiji Seika Co (Japan; http://www.meiji-seika-pharma.co.jp/english/corporate/index.html). The commercial preparations have been reported to lead to a higher glucose yield than Accellerase 1000 (a *T. reesei* preparation from Genencor International Inc.), likely because they are—like the *P. occitanis* preparations—less sensitive to inhibitors, such as lignin (Fujii et al., 2009). *A. cellulolyticus* has been reported to secrete up to 20 g/L protein in shake flasks (Gusakov, 2011).

Table 9.1. Comparison of the hydrolytic performances of cellulases from various *Penicillium* species and *T. reesei*.[a]

Organism	Substrate Concentration	Cellulase Loading Conditions	Increase in Hydrolytic Performance
Penicillium sp.	Short fiber cellulose wastes, 100 g/L;	2 FPU/mL, 50°C, pH 4.5, 24 h	1.35–2.13 fold; mainly β-glucosidase caused
P. verruculosum	Unbleached eucalypt cellulose, 18 g/L	10 FPU/g substrate, 50°C, pH 5.0, 6 h	1.5–3 fold
P. pinophilum IBT 4186, *P. persicinum* IBT 13226, *P. brasilianum* IBT 20888	Steam-pretreated spruce, 38.5 g/L	25 FPU/g cellulose, 40°C, pH 4.8, 24 h	Threefold over Celluclast, mainly β-glucosidase caused
Penicillium sp.	Hardwoods and softwoods pretreated by steam explosion or organosolv, 50 g/L	10 FPU/g substrate, 50°C, pH 5.0, 12 h	1.4–3.6 fold
P. verruculosum, *P. funiculosum*	Conifer wood pretreated with organosolv, 50 g/L	10 FPU/g substrate, 50°C, pH 5.0, 12 h	1.5–3.5 fold
P. pinophilum	α-Cellulose from bagasse, pulverized cellulose (CP-123), Sigmacell, Solka Floc (ball milled), 100 g/L	5–20 FPU/g substrate, 50°C, pH 4.8, 16–96 h	Notable higher than Accelerase 1000

Adapted from Gusakov (2011).
[a]Unless otherwise stated, the comparison was done with cellulase preparations from *T. reesei* laboratory cultivations; for references to individual experiments, see Gusakov (2011).

One of the currently most attractive alternatives to *T. reesei* as a cellulase producer is *Myceliophthora thermophila* (formerly called *Chrysosporium lucknowense,* which is already commercially used by Codexis Inc. (www.codexis.com) and Abengoa Bioenergy (http://www.abengoabioenergy.com). It was isolated in a broad screening for neutral cellulases from forest alkaline soil at the peninsula of Kola in northeastern Russia. On the basis of morphological characteristics, the isolate was identified as *C. lucknowense*. Recently, Visser et al. (2011) sequenced the internal transcribed spacer regions of the rRNA cluster of this fungus, which is widely used as a marker for molecular taxonomy. On the basis of these data, *C. lucknowense* had to be revised as *M. thermophila*, a member of the Sordariomycetes. This is interesting because *M. thermophila* had been subject to several misidentifications, including *C. thermophilum, Sporotrichum thermophile, Thielavia heterothallica,* and *Corynascus heterothallicus,* all of which had been described as potent cellulase and hemicellulase producers. Producer strains of *M. thermophila* were then developed by Dyadic International Inc's as a proprietary fungal enzyme producer under the name "C1" (Gusakov, 2011). *M. thermophila* has the additional benefit that during mutagenesis toward higher cellulase formation, it acquired a beneficial, distinct fragmented mycelial morphology, which reduced the viscosity of its submerged cultivations about 50-fold to a level that is also much lower than that usually obtained with *T. reesei* or other filamentous fungi. Consequently the fungus can be grown at much higher biomass densities (and thus reaching higher protein concentrations in the medium) and lower energy input for adequate mixing. C1 is reported to produce 50–70 g/L total extracellular protein in laboratory scale fed-batch fermentations and up to 100 g/L total extracellular protein in larger scale fermentations (http://www.dyadic.com/pdf/posterDNLBiotransNiek.pdf), which are values comparable to *T. reesei* strains used today. Gusakov (2011) reported that the hydrolytic potential of the most recent presented ArternaFuel C1 enzyme preparations from Dyadic International Inc. is comparable (or even significantly higher under certain conditions) to the performance of modern commercial cellulases, such as Cellic CTec or Accellerase 1000 and 1500 produced with current industrial strains of *T. reesei*.

Dyadic also developed a complete toolbox for genetic engineering of C1 and sequenced its 38 Mbp genome (Visser et al., 2011). The latter has been claimed to contain a high number of cellulases- and hemicellulase-encoding genes, and particularly the number of pectinolytic and arabinolytic enzymes (58 and 14, respectively) is striking (Hinz et al., 2009), and there is also an impressing number of GH61 cellulase-enhancing proteins (24) and cellulose-binding domains (46), which are the highest found in fungi so far (http://www.dyadic.com/pdf/posterDNLBiotransNiek.pdf), thus making it an excellent platform for the degradation of pectin-rich biomass. As a consequence of these accessory proteins, the individual cellulase proteins of C1 exhibited comparable activities on cellulose, yet synthetically composed multienzyme mixtures of C1 displayed a much higher performance than those from *T. reesei*. A drawback for the research community, however, is that much of the original research done with C1 has not been published, and its genome sequence is not available to the public (cf. Visser et al., 2011).

Neurospora crassa is used as a model organism for genetics, biochemistry, and fungal biology since more than 90 years (Davis and Perkins, 2002), and George Wells Beadle and Edward Lawrie Tatum were together awarded the Nobel Prize for Medicine in 1958 because of their fundamental work in this direction in the forties (Beadle, 1974). This species has recently attracted strong interest by US researchers, because *N. crassa* can frequently be isolated from sugar cane, which is closely related to *Miscanthus*, one of the proposed crops for biofuel production (Tian et al., 2009). An advantage of the use of *N. crassa* is the presence of a large, worldwide research community, and the therefore availability of several tools for high-throughput research (such as the free access to defined mutants for >70% of its genes; Dunlap et al., 2007). A drawback of *N. crassa*, however, is the fact that its conidia become activated by heat to germinate, thus making containment of the organism more difficult. However, *N. crassa* also belongs—like *T. reesei*—to the Sordariomycetes and both fungi share orthologous genes in their genomes (Martinez et al., 2008). Result from research with *N. crassa* may thus be directly used for strain improvement in *T. reesei*.

9.1.3 Thermophilic Fungi

Thermophilic or thermotolerant fungi deserve special attention: since these organisms are able to grow and metabolize at elevated temperatures, it is usually assumed that their extracellular enzymes should also be more stable or active at these temperatures. Such enzymes would be attractive because—apart from increasing the reaction rate and so decreasing the retention time in the reactor or the enzyme concentrations to be used—there are also engineering reasons why lignocellulose hydrolysis should be performed at higher temperatures (see Chapter 10).

Some filamentous fungi produce cellulases that retain relatively high cellulose-degrading activity at temperatures of 50–70°C, particularly species such as *Thielavia terrestris*, *Thermoascus aurantiacus*, and *Chaetomium thermophilum* (for review, see Rosgaard et al., 2006).

C1 (*Myceliophthora thermophila*) also belongs to this group, and strains published under the earlier names *Sporotrichum thermophilum* and *Corynascus thermophilus* were also shown to produce thermostable cellulases and hemicellulases.

9.2 Strain Improvement

9.2.1 Strain Breeding by Classical Mutagenesis

Wild-type isolates, while showing potentially useful properties, inevitably have to be improved toward high production. Because of the fact that most of the fungi that are used for enzyme production are reproducing asexually, improvement by crossing—as standard in plant breeding—is not possible. However, along with the sexual cycle, most ascomycetes have a "parasexual cycle" (Elliott, 1994), a process in which two vegetative haploid nuclei fuse into a vegetative diploid nucleus that can give rise to a vegetative mycelium. Repeated nondisjunction during mitosis in a diploid nucleus can revert it to haploidy by random loss of whole chromosomes, leading to recombination at the chromosome level. However, parasexuality has been used for strain improvement only at a low level.

With the identification that *T. reesei* is actually a sexually propagating fungus (Kuhls et al., 1996; Druzhinina et al., 2010), it was possible to develop conditions for crossing *T. reesei* in the lab, a tool which is not available for the other fungi used in enzyme production (Seidl et al., 2009).

As a result, most of the strain improvement with enzyme-producing fungi has been performed by classical mutagenesis. This approach has a long and successful history with almost all fungal strains that had been introduced to industrial production in the second half of the last century. A genealogy of the cellulase-producing mutant strains of *T. reesei* prepared from QM6a by various laboratories has been summarized by El-Gogary et al. (1990). Basically, this approach consists of subjecting the fungus to a nonlethal dose of a mutagen (frequent irradiation by UV light or addition of nitrosomethyl guanidine; but for a complete description of techniques used, see Lawrence, 2002) and subsequent screening of the survivors for improved production of the desired product. Obviously, this approach is only as good as the selection method. For cellulases and other extracellular enzymes, this method usually involves pregrowth of the fungus on agar plates and then overlaying the colonies with the substrate of the respective enzymes, for example, in the case of screening for cellulases with amorphous cellulose. Since the secreted enzymes diffuse into the agar, they will hydrolyze the cellulose at the place of contact and produce "clearing halos." In order to prevent growth of the mutant colonies, this is usually done at an elevated temperature (e.g., 50°C), which does not kill the cells and also enhances the rate of the enzymatic reactions. Larger halos then indicate larger enzyme production. To facilitate visualization of the halo, polysaccharides can either be stained with Congo Red (Carder, 1986) or be replaced by predyed polysaccharides (Biely et al., 1985).

One difficulty of this technique is that it is essential that the fungi grow in the form of small, defined colonies, which is not the case with all fungi. Therefore, many researchers added "colony restrictors" to the selection medium (usually a surfactant such a Triton X100), which causes the fungus to grow in the desired small colonies. Other researchers restricted the fungus by physical restriction (e.g., covering the agar plate with a metal grid) that enabled only a small diameter for fungal development (Durand et al., 1988) or by forcing the fungus to grow through an overlay of cellulose substrate (Toyama et al., 2008).

While most of the mutant selection procedures merely involved screening for improved product formation, it is of course also possible to screen for other desirable properties such a resistance to an inhibitor or a particular phenotype. Since cellulase production by *T. reesei* did not occur on glucose and was inhibited by glucose in the medium (for more details, see Chapter 7), Eveleigh and Montenecourt (1979) screened for *T. reesei* mutants that were resistant to 2-desoxyglucose when growing on glycerol. The rationale of this screening system, which had been established in yeast (Zimmermann and Scheel, 1977), is that 2-desoxyglucose can be phosphorylated by hexokinase but not be isomerized by phosphoglucose isomerase and thus leads to an accumulation of 2-desoxyglucose-6-phosphate and depletion of cellular ATP levels. While this selection can thus be a result of various different mutations, strains that tolerate a high intracellular 2-desoxyglucose-6-phosphate concentration may be less susceptible to carbon catabolite repression that is believed to be triggered by glucose-6-phosphate (see Chapter 7). This strategy was indeed successful and led to the isolation of *T. reesei* RUT C30, one of the best cellulase producers that are currently available to the public. Besides the fact that it produces some cellulases even in the presence of glucose, it also has increased cellulase production, which is optimal at increased cellulose or lactose concentrations (Allen and Roche, 1989). Ilmen et al. (1996) were the first to show that this mutant strain indeed has a truncated, nonfunctional allele of the carbon catabolite repressor *cre1*. Also, it bears a mutation in the glycoprotein processing β-glucosidase II (see Chapter 8; Geysens et al., 2005). Le Crom et al. (2009) sequenced the genome of *T. reesei* RUT C30 and its progenitor NG14, and showed that the two differed by mutation in the open reading frame of 46 genes whose potential relevance to cellulase production improvement is not yet known (Table 9.2).

9.2.2 Strain Improvement by Targeted Gene Manipulation

In 1983, first systematic attempts to introduce into filamentous fungi (*N. crassa, A. nidulans*) were published (reviewed by Cullen et al., 1988 and Timberlake and Marshall, 1989), thus opening the possibility to directly manipulate the genome of fungi by recombinant techniques. Soon later, the elementary principles were developed for all other fungi, for which research in relation to lignocelluloses degradation was performed, including *T. reesei*, *Humicola insolens,* and *P. chrysosporium* (for a summary, see Shishido, 1994). Gene transfer into multicellular fungi traditionally involved the addition of plasmid vectors or vector fragments to protoplasts under conditions that facilitated uptake such as precipitating the plasmid as Ca^{2+} salts on the cell surface and gluing protoplasts together by the addition of polyethylene glycol in order to force uptake by endocytosis and regeneration of the protoplasts in the presence of a selection marker present on the same or concomitantly added plasmid to facilitate detection of positives (Turgeon et al., 2010). One of the drawbacks of nonyeast fungi is the absence of plasmids that could replicate independently and thus be easily recovered from the transformed cells (Kroll et al., 2010), and this could only in some fungi be partially overcome by identification and use of autonomously replicating sequences (Verdoes et al., 1994; Fierro et al., 1996). Another drawback is the low number of transformants, which at best make up several 100/μg DNA but can be as low as <1/μg DNA in some basidiomycetes, probably due to degradation of the introduced DNA (Shishido, 1994). The introduction of improved techniques for transformation, such as electroporation (Kapoor, 1995), particle bombardment (also known as "biolistic transformation" or "gene gun"; Harrier and Millam, 2001) or the use of *Agrobacterium tumefaciens*-mediated DNA transfer (de Groot et al., 1998) reduced these bottlenecks in most albeit not all fungi. Nevertheless, gene manipulation is

Table 9.2. Additional mutations in *T. reesei* RUT C30 compared to QM6a.[a]

Protein ID	Annotation
Transcription factors	
70071	Zn2c6 transcription factor
109619	CBF/NF-Y family transcription factor
77513	Zn2C6 transcription factor
106009	Zn2C6 transcription factor
52368	Zn2C6 transcription factor
Secretion	
28409	MSF family. Related to vesicle transporters
112231	Kinesin-like Kif21a protein, implicated in vesicle trafficking
108962	Bears TLDC domains: related to yeast Ypt6 and Ypt7 (GTPase activator proteins of Rab-like small GTPases)
23001	Golgi complex component (cog1)
Metabolism	
54157	Weak similarity diene lactone hydrolase family protein
58073	Uroporphyrinogen decarboxylase
58790	Glycerol-3-phosphate phosphatase (GPP1)
59388	Maltose permease
59952	Choline transporter
65106	Nitrilase or amidohydrolyase
123786	Peptaibol (hypojecorin) synthase
123126	Unknown esterase, lipase, or thioesterase
Nuclear transport	
112346	Ataxin-7 ortholog
22294	Nuclear transport factor 2
78158	Importin β-3
RNA metabolism	
66895	Component of exosome complex 3′-5′ ribonuclease
3400	Protein with RNA recognition motif[b]
110423	CCR4-associated factor (CAF1)
63935	Isoleucyl-tRNA synthetase
Vacuole	
65104	VPS13, vacuolar sorting associated protein
43599	VSP1 vacuolar ATPase, may be Golgi associated
79014	16 kD proteolipid subunit of vacuolar ATPase
Other	
64375	GH5 exo-β-1,3-glucanase
120806	Putative kinase,[c] transcript present, orthologs expressed in plant pathogens during infection
74570	ABC-1 like domain protein (protein kinase?)
57609	Low similarity to Src1, sister chromatid separation protein
119999	Myb DNA binding
79304	Rad2p-like
60458	Small nonribosomal peptide synthase

Data taken from Le Crom et al. (2009).
[a]Protein IDs refer to the protein identifications in the *T. reesei* genome database (http://genome.jgi-psf.org/Trire2/Trire2.home.html). Unless otherwise stated all genes bear single SNPs.
[b]Frameshift (4 nts).
[c]Two SNPs, one deletion.

now possible with most if not all fungi that are used in lignocellulose hydrolysis, and the meanwhile available genome sequences for many of them further facilitates the application of these techniques. A continuously updated summary of fungi with sequenced genomes is found at http://www.ncbi.nlm.nih.gov/sutils/genom_table.cgi?organism=fungi. Major fungal databases are found at the Joint Genome Institute (http://genome.jgi-psf.org/) and the Broad Institute (http://www.broadinstitute.org/scientific-community/science/projects/fungal-genome-initiative/fungal-genome-initiative).

Yet a few recent innovations, which have only been applied to a few fungi, need special mentioning: one is associated with transformation markers. While markers based on complementation of an auxotrophy (such as for an amino acid, a nucleotide base, or a vitamin) are of little use to industrial strains because introduction of an auxotrophy into a production strain frequently leads to a drop in production, dominant selection markers are scarce and their number has only recently been increased (Table 9.3). Moreover, the inherent resistance of individual fungi toward the markers used to select for integration of the gene of interest can change even between different mutant strains of the same fungi. The number of truly applicable markers for

Table 9.3. Selection makers for filamentous fungi.

Auxotrophy			
Arginine	*argB*	Ornithine transcarbamylase	Buxton et al., 1985
Biotin	*bioDA*	7,8-Diaminopelargonic acid synthase and dethiobiotin synthetase	Magliano et al., 2011
Nitrate	*niaD*	Nitrate reductase	Daboussi et al., 1989
Tryptophan	*trpC*	Tryptophan synthase	Yelton et al., 1984
Uridine	*pyr4/pyrG*	Orotidine-5-phosphate decarboxylase	Ballance et al., 1983
Nutrient utilization			
Acetamide/acrylamide	*amdS*	Acetamidase	Tilburn et al., 1983
Acetate	*facA*	Acetyl-CoA synthetase	Gouka et al., 1993
Sulfur	*aasC*	ATP-sulfurylase	Buxton et al., 1985
Resistance			
Benomyl	*benA*	β-Tubulin	Orbach et al., 1986
Bialaphos/phosphotrinocin	*bar*	Glutamine synthase	Avalos et al., 1989
Blasticidin S	*BSD*	Blasticidin S deaminase	Kimura et al., 1994
Carboxin	cbx^R	Succinate dehydrogenase subunit	Shima et al., 2009
Cyclosporine	*csr1*	Cyclosporine C-binding protein	Bardiya and Shiu, 2007
Hygromycin B	*hpg*	Phosphotransferase	Cullen et al., 1987
Nourseothricin	*nat*	Nourseothricin acetyltransferase	Alshahni et al., 2010
Oligomycin	*oliC*	ATP-synthase subunit 9	Ward et al., 1986
Phleomycin	*ble*	Phleomycin-binding protein	Drocourt et al., 1990
Sulfonyl urea	*als1*	Acetolactate synthase	Lin et al., 2011
Terbinafine	*ergA*	Squalene epoxidase	Sigl et al., 2010

a given fungus is thus limited and can present an obstacle when a large number of genes are to be manipulated in a strain. To circumvent this problem, transformation systems independent of the number of available markers were recently developed, which are based on the removal of the marker gene after each transformation step (reviewed by Seidl and Seiboth, 2010). One system that makes use of recyclable markers is the blaster cassette system, which is dependent on marker genes that can be bidirectionally selected (i.e., selection is possible for both presence as well as absence). One such example, used successfully with many fungi including *T. reesei* makes use of the *pyr4* (encoding orotidine 5′-phosphate carboxylase) gene, which complements uridine auxotrophy, whereas its absence results in resistance to 5-fluoroorotic acid (5-FOA). Transformants can therefore be selected for *pyr4* integration by growing on a medium without uridine, whereas excision of the *pyr4* marker can be induced and detected by growth in the presence of 5-FOA, which is otherwise toxic for the fungus (Figure 9.1).

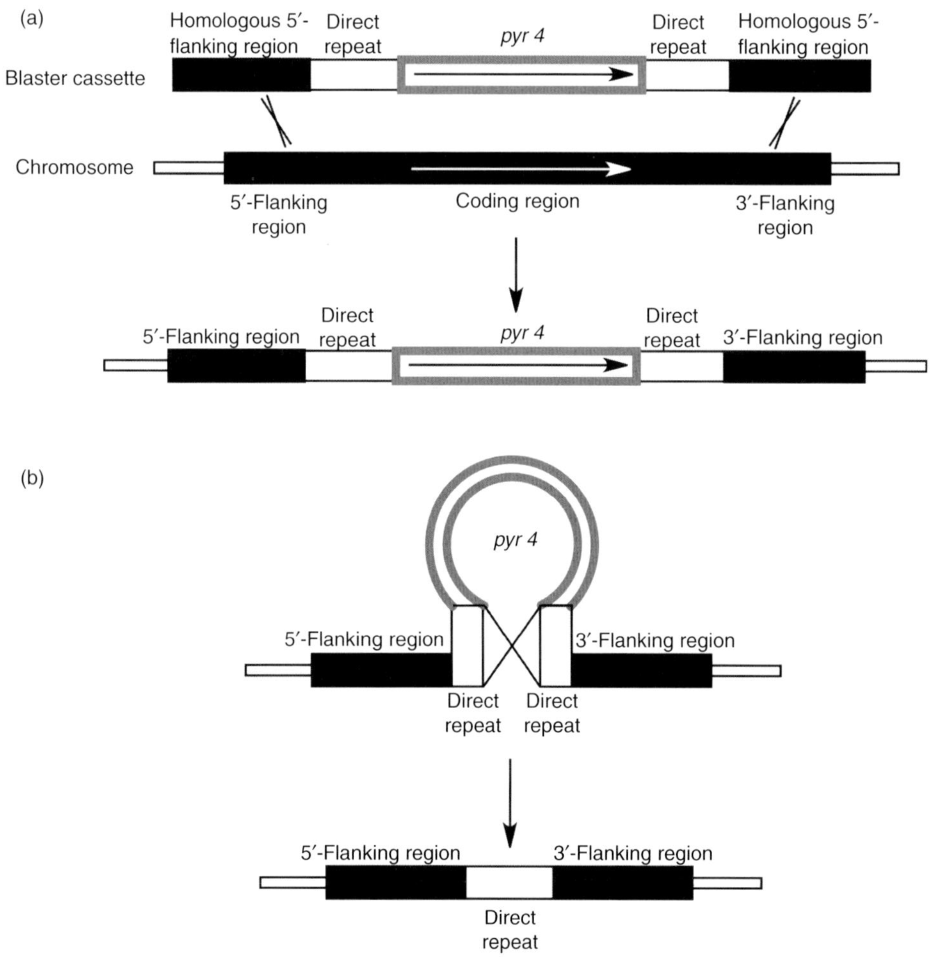

Figure 9.1. Gene deletion using a blaster cassette, using *T. reesei pyr4* (Hartl and Seiboth, 2005). (a) Indicates the status of integration of the whole plasmid into the genome; (b) the excision of *pyr4*, driven by selection in the presence of 5-fluoroorotic acid.

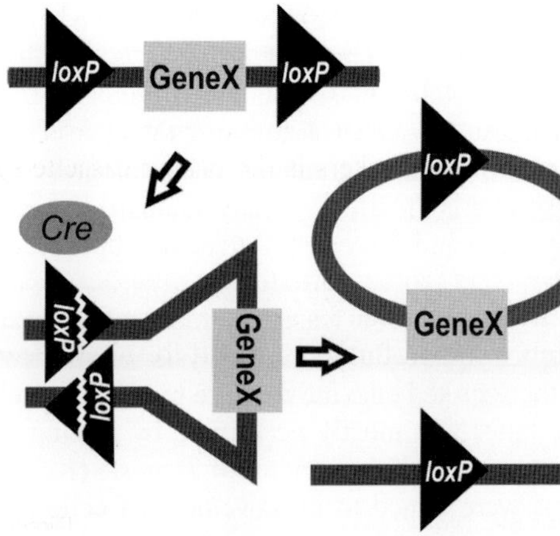

Figure 9.2. Marker gene removal by the Cre/Lox system. For details see text.

To facilitate homologous recombination that is necessary for the removal of the *pyr4* blaster cassette, the marker gene is located between two direct repeats. As a consequence, this blaster cassette can be reused for multiple rounds of gene deletions (Hartl and Seiboth, 2005).

Alternative strategies for marker recycling by induced recombination are the bipartite Cre/loxP recombination system (Figure 9.2) and the yeast flipper system. The former system makes use of a site-specific bacteriophage DNA recombinase, termed Cre, which is able to catalyze the recombination of DNA between specific 34 bp-long DNA sequences to which Cre specifically binds (the so-called loxP sites), and that surround the sequence that should be removed by recombination. Obviously, the recombinase should only act when desired. To achieve this, it is placed it under an inducible promoter that must be tightly turned off under noninducing conditions. This system was originally developed for plants but has later on been shown to work in yeast and in filamentous fungi, including taxa important in lignocellulose degradation like *T. reesei* (Iwaki and Takegawa, 2004; Forment et al., 2006; Honda and Selker, 2009; Steiger et al., 2011).

The flipper (FLP/*FRT*) recombination system is an alternative to the Cre/loxP system, which is based by the 2 μm (6.4 kb) plasmid found in most isolates of *S. cerevisiae* and which can attain a copy number of up to 100 copies per cell. A recombinase present on the plasmid catalyzes recombination at two 599-bp target sites that are located in an inverted orientation, which 34 bp alone are functional. A total of four FLP recombinases are needed for recombination, of which two bind to a single target site (Kilby et al., 1993; Morschhäuser et al., 1999). This system has recently been optimized for *Penicillium chrysogenum* (Kopke et al., 2010) and shown to work for *Podospora anserina* (Kopke et al., 2010) and *T. reesei* (B. Metz, B. Seiboth and C.P. Kubicek, unpublished data).

Another recent breakthrough concerns the frequency with which transformed DNA is integrated into the homologous locus, an essential requirement for gene disruption and gene replacement. The frequency of homologous integration in fungi is usually small (5–10%) and depends on the genomic locus that is targeted. This implies the screening of a high number of transformants to identify strains in which the desired gene deletion has actually occurred. The molecular reason for this nasty behavior is the dominant operation of the nonhomologous end-joining (NHEJ) pathway of double-strand break repair in fungi (Ninomiya et al., 2004).

Thus, impairment of the NHEJ pathway can be expected to lead to a higher number of transformants bearing homologous integrations. The NHEJ process is mediated by different proteins, including the DNA-dependent protein kinase catalytic subunit DNA-PKcs, the Ku70/Ku80 heterodimer, and the DNA ligase IV–XRCC4 complex (Walker et al., 2001). Disruption of ku70 or ku80 was indeed shown to lead to enhanced homologous integration and, in some cases, even to 100%. This system is therefore already standard for gene disruption in various fungi, including several Aspergilli and *T. reesei* (Weld et al., 2006; Kück and Hoff, 2010; Seidl and Seiboth, 2010).

With all these technique in hand, it is now feasible to manipulate most of the producer strains used in enzyme production for lignocelluloses hydrolysis. First attempts into this direction were dedicated to altering the secreted enzyme mixture by deleting selected components and overexpressing others. The latter was mostly performed by putting the target gene under a stronger promoter, for example, the *cel7A* promoter of *T. reesei* (Keränen and Penttilä, 1995). Most of these modifications were aimed to improvement of cellulase compositions for the textile industries, though.

One of the simplest examples for improvement of *T. reesei* cellulases for biofuel production is the introduction of a heterologous *A. oryzae* β-glucosidase gene (Nakazawa et al., 2011; Wang and Xia, 2011). This eliminated the shortage of *T. reesei* cellulase preparations in this enzyme, which is due to binding of more than two-third of the total β-glucosidase to the cell walls of *T. reesei*. However, the situation with β-glucosidase secretion by *T. reesei* has dramatically improved recently. Modern commercial cellulase preparations from leading enzyme producing companies, such as Cellic CTec and Cellic CTec2 from Novozymes (http://www.novozymes.com) or product series under the registered name Accellerase from Genencor International Inc. (http://www.genencor.com), are based on genetically modified strains of *T. reesei* and they possess rather high β-glucosidase activity (Merino and Cherry, 2007).

9.2.3 Strain Improvement by Complementation with Heterologous Enzymes

Various researchers have also searched for novel and more active cellobiohydrolases, which are believed to be rate-limiting enzymes in crystalline cellulose degradation. Recombinant GH7 cellobiohydrolases from *A. thermophilum*, *Chaetomium thermophilum*, and *Thermoascus aurantiacus*, when expressed in *T. reesei*, displayed higher thermostability (by 4–10°C) and much higher specific activity in the hydrolysis of Avicel at 45°C than did the CEL7A of *T. reesei* (Merino and Cherry, 2007).

Another method of improving the cellulase productivity is through increasing the expression of accessory proteins that synergistically raise the activity of the cellulases. These comprise swollenin (Huang et al., 2011), CIP1 (Foreman et al., 2010), and the most notable GH61 proteins (Schnorr et al., 2007; see Chapter 3). All these proteins do not exhibit any enzymatic activity but boost hydrolysis when supplemented to cellulase preparations and have resulted in a significant rise in the specific activity (Figure 9.3).

Another and very attractive approach for discovery of new enzymes that complement the lignocelluloses hydrolyzing enzyme system of cellulase producers is the metagenomic screening for new enzymes from unknown microorganisms. The rationale for this is that although approximately 4–6 × 10^{30} prokaryotes inhabit the Earth and constitute the world's major reserve of genetic diversity (Whitman et al., 1998), only about 95–99.9% of microorganisms have not (yet) been cultured by standard laboratory techniques (Epstein, 2009). One

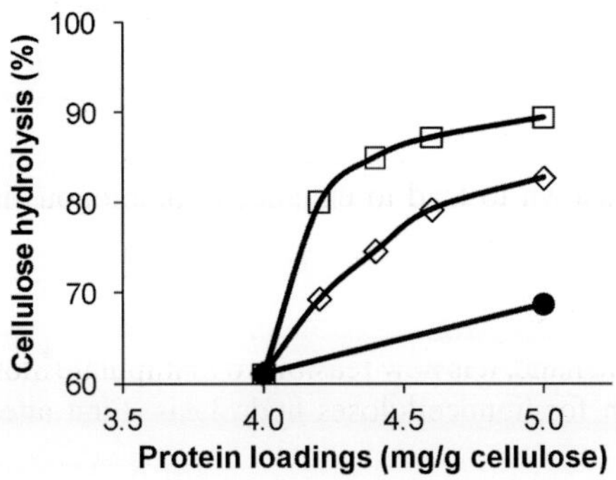

Figure 9.3. Effect of cellulose-enhancing proteins on the time course of lignocelluloses hydrolysis by cellulase preparations. Release of soluble sugars by cellulases alone (•) or with addition of increasing amounts of recombinant GH61E protein from *Thielavia terrestris* (◊) or GH61A protein from *Thermomyces aurantiacus* (□). (Reprinted with permission from Harris et al., 2010. Stimulation of Lignocellulosic Biomass Hydrolysis by Proteins of Glycoside Hydrolase Family 61: Structure and Function of a Large, Enigmatic Family. Biochemistry. Copyright 2012 American Chemical Society.)

way to detect all microorganisms in a given habitat is to extract their genomes and amplify and sequence a conserved gene of phylogenetic information, such as 16S/18S-rRNA genes. The term "metagenome" was introduced to specify such collections of genomes and/or genes representing all microorganisms in a sample). More recently, the "metatranscriptomes" have also been targeted, which consist of a collection of cDNAs of all mRNAs expressed in this habitat and which is supposed to lead to an enrichment of those genes whose expression is particularly triggered by the physicochemical conditions in the respective habitat (for review, see Cowan et al., 2005). Yet, the high number of microorganisms in natural habitats (e.g., up to 4,000 species in 1 g of soil; Curtis et al., 2002) limited these approaches at the beginning, and only the improvement of sequencing technologies during the last 5 years (such as the 454 pyrosequencing; Life Science, Roche) made metagenome shot-gun sequencing of environmental samples feasible.

The sequenced metagenomes theoretically represent all genes from all organisms in this habitat and can be used for mining of new enzymes by two approaches: function-based screening of expression libraries and sequence-based gene searches (Cowan et al., 2005). In the former, metagenomic expression libraries are constructed from metagenomic DNA and screened for target enzyme activities by (robot-based) high-throughput methods. For the latter, metagenome sequence databases are screened by bioinformatics tools and interesting candidates thereafter cloned and characterized. The power of these approaches has been reviewed by Li et al. (2011): they reported that the (by that time) 46 completed metagenomes present in GOLD (www.genomeonline.org) contained 7,338 putative GHase homologs (Table 9.4). Not surprising, metagenomes from biotopes characterized by a continuous turn over of plant cell wall biomass display an increased abundance of putative GHases. As an example, the metagenomes derived from termite, human, and mouse guts displayed more putative GHase homologs (approximately 1.5% total gene count) than those from other samples. However, as far as lignocelluloses hydrolysis is concerned, GH13 and GH 23 members were most abundant and they represent amylases and other α-glucanases and lysozyme, respectively. However, GH2 members, which contain various β-glycosidases, ranked third in abundance.

Table 9.4. Lignocellulose-biomass-related glycosyl hydrolase families detected in nonenriched environmental samples.[a]

	Total GHs	GH1	GH2	GH5	GH16	GH28	GH31	GH43
Acid mine drainage	73					9		
Human gut community	705	7	9				5	
Hypersaline mat	786		9					
Lake Washington formaldehyde enrichment (^{13}C-labeled formaldehyde)	397		8					
Lake Washington formate enrichment (^{13}C-labeled formate)	114		7			4		
Lake Washington methanol enrichment (^{13}C-labeled methanol)	373		5					
Marine archaeal anaerobic methane oxidation community	3		33					
Mouse gut community (lean mouse)	119							9
Mouse gut community (obese mouse)	58		10					10
Olavius algarvensis microbiome	41		5	5		5		5
Singapore air sample	514					5		
Sludge US Jazz assembly	126		6					
Sludge US Phrap assembly	260				7			
Farm silage surface soil	1078		5					
Termite gut	1,267		6	9				
Human oral microflora	14	21				7		
Uranium-contaminated groundwater	71					6		
Whalefall sample	433		6					

Data modified from Li et al. (2011).
[a]Values for individual GHs are percentage values of total GH.

This can be explained by the fact that most of these studies did not specifically focus on biotopes that were enriched in plant biomass.

Toward this goal, Li et al. (2011) sequenced the metagenome of a mesophilic, anaerobic microbial community that actively decays poplar woodchips, approximately 675 Mb of sequence was generated and, after assembly, resulted in 44,600 contigs and 1.42 M singletons totaling 382 Mb. Using tiled blastx, the sequences were searched against the CAZy database and approximately 4,000 glycoside hydrolase homologs were thereby identified. Using a shotgun expression library, the encoded enzymes were screened for GHase activities. Four novel GHs showed significant activity on *p*-nitrophenyl-α-L-arabinofuranoside, one enzyme had significant activity against *p*-nitrophenyl-β-D-glucopyranoside and one enzyme showed significant activity against *p*-nitrophenyl-β-D-xylopyranoside. The enzymes remained active in the presence of up to 20% of most ionic liquids tested, thus recommending their use for minimal solvent processing of biomass (see Chapter 10).

9.2.4 Strain Improvement by Protein Engineering

While the approaches described above consist of introducing and expression of copies of native genes into the respective genomes of fungal producer strains, there have also been several attempts to engineer these proteins toward an improved performance. Improvement of the thermostability and specific activity of the individual protein are the main targets, but reduced product inhibition and inhibition by lignin are also pursued. Two main approaches have and are being used to engineer cellulases toward this goal: directed evolution and rational design (Percival Zhang et al., 2006; Levine et al., 2011).

Directed evolution of cellulases with improved activity on crystalline cellulose involves the random exchange of either some or even all of the amino acids in a protein and therefore requires that all the arising mutant cellulases be screened. This is therefore mostly done in *E. coli* or yeast strains to enable high throughput. However, to obtain results that also hold true in the subsequent application, the native enzyme, not only the catalytic domain, needs to be tested, and any potentially improved enzyme must then be tested in the appropriate synergistic mixture on the actual substrate for the final process. In the past, the initial screening was mostly done by the aid of soluble, artificial substrates. However, Gao et al. (2010) have recently proposed a high-throughput 96-well microplate technique with crystalline cellulose and ammonia fiber expansion pretreated corn stover as substrate. As an alternative, Jäger et al. (2011) employed the BioLector technology (Gasdip AG; http://www.dasgip.com), which consists of a benchtop micro (0.2–1.5 mL) fermentation system that allows to perform 48–96 fermentations simultaneously with continuous online monitoring of critical process parameters. In their study, cellulase activities were examined by light scattering and used to evaluate the pH optimum of cellulases and nonhydrolytic deagglomeration of cellulose particles.

A major bottleneck of directed evolution is that it can be used only for screening of single amino acid exchanges. Massive screening may be able to identify even double mutations, but the screening for higher numbers of mutations is prevented by the mutant library size. Nevertheless, engineering for improved thermostability has been relatively straightforward, and variants of *T. reesei* CEL7A and CEL6A with improved thermostability have both been obtained (Aehle et al., 2006; Day et al., 2007; Wu et al., 2008). In particular, a mutation of S413P in *T. reesei* CEL6A or substitution of the amino acid at the equivalent to position 413 with a P in other GH6 cellulases leads to a mutant protein with increased thermostability (Wu et al., 2008).

Likewise, mutations at the equivalent of positions 103, 134, 136, 186, 365, and 410 within the catalytic domain of *T. reesei* CEL6A and other GH6 cellulases have been shown to lead to reduced inhibition by glucose (Lavigne et al., 2009). Directed evolution of *T. reesei* Cel12A for enhanced thermostability, and of a *Bacillus subtilis* GH5 endoglucanase toward higher activity on CMC and no activity on crystalline cellulose has been reported (reviewed by Wilson, 2009). Heinzelman et al. (2009) used SCHEMA structure-guided recombination (Meyer et al., 2003; Patrick and Firth, 2005), which allows the prediction of targets for producing functional variants, to create chimeras of *H. jecorina* CEL6A and two thermostable CEL6A orthologs from *Humicola insolens* and *Chaetomium thermophilum*. Five of the 6,561 possible chimeras had half-lives of thermal inactivation at 63°C, which were even higher than the most stable parent enzyme from *Humicola insolen*s, and thus the chimera collection may have contained hundreds of cellulases with improved stability over that of *T. reesei*. About 17 thermostable CBH II enzymes differed from their closest natural homologs at up to 63 amino acid positions. Using these data, the same authors constructed a quantitative CBH II chimera stability model that allowed pinpointing a single amino acid substitution (C313S) as being responsible for the large stabilizing contribution of one of the SCHEMA blocks. Introducing this mutation into the wild-type CEL6A increased temperature tolerance, raised the maximum temperature for Avicel hydrolysis and improved the performance in long-time hydrolysis experiments.

Interestingly, a significant diversity in enzyme stability was seen among the GH12 homologs even from closely related organisms: as an example, CEL12A from *T. citrinoviride* differs from the *T. reesei* ortholog at only 14 residues, but exhibits a significantly less thermal stability (Goedegebuur et al., 2002). Sandgren et al. (2003b) systematically introduced these 14 differences into the *T. reesei* Cel12A enzyme, and one of them (A35V) was indeed found to stabilize the enzyme.

Likewise, the differences in thermal stability between *H. grisea* and *T. reesei* (apparent denaturation midpoints, T_m, of 68.7 vs. 54.4°C, respectively) were used to identify the responsible amino acid residues (Sandgren et al., 2003b). There are three cysteines in the *H. grisea* Cel12A, which are not involved in disulfide bond formation and which are not present in the *T. reesei* enzyme. Substitutions at any of these three positions affected stability in *H. grisea*, while introducing them into *T. reesei* CEL12A increased in stability, the most significant effect seen with a P201C mutation that increased T_m by 3.9°C.

In a third study, the stability and specific activity of *T. reesei* CEL12A was optimized by directed evolution. The best clone harbored six mutations (G41E/T110P/K173M/Y195F/P201S/N218I) and showed broad pH stability (4.4–8.8) and strongly increased thermotolerance (100% activity at 55°C for 30 minutes). The k_{cat} for carboxymethyl cellulose was increased 1.4-fold, while K_m was reduced by 50% (Nakazawa et al., 2009).

At the time of this writing, there are no published reports of successful engineering of cellulases toward significantly (greater than 1.5-fold) increased activity on crystalline cellulose.

Scott et al. (2010) developed a *T. reesei* strain in which the hinge domain that links the catalytic domain and the carbohydrate-binding module of CEL7A was mutagenized to contain at least 50% P, S, or T, which resulted in a decrease in the calculated isoelectric point of the hinge domain. This modification led to a reduced inactivation by and binding to lignin during the hydrolysis process (see Chapter 10).

9.3 Cellulase Production

9.3.1 Cellulase Fermentation

Today, fungal cellulase production is mostly carried out in fermenters of 95–285 kL scale and resemble other fungal fermentations in the basic unit operations: the cultures of the producer strain are propagated in a series of smaller, yet stepwise increasing volumes, which typically serve as an inoculum of 10–15% of the final volume of the next fermentation step. Most of the fermentation improvement for cellulase production has already been done in the 80s of the last century, and only a few younger scientific articles are available on this topic. However, even from this time, practically no data are available on studies that have been carried out at least in a pilot plant or even higher scale. Three valuable exceptions are the reviews by Nystrom and Allen (1976), Esterbauer et al. (1991), and Tolan and Foody (1999).

The submerged batch process is the classical variant: in this performance, cellulase formation is correlated with growth rate and the biomass yield. Hence, high growth rate and biomass density are favorable. Since the fungi used for cellulase fermentation are aerobic organisms, this obviously results in a high demand for oxygen and concomitantly significant heat production (i.e., need for efficient cooling). Oxygen is usually supplied through a sparger at a rate of 0.3–1.2 vvm and should result in dissolved oxygen concentrations of >10%. Above this value, the oxygen concentration has no effect on cellulase production. The heat generated is removed through a cooling jacket or coil. Efficient mixing is achieved by appropriately designed impellers and eventually also baffles next to the wall.

There are several essential issues that determine the cellulase yield. One is the necessity of presence of an organic nitrogen sources (usually peptone in laboratory fermentations, but distillers spent grain or other nitrogen-rich waste products in industrial fermentations), which enables *T. reesei* to rapidly increase in biomass (the fungus utilizes the organic nitrogen as a carbon source, as indicated by a transient rise in pH) and subsequently attack cellulose without time gap.

Another issue is sterility, because the organic nitrogen and the cellulase and hemicellulase proteins that are secreted into the medium during the fermentation are ideal substrates for bacteria and yeasts. Therefore, the fermentor and support equipment must be sterilized before inoculation, typically by steam sterilization at 121°C for 20 minutes. Aeration is filtered for the same reason.

A further point is foaming: both the proteins present as nutrients at the beginning of the fermentation as well as the extracellular enzymes formed result in enormous foaming due to the stirring and air input. Consequently, antifoam agents such as polyethylene glycol or plant oils (such as soybean oil or palm oil) must be added. Alternatively, mechanical foam disruptors such as table centrifuges or a Waldhof draught tube must be implemented into the fermenter (Nystrom and Allen, 1976).

The time needed for a simple batch fermentation with *T. reesei* is dependent on the type and concentration of the carbon source and can take up to 4 days. The inducing carbon sources are proprietary to each manufacturer but obviously contain an inexpensive mixture of soluble and/or insoluble sugars such as a cellulosic material (Solca floc, paper mill waste streams) or lactose/milk whey. Their concentration is strongly strain dependent and varies between 1% and 8% (w/v), the higher concentrations usually only being feasible with the soluble carbon sources. Toward application of the resulting cellulases for biorefineries, there is a tendency of cultivating *T. reesei* already on the same pretreated substrate that is going to be subsequent process.

Other medium components include inorganic sources for phosphate, sulfur, potassium, and trace elements. The nitrogen source and its concentration are critical, given that the product (enzyme protein) contains nitrogen at about a sixth of its weight. In addition, the supply of an organic nitrogen source, from which the amino acids are consumed as a carbon source prior to the cellulose, has been proven to be essential for a high enzyme yield.

Tween 80 and related surfactants are also added. The reason for this is not completely clear: they may aid in removing the enzymes from the insoluble substrate or may be used by the fungus to build up the endogenous membrane system needed for protein secretion (cf. Chapter 8). In any case, their addition already at the beginning of the fermentation would amplify the foaming caused by the protein in the medium. The surfactants are therefore usually added stepwise or continuous after the end of the early phase of the fermentation.

The pH of the fermentation has been shown to strongly influence the final enzyme titer. A pH of 5.0 is optimal for growth, but production is usually better at a lower pH (down to 3), which gave rise to pH cycling strategies for controlling the fermentation. Although cellulases do not display an optimal activity at pH 3, they are stable at this pH for the duration of fermentation. The temperature is also strain dependent, but is generally between 25°C and 30°C.

The morphology of filamentous mycelia is related to the protein secretion (see Chapter 8) and affected by the agitation conditions, which in turn impacts the rheological properties of the system (Riley et al., 2000). The mycelial morphology of fungi in fermenters occurs either in a dispersed or pelleted form. The former leads to a higher viscosity but also higher specific productivity than pellets. Different fungi have different tendencies toward one of these two alternatives: during the production of cellulase, *T. reesei* grows mainly in the dispersed form (Lejeune et al., 1995). In contrast, *A. niger* is difficult to control not to produce pellets. Various semiautomatic and fully automatic image analysis methods for analyzing dispersed filamentous fungi have been developed. Lejeune et al. (1997) studied the morphology of *T. reesei* QM9414 in submerged culture, but their analysis was limited to freely dispersed mycelia. Choy et al. (2011), using image analysis, showed that enzyme production of *T. reesei* was dependent on the agitation rate, and this dependency was reflected in the morphology: highest productivity was obtained at 400 rpm, and higher shear was shown to lead to higher fragmentation and more dead tips that do not produce proteins.

Esterbauer et al. (1991) performed a mass balance of cellulase production with the strain *T. reesei* MCG77 in pilot plant fermenters on cellulose and lactose as a carbon source. They arrived at values of 0.26 g fungal biomass and 0.25–0.27 g cellulase protein from 1 g of carbon source.

As an alternative to submerged fermentation, solid-state fermentation (SSF) offers a low-cost alternative for producing cellulases using natural polymers derived from agroindustrial residues (Singhania et al., 2010; Brijwani and Vadlani, 2011). It has originally been used particularly in South and East Asia (due to the traditional Koji process) and used for cellulase production by Japanese manufacturers. SSF makes use of the insoluble cellulosic materials to produce a solid phase, which is wettened with soluble accessory nutrients, on which the fungus grows. The availability of space between the substrate particles ensures availability of oxygen (Brijwani and Vadlani, 2011). This process has the advantage that it is less susceptible to infection and does not require sophisticated instrumentation. It is therefore an alternative for cheap and small-scale local production.

Obviously, the currently reported industrial cellulase titers of >120 g protein per liter cannot be obtained in a batch fermentation because of the too high and growth inhibitory concentrations of nitrogen and carbon that would be needed. Such yields are therefore only possible with fed-batch strategies. The rationale is to maintain the organism as long as possible in the production stage. Therefore, the culture is typically first allowed to grow until the stage of cellulase production is reached. Then, further substrate is added in appropriate portions/concentrations that maintain the organism to continue producing further cellulases without too much accompanying biomass growth. Only little ammonium has to be included in the fed, because the average content of N in protein is a seventh of carbon (g/g). Fed-batch fermentations with *T. reesei* usually take up to 300–400 hours of cultivation and are mostly done with lactose.

There has also been numerous literature about producing cellulases with *T. reesei* in continuous cultivation with free or immobilized cells; however, none of them has been applied on an industrial level so far (Esterbauer et al., 1991; Tolan and Foody, 1999).

9.3.2 Cellulase Downstream Processing

After termination of the fermentation, the enzyme-containing extracellular culture is separated from cell mass and eventually residual insolubles are separated by filtration, typically using a rotary vacuum filter or filter belt press. For confection, the broth, which typically contains

8–14% of protein, is either concentrated to about 30% by ultrafiltration, using membranes of 10,000 molecular weight cut off. To stabilize the preparations, lactose, glycerol, or polyols can be added. Alternatively, solid preparations can be obtained by spray drying, using an inlet temperature below 150°C to prevent inactivating the enzymes (Esterbauer et al., 1991). For the latter treatment, salts and other low molecular weight substances from the medium can be removed by diafiltration. In addition, precipitation with acetone or isopropanol (typically using the double or threefold volume of the solvent below 4°C) is sometimes done, which results in white powder consisting of >98% protein.

In order to reduce enzyme costs, on-site or near-site production of the enzymes is considered to avoid transportation and formulation costs. The least expensive alternative in this situation involves the direct use of whole fermentation broth (including cell mass) to circumvent expensive cell removal and enzyme formulation steps (Sørensen et al., 2011). However, strain containment is a major issue in these considerations.

Enzyme formulation can be carried out at the fermentation plant in a separate formulation facility or at the end users.

For solid formulations, the spray-dried enzyme can be compacted or granulated (Tolan and Foody, 1999). In order to reduce inactivation of the enzymes, a stabilizer must be used: in liquid formulations, stabilizers of protein hydration (to keep it in solution) such as sorbitol, glycerol, or propylene glycol are used at concentrations of 10%–50% (w/w). In addition, preservatives such as sodium benzoate, potassium sorbate, or sodium chloride are added to prevent microbial contamination.

Chapter 10

Production of Fermentable Sugars from Lignocelluloses

As discussed at the beginning, in Chapter 1, evolution has driven the plant cell wall to serve as a barrier against attack by potential pathogenic organisms and to provide resistance against mechanic forces. Therefore, the plant cell wall is recalcitrant against enzymatic attacks. To this end, pretreatment of the plant biomass by various means (as outlined in this chapter) is essential to overcome this recalcitrance. The objective is to increase the surface area and porosity of the substrate, reduce the crystallinity of cellulose, and disrupt the hydrogen bonds that hold the heterogeneous structure of cellulosic materials together, and by all this render it more accessible for enzymatic hydrolysis. Reports have shown that pretreatment can improve sugar yields to higher than 90% theoretical yield for biomass such as wood, grasses, and corn (Brodeur et al., 2011).

Obviously, no general pretreatment procedure can be recommended because the various lignocellulosic materials, such as grasses, softwoods, and hardwoods, have different physical and chemical properties and, therefore, require different strategies to render them accessible to enzymes.

It must also be kept in mind that the method of pretreatment will have an important impact on the physicochemical properties of the pretreated biomass, which in turn will affect all the subsequent downstream processes such as selection of the most appropriate enzyme mixture and source, preconditioning, by-product utilization, waste residue handling, and product recovery. Pretreatment of the lignocellulosic biomass is therefore the central operation unit in various biorefinery products (Hendriks and Zeeman, 2009).

A number of pretreatment technologies have been developed and applied to various sources of lignocellulosic plant biomass (reviewed, among others, by Galbe and Zacchi, 2007; Hendriks and Zeeman, 2009; Alvira et al., 2010; Talebnia et al., 2010; Zhu et al., 2010; Brodeur et al., 2011). The pretreatments can roughly be classified into physical, physicochemical, chemical, and biological processes (Figure 10.1). The methods then actually applied make use of a combination of mechanical treatment together with thermal and chemical treatments, with the overall goal to achieve highly efficient sugar release, minimize the production of toxic

Fungi and Lignocellulosic Biomass, First Edition. Christian P. Kubicek.

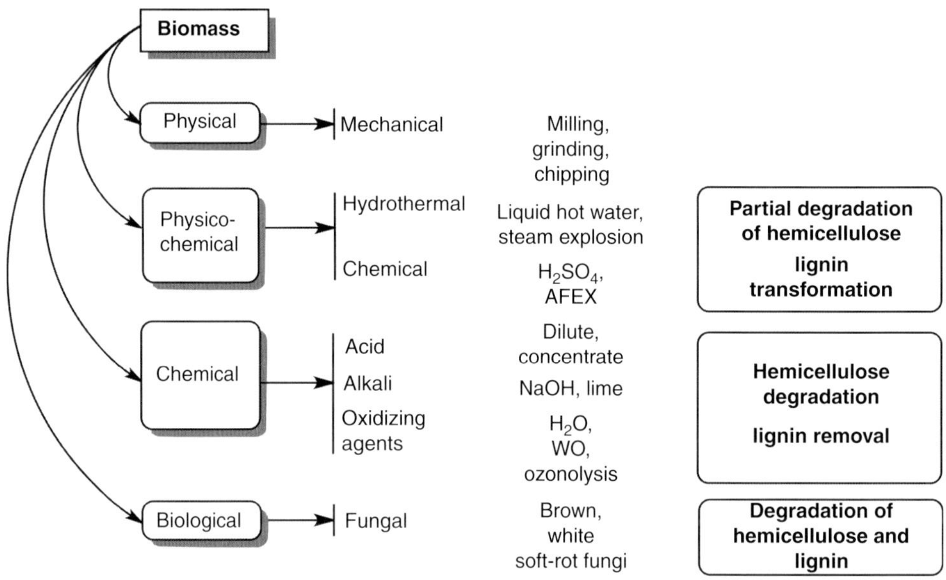

Figure 10.1. The most common pretreatment methods used on wheat straw and their possible effects.

by-products, avoiding the degradation of the pentose sugars, recovery of lignin for conversion into valuable coproducts, and to be cost effective by operating in reactors of moderate size and by minimizing heat and power requirements (Brodeur et al., 2011). In the following section, I will review these major pretreatment methods.

10.1 Pretreatment Technologies

10.1.1 Physical Pretreatment

Physical pretreatment of lignocelluloses biomass is mandatory in order to increase the ratio of surface area to volume, which alleviates mass and heat transfer problems and makes most other pretreatments more effective, including susceptibility of the untreated substrate to enzymatic hydrolysis. This can be obtained by dry, wet vibratory, and compression-based ball milling procedures. Also γ-irradiation has been shown to disrupt hydrogen bonds, thus leading to smaller particles, but it is unlikely that this method will find acceptance by the concept of "green biofuels." However, a drawback of physical pretreatments is the high energy input that is required, which can seriously impair the energy obtained by the biofuels produced. Therefore, the particle size of the original material and of the material after treatment need to be carefully calculated in order to keep the energy consumed by the process in a reasonable range.

10.1.2 Chemical Pretreatment

Chemical ways for biomass pretreatment comprises acidic, alkaline, and oxidative-based pretreatments. Depending on the type of chemical used, pretreatment will have different

Table 10.1. Summary of chemical pretreatments.

	Acidic Treatments	Alkaline Pretreatment	Oxidative Treatment
Agent	Sulfuric acid (4%, w/v)	Lime	Air or oxygen
T (°C)	140–200	85–150	>120
Duration range	Min to h	h	h
Disadvantage	Production of toxic degradation products	Formation of irrecoverable salts	Appropriate only for biomass with low lignin content

effects on lignocellulose structural components. Alkaline pretreatment, ozonolysis, peroxide, and wet oxidation pretreatments are more effective in the removal of lignin, whereas dilute acid pretreatment is more efficient in hemicellulose solubilization. Optimal operational conditions for some of the chemical pretreatments most frequently used are summarized in Table 10.1. The type of chemicals used results in different chemical reactions that cause the ultrastructural and chemical modifications of the lignocelluloses, which ultimately improve enzyme accessibility.

Acidic Treatments

Inorganic acids such as H_2SO_4 have long been used for pretreatment of lignocelluloses. Two types of processes are in use, that is, concentrated- and/or dilute-acid hydrolysis. In the former, the lignocellulose is treated with high concentrations of acids at ambient temperatures, whereas in the latter only low concentrations of acids (0.5–1% H_2SO_4) are used but at high temperatures (Pedersen et al., 2011).

Concentrated acid hydrolysis results in a high yield of sugars, which bears the advantage that the hemicelluloses hydrolysate can directly be used for further biorefineries, and no xylanases need to be added. However, this method also has the disadvantages of high acid and energy consumption, production of toxic by-products, equipment corrosion, longer reaction time, and the necessity of acid recovery after the treatment. In addition, the cellulose-rich debris needs to be extensively washed before it can be used for fermentation.

For these reasons, the dilute acid/high temperature process is generally preferred. Many of the disadvantages of the concentrated acid process can be bypassed, but the high temperature causes increased equipment corrosion and an increased rate of decomposition of the hemicellulose sugars. These degradation products include furfural, 5-hydroxymethylfurfural, phenolic acids and aldehydes, and levulinic acid (Figure 10.2), which are inhibitory to the cellulases as well as the microorganisms during fermentation and are the biggest drawback of the sulfuric acid hydrolysis process.

In addition, several other mineral and organic acids (nitric, carbonic, succinic, fumaric, maleic, and citric acid) have been used successfully as pretreatment catalysts (Talebnia et al., 2010), but their use in commercial processes has not yet been reported.

Alkaline Pretreatment

Alkaline chemicals, such as calcium oxide (lime), ammonia, and sodium hydroxide, can hydrolyze acetyl groups in hemicelluloses and lignin–carbohydrate ester linkages and thus aid in the solubilization and extraction of lignin from lignocellulose (da Costa Sousa et al.,

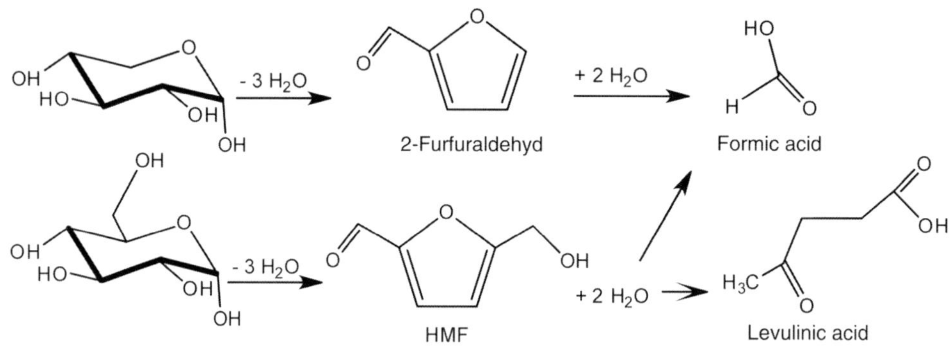

Figure 10.2. Acid-catalyzed degradation of xylose (top) and glucose (bottom) during pretreatment.

2009). At low temperature, the alkali (either NaOH alone or NaOH–urea mixture solution) can slightly remove lignin, hemicelluloses, and cellulose in the lignocellulosic materials, disrupt the connections between hemicelluloses, cellulose, and lignin, and cause a swelling of the pores in wood and thus lead to an increased surface beneficial for enzymatic attack. In addition, wood fiber bundles were broken down to small and loose lignocellulosic particles. Last not least, they display the further advantage to operate under lower temperatures and pressures than other pretreatment technologies. As lignin is a major target of alkaline treatment, it strongly influences the efficacy of this process: Zhao et al. (2008) showed the effectiveness of sodium hydroxide pretreatment for hardwoods, wheat straw, switchgrass, and softwoods with less than 26% lignin content, but there is almost no effect on softwood with higher lignin content. Typical conditions for alkaline pretreatment are, for example, 3% (w/v) NaOH/12% (w/v) urea that ultimately lead to 60% glucose conversion. However, NaOH is expensive, and thus the use of calcium hydroxide (lime) as a cheap alkaline agent (with also easy recovery and regeneration) is likely the alkali of choice for the technology (Kim and Holtzapple, 2005).

Oxidative Treatment

Oxidizing agents can also be used to remove lignin and hemicellulose from biomass. There are four variants of this technology: wet oxidation, alkaline oxidative treatment, alkaline wet oxidation, and ozonolysis.

In wet oxidation, the lignocellulosic biomass is treated with water in the presence of either high-pressure oxygen (120–480 psi; Palonen et al., 2004) or air at elevated temperatures (above 120°C). It is an effective pretreatment method for the fractionation of lignocelluloses biomass into a soluble hemicellulose fraction and a cellulose-rich solid fraction with high susceptibility to enzymatic hydrolysis (Klinke et al., 2002). Because of the oxidation of the hemicellulose components, acids that catalyze the subsequent hydrolysis of hemicelluloses to soluble monosaccharides are formed. Oxidative treatment also leads to significant lignin degradation because C–C bonds in it are very reactive under wet oxidation conditions. Thus, lignin is decomposed to CO_2, H_2O, and carboxylic acids (Klinke et al., 2002). However, wet oxidation is usually not very efficient with lignocelluloses of high lignin content (Galbe and Zacchi, 2007).

A variant of this process is alkaline oxidative pretreatment. This technology avoids the use of gases under high pressure and employs an oxidizing compound (usually hydrogen peroxide or peracetic acid) at low temperatures under alkaline (e.g., NaOH) conditions. Gould (1984) pioneered in demonstrating the use of H_2O_2 for delignification of agricultural residues: approximately one-half of the lignin and most of the hemicellulose present in agricultural residues such as wheat straw and corn stover were solubilized by an alkaline solution of hydrogen peroxide at 25°C. The delignification reaction was most efficient when the ratio of hydrogen peroxide to substrate was at least 0.25 (w/w) and the pH 11.5. He also showed that the supernatant after pretreatment could be reused at least six times, when supplied with further peroxide and pH readjustment, resulting in a substantial concentration of hemicellulose and soluble lignin degradation products (Gould, 1984). Hydrolysis of the insoluble fraction yielded glucose with almost 100% efficiency, indicating that alkaline peroxide pretreatment is a very easy and efficient method for pretreatment. A caveat, however, has to be paid to the fact that alkaline oxidation can also degrade hemicellulose and cellulose, and thus affect the final sugar hydrolysis yields. Additionally, a number of degradation products can also be formed (primarily aliphatic aldehydes and aliphatic organic acids) that will inhibit the enzymatic hydrolysis of cellulose.

Alkaline wet oxidation is actually a modification of alkaline oxidation treatment that uses sodium carbonate as alkali. Sodium carbonate has the advantage that it minimizes the formation of furan-based degradation products that would inhibit enzymes.

In ozonolysis, ozone is used to make lignin and a small fraction of hemicellulose from plant biomass soluble (García-Cubero et al., 2009). It is performed at room temperature and can effectively remove the lignin without producing any toxic residues. The main drawback of this process is the large amount of ozone utilization that makes the process too expensive for industrial application.

10.1.3 Physicothermal Pretreatment

Steam explosion is one of the most cost-effective and widely used pretreatment methods that is successfully applied to several types of lignocellulosic biomass (e.g., poplar chips, olive tree residues, corn stover residues, and wheat straw). It involves the subjection of biomass of reduced particle size to a high-pressure steam at a high temperature (160–290°C) for a short time (ranging from a few seconds to several minutes), followed by sudden reduction of the pressure. This results in an explosive decompression and fragmentation of the lignocelluloses (Ballesteros et al., 2006). A two-step steam-explosion technology has also been developed, in which the first step uses temperatures of 180°C to solubilize and remove the hemicellulose fraction. The second step (which is performed at 210–240°C at high pressure) then hydrolyses the remaining cellulose fraction (Brodeur et al., 2011). Parameters influencing the efficiency of this technology are temperature, residence time, original particle size, and moisture content of the lignocellulose. Generally, the operation at higher temperature increases the susceptibility for enzymatic hydrolysis of the cellulose-rich fraction, but it is also accompanied by a higher loss of cellulose by degradation.

Steam explosion can be made even more efficient by treating the lignocellulose with different chemicals (e.g., sulfuric acid, sulfur dioxide, sodium hydroxide, and ammonia) before steam explosion. Dilute acids have the particular advantage of not only reducing the retention time and temperature needed, but they also allow the use of softwoods with steam explosion, which otherwise has been proven not to be economic (Brodeur et al., 2011). Decreasing the retention time and temperature also has the benefit of decreasing the formation of inhibitory compounds (Ballesteros et al., 2006).

Steam explosion is the pretreatment technology that is closest to commercialization. Galbe and Zacchi (2007) listed several companies in the United States of America, Canada, and Sweden that run demonstration plants including this technology.

Another popular technology for making lignocelluloses susceptible for enzymatic attack is liquid hot water pretreatment. This pretreatment applies a high pressure ($P > 5$ MPa) to maintain water in a liquid state at even high temperatures (Yu et al., 2010). This approach bears several similarities to steam explosion (i.e., the use of water at elevated temperatures and high pressure for disintegration and separation of the lignocelluloses matrix). Typically, temperatures of 160–240°C and residence times of a few minutes up to an hour are used (Yu et al., 2010). The adjustment of the most appropriate temperature and pH can thereby be used to manipulate the solubility and partitioning of the plant cell wall components, the hemicelluloses usually being the most thermochemically sensitive fraction. Xylan can be extracted even by water when the temperature is higher than 150°C. The method has the disadvantage, however, that it uses more water than the other technologies and thus results in a more diluted sugar solution.

Several process technologies have been applied to promote an effective contact between the lignocellulose and the liquid water, varying from cocurrent, counter-current, and flow-through operation (da Costa Sousa et al., 2009). A pilot plant (up to 100 kg/h) that uses a two-step hydrothermal pretreatment in continuous operation has been developed for wheat straw: it involves a first soaking of the straw at 80°C for 5–10 minutes, followed by heating at 195°C for 6–12 minutes. Under these conditions, approximately 70% of the hemicellulose is recovered, 93–94% of the cellulose is recovered in the fibers and 89% of the cellulose in the fibers could be hydrolyzed by commercial cellulase mixtures (Pedersen and Meyer, 2009).

Supercritical fluids, that is, chemicals that can exist as a liquid or a gas and, when maintained above a critical temperature and critical pressure, display a coexistence of both aggregate states, also have a high potential for lignocelluloses pretreatment. Their advantage lies in the fact that they display a liquid-like density but exhibit the transport properties, diffusion, and viscosity of a gas (Brodeur et al., 2011). Supercritical fluids thus can penetrate the crystalline structure of lignocelluloses, thereby disrupting their hydrogen bonding, and overcome the mass transfer limitations that are inherent to other pretreatments (McHardy and Sawan, 1998). In addition, the solubility of compounds in supercritical fluids can be influenced by small changes in temperature or pressure close to critical point and thus be adapted for product separation after the pretreatment. Supercritical carbon dioxide (CO_2) with a critical temperature (T_c) of 31°C and a critical pressure (P_c) of 7.4 MPa has today the best proven potential for biomass pretreatment (Schacht et al., 2008). While this technology has the obvious advantage of the low temperature for operation, which prevents the degradation of the sugars, it has so far not passed the criterion of economy. In addition, the high pressure involved is still an obstacle against implementation on a large scale.

Ammonia fiber explosion (AFEX) is yet a further and popular technology that makes use of aqueous solvents and high temperatures to solubilize hemicelluloses and render cellulose accessible for attack (Chundawat et al., 2007). It operates by exposing the biomass to liquid concentrated ammonia (around 1 kg/kg of dry weight biomass) at elevated temperature (60–140°C) and pressure for 5–45 minutes followed by a rapid pressure release. This low temperature (*vide supra* for comparison) is clearly an advantage over other processes. The rapid expansion of the ammonia gas causes a swelling of the lignocellulose and thereby also disrupts the linkages between lignin and the polysaccharides, hydrolyzes the hemicellulose, and partially affects the crystallinity of cellulose (Brodeur et al., 2011). The advantages of this method are that the ammonia (because of its volatility) can be recovered and reused, thus leaving the dried biomass ready for enzymatic hydrolysis. Also, this technology does not produce inhibitors for the downstream processes and it is thus possible to ferment and hydrolyze this substrate without any need for detoxification. Finally, AFEX can be applied without the necessity for a small particle size of the biomass (Bals et al., 2011). The disadvantage of this method, however, is that it is less efficient for lignocelluloses with higher lignin content because lignin is not removed but eventually adsorbs on the surface of the cellulose and consequently impedes the subsequent enzymatic hydrolysis. Also hemicellulose can become acetylated and degraded into insoluble oligomers (Galbe and Zacchi, 2007). Like other processes that make use of alkaline treatments, AFEX gives best results with agricultural materials: almost theoretical sugar yields have been reported for corn stover and switchgrass after AFEX treatment under optimum conditions (Teymouri et al., 2004; Alizadeh et al., 2005). However, it has not proven successful with wood of higher lignin content (Galbe and Zacchi, 2007).

Other ammonia-based methodologies (e.g., ammonia recycle percolation (ARP), soaking in aqueous ammonia, supercritical ammonia, and ammonia-hydrogen peroxide pretreatments (Weimer and Chou, 1986; Kim and Lee, 2005)) are available. The differences between these methods are mainly the thermodynamic state of the ammonia–water mixtures, and the concentration of ammonia. ARP deserves some further remarks, though: it consists of a packed-bed reactor containing the respective biomass, which is rinsed by 5–15% (w/v) ammonia at a flow rate of about 5 mL/min and at 140–210°C. Although the need for a higher temperature than in AFEX implies higher energy costs, this could be matched by the higher recovery of biomass components: ARP can remove most of the lignin (75–85%) and solubilize more than half of the hemicellulose (50–60%) while maintaining a high cellulose content (Kim and Lee, 2005).

10.1.4 Solvent Treatment

Another principle that is successfully used for the pretreatment of lignocelluloses biomass is the extraction of lignin and hydrolyzed hemicelluloses by organic solvents (known as the Organosolv process). This technology uses organic solvents, usually ethanol in the presence of an acid, to extract lignin from lignocellulosic biomass. As an example, Pan et al. (2005) fractionated poplar chips into a cellulose-rich solids fraction, an ethanolic lignin fraction, and a water-soluble fraction that contained hemicelluloses monomers, sugar breakdown products, degraded lignin, and other components. Optimal conditions were 180°C, 60 minutes, 1.25% (w/v) H_2SO_4, and 60% (w/v) ethanol, which resulted in the recovery of 88% of the cellulose that was originally present in the biomass before treatment, of which approximately 82% could be hydrolyzed by cellulase preparations. About 74% of the original lignin was recovered in the ethanolic phase. The solvent is recovered by distillation and reused.

Phosphoric acid fractionation is a modification of the Organosolv process, which allows a strong reduction of the operation temperatures (Zhang et al., 2009): this technology consists of the use of phosphoric acid as an acid catalyst in a series of extractions by acetone and water for fractionation of the plant cell wall. This treatment also decrystallizes the cellulose fibers. A bottleneck of phosphoric acid modification technology, however, is the need to recover and purify the solvents, particularly phosphoric acid for economic performance of the process, for which economic solutions have not yet been obtained.

Room temperature ionic liquids (RTILs) are salts, typically composed of a small anion and a large organic cation (such as 1-*n*-butyl-3-methylimidazolium chloride, 1-allyl-3-methylimidazolium chloride, or 1-ethyl-3-methylimidazolium acetate), which exist as liquids at room temperature and are characterized by very low vapor pressure, high thermal stability, high electrical conductivity, and a wide range of resistance against electrochemical reaction (Quijano et al., 2010). The chemistry of the anion and cation can be tuned to generate a wide variety of liquids with varying viscosity, polarity, and hydrogen bond basicity, which can be used to dissolve a number of biomass types such as corn stover, cotton, bagasse, switchgrass, wheat straw, and woods of different hardness (pine, poplar, eucalyptus, and oak) (Brodeur et al., 2011; Mora-Pale et al., 2011). The low vapor pressure of ionic liquids, which makes them recoverable by more than 99% (w/v), and the lack of formation of toxic products makes their use highly advantageous. The biomass solubilized by this treatment can subsequently be precipitated by the addition of water, methanol, and ethanol. For all these reason, RTILs receive increased interest and are considered emerging solvents for the pretreatment of lignocellulose biomass (for review, see Mäki-Arvela et al., 2010 and Mora-Pale et al., 2011).

In a pioneering study with various lignocellulose sources, including pulps, fibrous cellulose, and filter papers, Swatloski et al. (2002) tested a range of 1-alkyl-3-methylimidazolium cations and inorganic anions for their abilities to dissolve cellulose. They found 1-butyl-3-methylimidazolium chloride to be the most effective by dissolving up to 25% (w/w) cellulose. Using a high-throughput system, Zavrel et al. (2009) screened more than 20 RTILs for their ability to act as a solvent for microcrystalline cellulose. In addition to the chloride-based ionic liquids, 1-ethyl-3-methylimidazolium acetate was also found to be a good solvent for cellulose for up to 4% (w/w) at 80°C. A list of RTILs that have meanwhile been successfully used in lignocellulose solubilization is given in Table 10.2. Generally, RTILS act by rendering cellulose far less crystalline without undergoing solubilization, which results in the high recovery (>90%) of the cellulose. A number of ionic liquids have also been tested for dissolution of lignin (Pu et al., 2007; see Table 10.2). As with cellulose, the anion plays a critical role in dissolving lignin: noncoordinating anions exhibited no or very limited ability to dissolve residual softwood Kraft lignin, whereas methylsulfate-containing ionic liquids could significantly dissolve lignin at 50°C and partially at room temperature. On the basis of the solubility of lignin in RTILs, acid-catalyzed depolymerization reactions have been attempted in RTILs. A disadvantage, however that has so far not been solved, is the tendency for self-condensation of the depolymerized lignin (Mora-Pale et al., 2011). Imidazolium-based ionic liquids have also been used extensively for solvation of hardwoods and softwoods (Fort et al., 2007; Kilpelainen et al., 2007).

Samayam et al. (2011), using X-ray fractionation analysis, demonstrated that ionic liquids, when used at high temperatures, bring about a conversion of cellulose I to cellulose II, which is likely the effect that increases biomass digestibility (Figure 10.3). They explained this finding by proposing that lignin restricts the swelling of cellulose at low temperature. At higher temperatures, there would appear to be a change in lignin structure that allows a complete cellulose dissolution.

Despite these promising results, the application of ionic liquids still suffers from the expensiveness of these chemicals (although their recovery by distillation is possible) and a number of other factors. For example, the retention of some of the ionic liquids can contaminate the biomass after regeneration, which could in turn interfere with the enzymatic hydrolysis. Increasing the economy of the process will therefore require an optimization of large-scale pretreatment conditions, performing post-pretreatment steps in RTILs, reusing RTILs, recycling the RTILs with reduced energy consumption, and enhancing process efficiency (Mora-Pale et al., 2011). Also, the improvement of generation of high-value products from the extractable lignin that could replace the petroleum-based production will increase the economy of the RTIL technology.

Table 10.2. RTIL that have been applied to lignocelluloses pretreatment.

RTIL	Effect
1-Ethyl-3-methylimidazolium acetate	Dissolution of cellulose
	Extraction of lignin from maple wood flour
	Dissolution of a variety of carbohydrates such as sugars, starch, and cellulose
1-Ethyl-3-methylimidazolium chloride	Dissolution of cellulose
1-Allyl-3-methylimidazolium chloride	Dissolution of cellulose
	Dissolution of hardwood and softwoods
	Extraction of lignin from maple wood flour
1-Butyl-3-methylimidazolium chloride	Dissolution of cellulose
	Pretreatment of cellulose for enhancing enzymatic hydrolysis
	Dissolution of hardwood and softwoods
	Extraction of lignin from maple wood flour
1-Butyl-3-methylimidazolium bromide	Dissolution of cellulose
	Dissolution of lignin
1-Butyl-3-methylimidazolium acetate	Extraction of lignin from maple wood flour
1-Butyl-3-methylimidazolium tetrafluoroborate	Dissolution of cellulose
	Extraction of lignin from maple wood flour
1-Butyl-3-methylimidazolium hexafluorophosphate	Dissolution of cellulose
	Extraction of lignin from maple wood flour
1-Butyl-3-methylimidazolium methylsulfate	Extraction of lignin from maple wood flour
	Dissolution of lignin
1-Octyl-3-methylimidazolium chloride [OMIM]Cl	Dissolution of cellulose

Data extracted from Pedersen and Meyer (2010).

Interestingly, a non-RTIL solvent that also effectively dissolves cellulose and has a similar low vapor pressure as ionic liquids, but which is nontoxic and does not inhibit cellulose hydrolysis, is *N*-methyl morpholine *N*-oxide (NMMO), which is known as Lyocell solvent, used by the Lenzing company commercially in paper manufacture (Rosenau et al., 2001). Ramakrishnan et al. (2010) showed that NMMO can be used for *in situ* enzymatic hydrolysis without the need to recover regenerated cellulose. Their results showed superior rates of hydrolysis of cellulose and yields of reducing sugars in the presence of NMMO-water initially and comparable to that of regenerated cellulose when suspended in aqueous solutions. Thus, cellulases remain active in NMMO, which eliminates the need to recover regenerated cellulose.

10.1.5 Biological Pretreatment

Biological pretreatment is a "soft" alternative to the other methods outlined above. It uses microorganisms for selective degradation of lignin and hemicelluloses and—because it is not

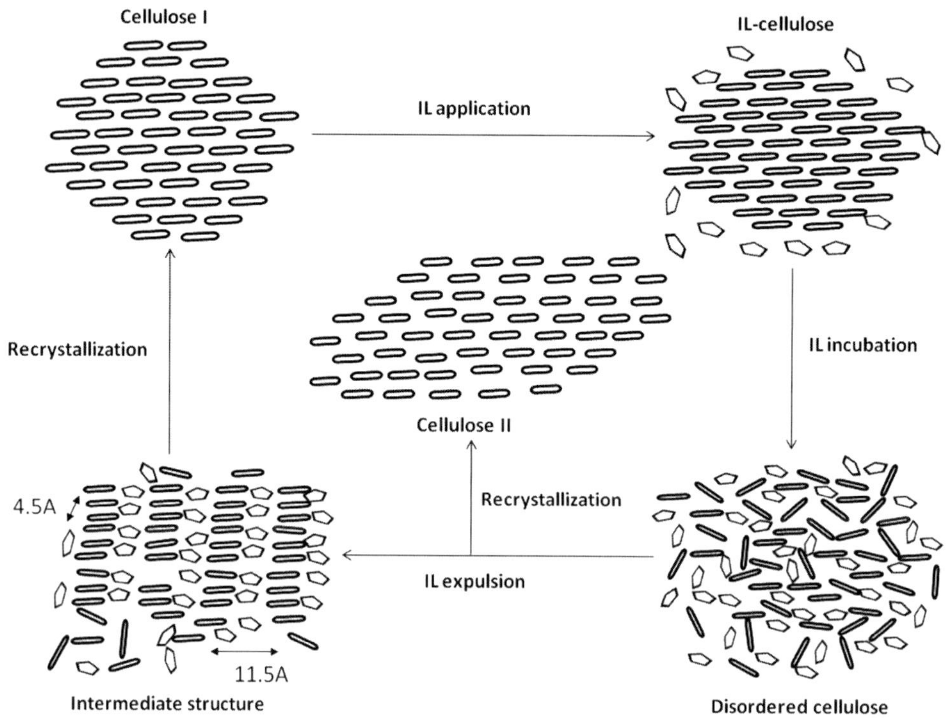

Figure 10.3. The effects of varying ionic liquid pretreatment parameters on various sources of lignocellulosic biomass have been studied using X-ray powder diffraction, X-ray fiber diffraction, and compositional analysis. Comparative enzymatic hydrolysis and sugar analysis were used to relate the observed changes in cellulose structure to biomass digestibility. In this study, the factor most clearly associated with enhanced biomass hydrolysis is the conversion of cellulose fibers from the cellulose I to the cellulose II crystal phase. (Reprinted with permission from Samayam et al., 2011. Ionic-Liquid Induced Changes in Cellulose Structure Associated with Enhanced Biomass Hydrolysis. Biomacromolecules. Copyright 2011 American Chemical Society.)

energy intensive—is an attractive alternative to the chemical processes. This way of pretreatment is normally performed using fungi (e.g., white rot basidiomycetes) and actinomycetes, which have a higher affinity for lignin and degrade it faster than carbohydrate components (Yu et al., 2009; Bak et al., 2010; Salvachúa et al., 2011). As described in Chapter 2, white-rot fungi can efficiently degrade lignin and are therefore the most preferred candidates for biological pretreatment of lignocellulose. However, different white rot fungi vary greatly in the relative rates at which they degrade lignin and carbohydrates in lignocelluloses, and only the selective lignin-degrading white rot fungi are considered to be good candidates for biological pretreatment (Itoh et al., 2003; Taniguchi et al., 2005). In several reports, the authors attributed the enhancement of enzymatic hydrolysis after biological pretreatment to the biodegradation of the lignin, but it has not yet proven that fungal delignification actually affects the physical properties of the substrate. Some studies suggested that biological pretreatment affects the cellulase–substrate interaction by altering substrate properties, leading to the enhancement of enzymatic hydrolysis. Yu et al. (2009) used a new strain of *Echinodontium taxodii* in biological pretreatment to enhance the enzymatic hydrolysis of two trees native to China: Chinese willow (hardwood) and China-fir (softwood). *E. taxodii* preferentially degraded the lignin during the pretreatment, and the pretreated woods showed significant increases in enzymatic hydrolysis ratios (4.7-fold for hardwood and 6.3-fold for softwood). They showed that in fact the enhancement of enzymatic hydrolysis was due to a decrease in the irreversible adsorption of cellulase to lignocelluloses, which may result from alteration in lignin structure after biological pretreatment. Salvachúa et al. (2011) reported that most of the basidiomycetes that they used for biological pretreatment indeed degraded lignin with variable selectivity degrees, but only eight of them improved sugar recovery compared to untreated samples. Glucose yield after 21 days of pretreatment with *Poria subvermispora* and *Irpex lacteus* reached 69% and 66% of cellulose available in the wheat straw, respectively. Conversions from glucose to ethanol reached around 90%, showing that no inhibitors were generated during this pretreatment.

Yet—although a considerable amount of research has been dedicated to biological pretreatment—it is currently considered to be industrially not feasible (Brodeur et al., 2011), the major drawback being the longer residence time (several hours to few days) of the plant biomass and the consumption of some part of the carbohydrates available in the biomass by the fungus used, thus lowering the sugar yield at the end of the process. In fact, most of the work in this area focused on biopulping in the pulp and paper industries rather than preparation of lignocelluloses for enzymatic hydrolysis.

10.1.6 Summarizing Considerations on Pretreatment

A summary of the advantages and disadvantages of the main technologies in use and described earlier is given in Table 10.3. Clearly, there is no technology that is universally suitable for all lignocelluloses feedstocks, and the actually used technique must be adapted and optimized. We must also note that the costs of the feedstocks are the highest variable in the biorefinery processes, which consequently influences the choice of the pretreatment method. Consequently, an evaluation of a pretreatment must take into account both the technological challenges and economical and environmental parameters. Maximizing sugar yield and concentration have the strongest impact on the economy of the process and influence also the volume of reactors. Utilization of chemicals and water, temperature, pressure, mixing, and downstream processing (including removal of inhibitory substance) all are further parameters that influence not only the costs but also have environmental impacts and are related to energy requirements of the process. Pedersen and Meyer (2010) have recently compared the most critical variables in most of the pretreatment processed, that is, pH, temperature, and holding time. Their data illustrate that the pretreatment pH correlates with the hydrolysis yields (glucose and xylose), but not with the pretreatment temperature (90–200°C). Last not least, lignin recovery is an important issue as it can be used as a precursor for the manufacture of valuable chemicals (see Chapter 11), which is an option that could beneficially influence the economical feasibility of the whole biorefinery process. Moreover, some pretreatments have the potential to produce dry pretreated substrates, which could benefit distributed processing facilities because of lower transportation costs.

10.2 Hydrolysis

10.2.1 Enzymatic Hydrolysis

After pretreatment, the lignocellulolytic biomass is principally ready for hydrolysis. The enzymatic hydrolysis process can be designed in various ways: it can be run as a separate process

Table 10.3. Summary of advantages and disadvantages of different lignocellulose pretreatment methods.

	Advantages	Disadvantages	Applied to
Lime	Mild temperatures, low inhibitor production, partial removal of lignin	Slow action of lime causing long pretreatment time, salt production	Corn stover, wheat straw, hardwood and other agricultural residues
AFEX	Reduces lignin and hemicellulose content, swelling and decrystallization of cellulose, low inhibitor formation, low temperature	Expenses on ammonia, not effective when high lignin content, not suitable for softwood	Wheat straw, corn stover, and other agricultural residues
AWO	Increased digestibility of biomass, neutral after treatment	Costs on oxygen, temperature, alkaline catalyst	Wheat straw, softwood, and agricultural residues
IBUS	High DM, large-scale, no additives	Equipment requirement	Wheat straw, corn stover
WO	Easy to filter, minimal formation of inhibitors, exothermic process	Tank requirement, pressure, temperature, cost of oxygen	Wheat straw, softwood, and other agricultural residues
SE	High yield of glucose, hemicellulose degradation, lignin transformation	Pressure, temperature, formation of inhibitors	Wheat straw, wood, and other agricultural residues
MIC	Low energy requirement, no corrosive resistant reactor needed, degrades lignin and hemicellulose	Time, microorganisms digest some of the glucose	Wheat straw, wood, and agricultural wastes
2-step	Most lignin removed, minor loss of fermentable glucose and xylose, low formation of inhibitors, low temperature and pressure	Optimization needed toward continuous pretreatment	Wheat straw
ASE	High glucose release, removal of hemicelluloses	Uses SO_2 or H_2SO_4, requires high temperatures	Wheat straw, hardwood, and other agricultural residues
AH	High glucose yield, alters lignin structure, ambient temperatures	Tank requirements, corrosion, costs of acid and for acid recovery, degradation products	Agricultural residues
DAH	Removal of hemicellulose, tends to remove some lignin	Degradation products formed, corrosion	Agricultural residues

Data adapted from Pedersen and Meyer (2010).
Lime, lime pretreatment; AFEX, ammonia fiber explosion; AWO, alkaline wet oxidation; IBUS, integrated biomass utilization system; WO, wet oxidation; SE, steam explosion; MIC, microbial; 2-step, 2-step pretreatment first with acid then alkaline catalyst; ASE, acidic steam explosion; AH, acid hydrolysis; DAH, diluted acid hydrolysis.

(separate hydrolysis, SH) or it can be run simultaneous with the subsequent fermentation utilizing the arising soluble sugars. The latter option has mainly been attempted for ethanol production (simultaneous saccharification and fermentation, SSF). SH and SSF yield comparable process economics with hardwood as raw material but not with softwood, especially spruce, depending on the pretreatment regime (Galbe and Zacchi, 2007). Clearly, SH offers the advantage to carry out each step under optimal conditions (e.g., hydrolysis and fermentation have different temperature optima), but it has the disadvantage that the sugars released will inhibit the enzymes during the course of hydrolysis (Galbe and Zacchi, 2007). This would not occur in SSF, because the glucose produced is immediately consumed by the fermenting microorganism. Also, although the ethanol produced can act as an inhibitor of the enzymes used for hydrolysis, this is negligible at the concentrations produced currently (3–5%, w/v). Therefore, SSF would be preferred because it reduces the number of unit operations and reactors. A drawback of SSF is, however, that the mixture of lignin and yeast biomass that will remain after the process will make recycling of the yeast—a usual requirement for the economy of ethanolic fermentation—difficult. Ethanol fermentation and SSF processes will be described in Chapter 11.

Most cellulase enzymes show an optimum activity at temperatures and pH in the range of 45–55°C and 4–5, respectively (Galbe and Zacchi, 2007). Technically, this is therefore performed in a batch process at around 50°C and a pH around 5.0 (although these parameters are dependent on the residence time) with constant stirring to maintain a homogenous suspension. While this sounds simple, there are a number of parameters that strongly affect this process and also can cause costs that render it uneconomic.

β-Glucosidase plays a significant role in the hydrolysis process, since cellobiose is an end-product inhibitor of many cellulases, including both exo- and endoglucanases (Galbe and Zacchi, 2007). β-Glucosidase, in turn, is inhibited by glucose and enzymatic hydrolysis is thus sensitive to the substrate concentration. Also, β-glucosidase can be inactivated by prolonged residence time and agitation.

The removal of hemicellulose positively correlates with the efficiency of the hydrolysis (Palonen and Viikari, 2004). Even low amounts of residual xylan can limit the extent and rate of the hydrolysis, and the presence of sufficient amounts (activities) of xylanases in the enzyme mixture need to be present to overcome this; *Trichoderma reesei* preparations usually have adequately high xylanase activities to overcome this limitation on xylan-containing substrates (Viikari et al., 2007).

Current commercial preparations, mainly derived from fermentation of the filamentous fungus *T. reesei*, contain more than 80 proteins (Banerjee et al., 2010a), and although their identities are largely known (Herpoël-Gimbert et al., 2008), their necessity for hydrolysis of the pretreated biomass is not fully known. Reducing the levels of nonessential enzymes and enhancing the relative proportions of the critical enzymes are therefore potential strategies to increase the specific activities, and thereby lower the cost, of biomass-converting enzymes. Banerjee et al. (2010b) developed a working platform to synthesize and analyze individual cellulase enzymes from *T. reesei* and to test their combinations for the effect in the hydrolysis of AFEX-pretreated corn stover. Starting with a core of six enzymes (the cellobiohydrolases CEL7A and CEL6A, endo-β-1,4-glucanase CEL7B, β-glucosidase BGL1, endo-β-1,4-xylanase XYN3, and β-xylosidase BXL1), they showed that the hydrolytic efficiency of this core set can be significantly improved by the addition of other "accessory" enzymes. The specific activity of an optimized 11-component mixture equaled that of contemporary commercial enzymes (Accellerase 1000 or Spezyme CP) with respect to the yield in glucose and produced even higher yields of xylose.

Addition of surfactants has been reported to improve the enzymatic cellulose conversion into monomeric sugars. Nonionic surfactants such as Tween 20 were shown to be most effective for enhancing the enzymatic hydrolysis. The reason for this is not clear but it could be due to a modification of the surface properties of cellulose, a reduction of irreversible binding of cellulases on the cellulose, or decreased binding of enzymes to lignin. The optimum surfactant concentration is usually around 0.05% (g/g biomass) and is independent of the pretreatment type.

When scaling up lignocellulose hydrolysis for biorefineries such as ethanol production, the final ethanol concentration after the fermentation should optimally be as high as in the sugar-based processes (10–12% or higher). This is primarily desirable because of the economy of the unit operations involved in ethanol recovery and purification. Currently achieved concentrations are far below these values and more than 4% would already be a considerable improvement, because distillation of ethanol is considered economic only with a fermentation broth that contains more than 4% (w/w) ethanol (Zacchi and Axelsson, 1989). Since cellulose (containing the fermentable glucose) makes up 30–50% of the lignocelluloses biomass, and the theoretical ethanol yield is 50% (g ethanol/g glucose), this implies hydrolysis of a 16–24% (w/v) biomass suspension. For most technical substrates, the use of such high suspensions creates an environment in which practically no free water exists and which has negative impacts hydrolysis and also leads to high energy demands for efficient mixing (Palmqvist et al., 2011). Yet Jorgensen et al. (2007) and Roche et al. (2009), using biomass suspensions between 20% and 40%, w/v respectively, reported that efficient initial mixing is important for equal distribution of the enzymes. However, once the enzymes are distributed, rotation speed becomes less important and can be reduced. Also, recent progress in replacing the conventional stirring systems by gravimetric mixing have also helped to overcome this problem and made the liquefaction of pretreated lignocellulose possible at up to 40% initial DM content (Szijártó et al., 2011).

As for the viscosity of the suspension, one solution to combat its increase when working with high lignocellulose concentrations would be to use higher operation temperatures, and thus thermostable or thermophilic enzymes, because elevated temperatures result in lower viscosities and have the additional benefit of higher reaction rates. Also, a faster action of the enzymes on the lignocelluloses will reduce the viscosity during the hydrolysis process. To this end, Szijártó et al. (2011) compared novel thermostable enzymes (Cel7A from *Acremonium thermophilum, Chaetomium thermophilum,* and *Thermoascus aurantiacus;* Cel5A from *T. aurantiacus* and *A. thermophilum;* Xyn10A from *T. aurantiacus;* Xyn11A from *Nonomuraea flexuosa; and* Cel3A from *A. thermophilum*) for their ability to hydrolyze and concomitantly reduce the viscosity of pretreated wheat straw lignocellulose at 15% (w/v) solids concentration at 55°C. They showed that endoglucanases in fact showed such a superior ability, *T. aurantiacus* Cel5A giving best results.

Biomass hydrolysis in batch operation is characterized by a continuous decline in the reaction rate: initially, the cellulases rapidly adsorb to the accessible cellulose and perform a fast rate of hydrolysis (Figure 10.4). However, this rate of hydrolysis exhibits a significant decline at a point where about 50–70% of the original substrate has been hydrolyzed, giving rise to a steady but slow hydrolysis rate until the end of the process. Typically, very long hydrolysis times and/or high enzyme concentrations are required to achieve a near-complete conversion of cellulose. In addition, there are also cases where complete cellulose hydrolysis cannot be achieved. These long hydrolysis time associated with achieving complete cellulose saccharification adds significantly to the operating costs of the enzymatic hydrolysis step and, consequently, to those of the overall biomass to ethanol bioconversion process (Gregg and Saddler, 1996).

One of the most important factors responsible for the decline of the hydrolysis rate has been suggested to be the accessible surface area of the cellulose because a significant amount of the cellulose is buried within the highly ordered and tightly packed fibrillar architecture of the cellulose microfibrils (Arantes and Saddler, 2010a, 2011b). Cellulosic materials are typically not smooth but rather heterogeneous porous substrates, and the available surface area consists of exterior and interior surfaces. The latter can consist of internal pores, fissures, and

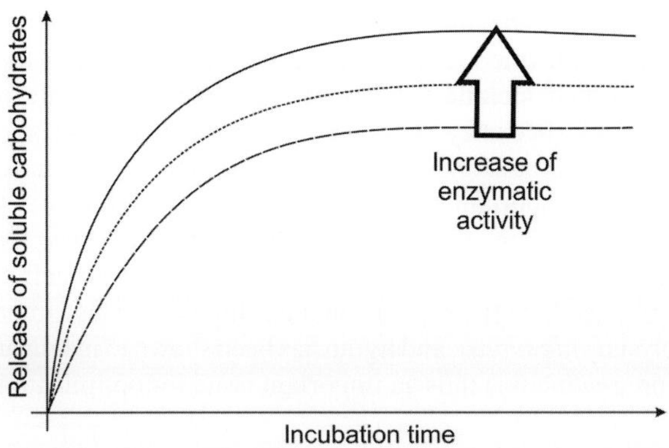

Figure 10.4. Typical time course of lignocelluloses hydrolysis by fungal cellulase mixtures. The arrow indicates increasing concentrations of the enzyme used.

microcracks, which typically arise from "discontinuities" of the molecular packing built into the cellulose during biosynthesis or surface openings/internal slits, voids, or spaces created by the removal of noncellulosic cell wall polymers during the pretreatment (Arantes and Saddler, 2011b). Grethlein (1984) reported that the initial hydrolysis rate of pretreated biomass was directly proportional to the pore size accessible to a molecule of 5.1 nm (= the diameter of an average 50 kDa cellulase).

Arantes and Saddler (2011b) investigated the parameters that govern the minimum protein loading for efficient hydrolysis after about 50% decline of the hydrolysis rate with a variety of pretreated lignocelluloses substrates. The optimized minimum protein loadings to achieve efficient hydrolysis of seven similarly pretreated substrates ranged between 18 and 63 mg protein per gram of cellulose, agricultural residues (corn stover and corn fiber) thereby requiring significantly lower protein loadings than the woody biomass (poplar, douglas fir, and lodgepole pine). On the basis of their results, they concluded that the rate-limiting step during hydrolysis is not the catalytic cleavage of the cellulose chains *per se*, but rather the limited accessibility of the enzymes to the cellulose chains.

Murphy et al. (2011b) have recently used isothermal calorimetry to monitor the early time course of the hydrolysis of insoluble cellulose by the three main endoglucanases from *T. reesei*, namely CEL7B, CEL5A, and CEL12A. They exhibited the typical initial burst in activity after which the rate significantly declined. Using a continuous monitoring of endoglucanase activity revealed that the slowdown is linked to a transient inactivation of enzyme on the cellulose surface. On the basis of these data, Murphy et al. (2011b) proposed that the frequency of the structures on the cellulose surface that cause this transient inactivation determine the extent of the burst phase.

An attractive strategy to reduce the cost of enzymatic hydrolysis would be to recover and recycle the enzymes. However, this is complicated by the two facts: first, part of the enzymes binds tightly to the residual substrate, and therefore the residual biomass substrate would have to be transferred to the fresh solid substrate, to which the bound enzymes will eventually migrate. Such a process, however, would lead to an enrichment of the solid residue with lignin, which would adversely affect the hydrolytic activity. The second complication is β-glucosidase, which does not adsorb to lignocellulose and thus would have to be supplemented at the beginning of each round of hydrolysis again.

Therefore, desorption of the enzymes adsorbed to the recalcitrant biomass is theoretically the only feasible solution for enzyme reutilization. A number of methods for desorbing enzyme have been developed, which include the use of surfactants, of alkali, urea, or regimes of varying the pH. The so desorbed enzymes can be recovered by ultrafiltration, which is capable of recovering all of the cellulase enzyme components, including β-glucosidase.

Probably an even more important factor that significantly influences the rate of biomass hydrolysis is the content in lignin. This is mainly due to three reasons: lignin forms a barrier to the enzymes to access the cellulose fraction in the biomass to cellulase; lignin absorbs cellulase and thus prohibits them from acting on the cellulose and also interferes with cellulase recovery after hydrolysis (*vide supra*); and lignin has been shown to inactivate cellulases. Lignin removal during the pretreatment is thus an important issue for optimization of lignocelluloses hydrolysis.

10.2.2 Chemical Hydrolysis

Because of the inherent problems with the enzymatic hydrolysis of plant lignocellulosic biomass, acid hydrolysis—a method used since 1819 (cited in Galbe and Zacchi, 2007)—is still in operation. Acid hydrolysis can be performed with several types of acids, including sulfurous, sulfuric, hydrochloric, hydrofluoric, phosphoric, nitric, and formic acid. These acids may be used either concentrated or diluted. With respect to the concentrated acid process, yields of glucose are very good (around 90%), this process has the problems of equipment corrosion and the high energy demand needed for recovery of the acid. Furthermore, when sulfuric acid is used, the neutralization process produces a large amounts of gypsum for which (because of its contamination with other products from the plant and the process) cannot be used for other applications and must therefore be deposited. Although improved and more economic methods for acid recovery have recently been developed by some companies, I am unaware of any attempts to revitalize this process.

The dilute acid hydrolysis process would have the advantage of low acid consumption and thus also reduction of recovery costs. However, this process must operate at high temperatures and thus also leads to corrosion and in addition also causes losses of hemicelluloses because of degradation to furfural, hydroxymethylfurfural, and other degradation products. This not only reduces the sugar yield but also introduces inhibitory substances into the hydrolysate. Glucose yields are consequently only between 50% and 60% of the theoretical value (Wyman et al., 2005).

Recently, Binder and Raines (2010) reported a high-yielding chemical process for the hydrolysis of biomass into monosaccharides. They gradually added water to a chloride ionic liquid-containing catalytic acid that achieved a nearly 90% yield of glucose from cellulose and 70–80% yield of sugars from untreated corn stover. Ion-exclusion chromatography allowed recovery of the ionic liquid and delivered sugar feedstocks that were used by ethanol-producing microorganisms. The authors concluded that this simple chemical process could enable crude biomass to be the sole source of carbon for a scalable biorefinery. The economics of this method still needs to be assessed.

Chapter 11
Lignocellulose Biorefinery

Biorefinery is a term that describes integrated bioconversion processes of plant biomass to value-added products. The concept was advocated by the International Energy Agency in the Bioenergy Task 42 in analogy to the petroleum refinery concept. Two different ways exist in the biorefinery of lignocellulose biomass. The *thermochemical* strategy uses the whole biomass without fractionation and subjects it to cracking and gasification for the production of synthetic biofuel via the Fischer–Tropsch method (Dodds and Gross, 2007). The *biological pathway* is based on biotechnological processes that transform fractions of the biomass into building-block chemicals by fermentation or by enzymatic and chemical reactions. The main difference between these two strategies lies in the first step of biomass pretreatment, in which lignin is separated from cellulose, which thus has a significant impact on the final yield in lignocelluloses biorefinery.

The lignocellulosic feedstock biorefinery scheme (Figure 11.1) must be applicable to a wide range of the respective feedstocks (straw, reeds, grass, wood, paper waste, etc.). Techniques for the fractionation of the biomass into its major constituents are of crucial importance to this, but are still in their infancy, particularly with respect to their technical and economical feasibility (FitzPatrick et al., 2010). Basic and applied research in this area is, therefore, a subject of intensive ongoing research. Acid-based and ionic liquid-based methods are currently serving best and they have been described in Chapter 10.

The resulting chemicals that can be directly produced from the fractionated, extracted, and/or hydrolyzed biomass can either be used as they are or represent key building blocks for production of further products. These are mainly alcohols (ethanol, glycerol, sorbitol) and carboxylic acids of C3–C6 length. All these compounds can also be produced from other, more easily accessible substrates, and it is thus important to mention that the current costs of many of the biomass-derived carbohydrates is already comparable to that of petrochemicals and solvents (Ragauskas et al., 2006). The US Department of Energy recently published a comparative study on the top 12 chemicals from carbohydrate biomass, identifying several

Fungi and Lignocellulosic Biomass, First Edition. Christian P. Kubicek.

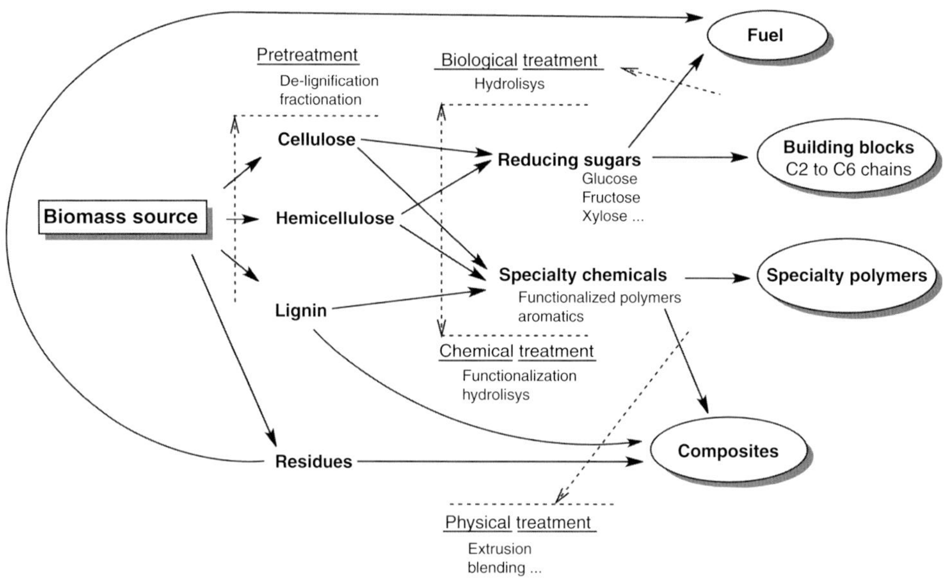

Figure 11.1. Overview of the biorefinery concept.

particularly promising compounds, including sorbitol, levulinic acid, and glycerol (Ragauskas et al., 2006).

11.1 Ethanol

The production of ethanol as a biofuel from renewable materials has provided the most significant impetus to lignocellulose utilization and biorefinery research and therefore deserves a more detailed treatment in this book. Ethanol is a classical fermentation product that can today be produced in high yields and with excellent economy from sucrose and starch hydrolysates (Berg, 2002). However, in contrast to these processes, lignocelluloses hydrolysis results in the production of a mixed-sugar solution.

The major problem in the production of bioethanol from lignocelluloses hydrolysates is that the pentose sugars xylose and arabinose are not fermented to ethanol by the *Saccharomyces cerevisiae*, which is traditionally used for ethanol formation from sucrose or glucose. As has been described in Chapter 1, plant biomass—in particular hardwood and agricultural raw materials—can contain 5–20% pentose sugars. Xylose is usually thereby by far the most abundant sugar, but arabinose can constitute as much as 14–15% in selected biomasses such as corncob hulls and wheat bran, respectively. Consequently, the development of microorganisms that efficiently ferment xylose to ethanol has been a major research effort in this area. Principally, such microorganisms can be found in bacteria, yeast, as well as filamentous fungi; however, while anaerobic bacteria ferment pentoses well, they are inhibited by already low sugar and ethanol concentrations and produce a considerable concentration of by-products, which reduce the ethanol yield and would warrant additional purification steps (Hahn-Hägerdal et al., 2006). Filamentous fungi, on the other hand, tolerate higher concentrations of inhibitors, but their ethanol production rate is too slow for a competitive industrial process, and they also tolerate only much less ethanol concentrations as, for example, yeasts (Panagiotou et al., 2005), thus leading to lower final ethanol concentrations. Yeasts such as *Pichia stipitis* can ferment xylose to ethanol with reasonable yield and productivity, but are very sensitive to the presence of inhibitors (Hahn-Hägerdal et al., 2007). Some other yeasts have been reported to ferment xylose into ethanol (reviewed by Schneider, 1989), but their rate and yield of ethanol production from xylose was considerably lower than from glucose. Therefore, no suitable natural organisms are available for pentose fermentation, and efforts have therefore been made to construct recombinant strains of bacteria and yeast that fulfill the requirements of industrial lignocellulose fermentation (see below; for review on bacteria, see Hahn-Hägerdal et al., 2006; and Peralta-Yahya and Keasling, 2010).

11.1.1 Yeast Strain Improvement

Despite the attempts to produce bioethanol with bacteria and other yeasts, *S. cerevisiae* has still remained the prime organism for this purpose. Therefore, attempts to overcome the inability of it to utilize xylose for growth or fermentation have been a major research subject. A starting point for this was the finding that *S. cerevisiae* can metabolize D-xylulose, an intermediate of the fungal xylose catabolic pathway (see Chapter 6), thus offering genetic engineering strategies for development of xylose-fermentating strains of *S. cerevisiae.*

Engineering Yeast toward Xylose Utilization

The utilization of xylose was obviously the first step that was tackled by metabolic engineering. Because of the existence of two different metabolic pathways for xylose utilization in fungi and bacteria (see Chapter 6), two different strategies have been pursued, that is, the introduction of the bacterial pentose catabolic genes and the introduction of the fungal genes, respectively. While the latter would be a logical way (in view of the fact that yeast are also fungi), the "fungal" pathway has the drawback that the two redox reactions that catalyze the first two steps in xylose catabolism utilize different coenzymes: the reduction of D-xylose to xylitol is usually catalyzed by an NADPH-dependent xylose reductase (XR; EC 1.1.1.21), whereas the oxidation of xylitol to D-xylulose occurs by an NAD^+-dependent xylitol dehydrogenase (XDH; EC 1.1.1.9; see Chapter 6). Thus, the first two reactions of the fungal xylose catabolic pathway lead to a shift in the NADPH/($NADP^+$ + NADPH ratio) (anabolic reduction charge) on the expense of the NADH/(NAD^+ + NADH ratio) (catabolic reaction charge). Since *S. cerevisiae* does not possess a transhydrogenase for conversion of NADH to NADPH (Bruinenberg et al., 1984), this results in cytosolic NADH accumulation. Interestingly, although *S. cerevisiae* cannot utilize xylose, the genes encoding XR (*YHR104w*, *GRE3*), XDH (*YLR070c*, *ScXYL2*), and xylulose kinase (*XKS1*) are present in its genome. Yet their expression levels under normal conditions are too low to allow for xylose utilization, but even an overexpression under strong promoters leads only to very slow growth on xylose (Toivari et al., 2004), suggesting that the function of these genes has been lost since a considerable time of evolution.

Successful engineering of *S. cerevisiae* for anaerobic xylose fermentation was first demonstrated by the introduction and expression of the *P. stipitis XYL1* and *XYL2* genes encoding XR and XDH, respectively, in it (Rizzi et al., 1988, 1989). *P. stipitis* can ferment xylose to ethanol at yields close to the theoretical maximum under anaerobic conditions, but xylose utilization by *P. stipitis* requires oxygen and it is also characterized by a slower sugar consumption rate compared to *S. cerevisiae* (Van Vleet and Jeffries, 2009). Despite the successful expression of the XR- and XDH-encoding genes in *S. cerevisiae*, the respective strains produced a considerable amount of xylitol in addition to ethanol, which reduces the yield of the latter. This property was already known also from natural xylose-fermenting yeasts and is due to the aforementioned difference in coenzyme specificities between the XR and XDH. Therefore, a change in this coenzyme specificity could theoretically eliminate or at least reduce xylitol formation. In fact, this was demonstrated in *P. stipitis*, which is one of the few yeasts that also possess an NADH-specific XR in addition to the major NADPH-specific XR, and consequently excretes less xylitol during xylose fermentation. In addition, the rate of xylose utilization by these recombinant *S. cerevisiae* trains was generally only between 5% and 10% of that of glucose (Hahn-Hägerdal et al., 2007).

Consequently, additional metabolic engineering strategies have attempted to improve xylose fermentation in the XR- and XDH-expressing *S. cerevisiae* strains. One aspect was to increase the activity of XDH relative to XR, based on the rationale that a lower activity of the second enzyme may present a bottleneck, and this strategy was indeed successful (Matsushika et al., 2009). Another strategy was to delete the gene for the endogenous, strictly NADPH-specific Gre1p aldose reductase of *S. cerevisiae*. While this indeed led to decreased xylitol excretion, the respective strains also showed reduced biomass formation, probably because of a function of *GRE3* in stress tolerance. Deletion of the *ALD6* gene encoding an NADPH-dependent aldehyde dehydrogenase was more successful in improving ethanol yields and xylose fermentation rates (Sonderegger et al., 2004), likely because this enzyme can also convert xylose to xylitol.

Another promising approach for reducing xylitol excretion and enhancing ethanol yield targeted the coenzyme specificity of XR and/or XDH. Jeppsson et al. (2006) were the first that reported enhanced ethanol production and decreased xylitol formation in a XDH-containing strain that expressed a mutated version of XR whose affinity for NADPH was reduced by a K270M amino acid exchange. Similarly, the introduction of a K274R-N276D *Candida tropicalis* XR double mutant into *S. cerevisiae* showed enhanced ethanol production with simultaneously decreased xylitol formation (Petschacher and Nidetzky, 2008). In addition, mutants of *P. stipitis* XDH that displayed reversed coenzyme specificity ($NADP^+$ instead of NAD+) have been prepared. One of these quadruple mutants (D207A/I208R/F209S/N211R) showed more than 4,500-fold higher values of k_{cat}/K_m for $NADP^+$ than the wild-type enzyme and similar values of k_{cat}/K_m for NAD^+ as the wild-type and increased ethanol production from xylose on the expense of xylitol excretion.

The expression of a bacterial xylose isomerase (XI) gene in *S. cerevisiae* appeared as an obvious solution to bypass the xylitol production problem. XI requires no redox cofactors and therefore does not cause an intracellular redox imbalance during xylose fermentation. However, this strategy resulted in several unforeseen problems: expression of the XI genes from various bacteria generated either inactive proteins or proteins that were insoluble and catalytically inactive. Only XI genes from the archaebacterium *Thermus thermophilus* and from the anaerobic fungal genera *Piromyces* and *Orpinomyces* were functionally expressed in *S. cerevisiae* at high levels. However, while these strains accumulated much less xylitol, their growth on xylose and the concomitant ethanol production rate was very slow (Karhumaa et al., 2007).

Metabolic Fine Tuning of Xylose Utilization in Yeast

Since the introduction of xylose-to-xylulose conversion pathways in *S. cerevisiae* alone did not lead to strains that were able to ferment xylose to ethanol in high yields, other strategies

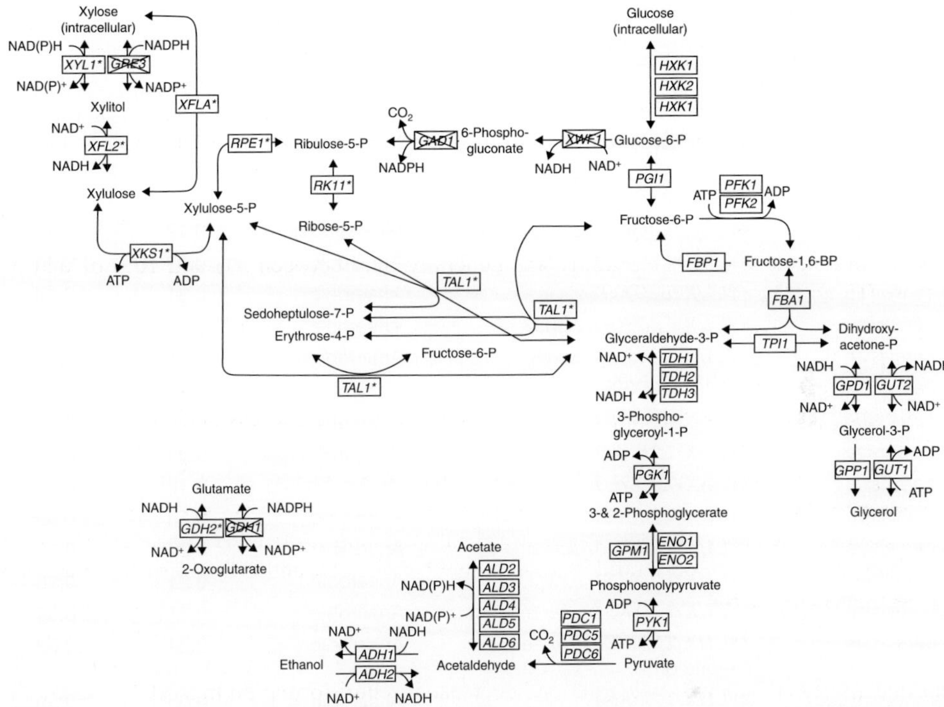

Figure 11.2. Overview of the metabolic pathway in *S. cerevisiae* engineered for improved xylose fermentation. The genes marked with an asterisk have been overexpressed; crossed genes have been deleted. *ADH1* and *ADH2*, alcohol dehydrogenase; *ALD2*, *ALD3*, *ALD4*, *ALD5*, and *ALD6*, aldehyde dehydrogenase; *ENO1* and *ENO2*, enolase; *FBA1*, fructose 1,6-bisphosphate aldolase; *FBP1*, fructose-1,6-bisphosphatase; *GDH1* and *GDH2*, glutamate dehydrogenase; *GDP1*, glycerol-3-phosphate dehydrogenase; *GLK1*, glucokinase; *GND1*, 6-phosphogluconate dehydrogenase; *GPM1*, phosphoglycerate mutase; *GPP1*, glycerol-3-phosphatase; *GRE3*, aldose reductase; *GUT1*, glycerol kinase; *GUT2*, glycerol-3-phosphate dehydrogenase; *HXK1* and *HXK2*, hexokinase; *PDC1*, *PDC5*, and *PDC6*, pyruvate decarboxylase; *PGI1*, phosphoglucose isomerase; *PGK1*, 3-phosphoglycerate kinase; *PFK1* and *PFK2*, phosphofructokinase; *PYK1*, pyruvate kinase; *RKI1*, ribose-5-phosphate ketol-isomerase; *RPE1*, ribulose-5-phosphate 3-epimerase; *TAL1*, transaldolase; *TDH1*, *TDH2*, and *TDH3*, glyceraldehyde-3-phosphate dehydrogenase; *TKL1*, transketolase; *TPI1*, triose phosphate isomerase; *XKS1*, xylulokinase; *XYL1*, xylose reductase; *XYL2*, xylitol dehydrogenase; *XYLA*, xylose isomerase; *ZWF1*, glucose-6-phosphate dehydrogenase. (Reprinted from Matsushikal, 2009, with permission from Springer.)

for metabolic engineering and genetic modification were sought, which would enhance xylose fermentation. These included engineering xylose transport, overexpression of native xylulokinase (XK), changing the intracellular redox balance, and enhancing the flux through pentose phosphate pathway (Figure 11.2). The ethanol and xylitol yields from xylose that resulted from these manipulations are summarized in Table 11.1.

Solute transport over the plasma membrane is often a limiting point in the utilization of carbohydrates (Rejenga et al., 2001), and xylose transport was thus speculated to be of possible importance for successful ethanol production from xylose. In fact, *S. cerevisiae* lacks a xylose-specific transporter and takes up xylose by facilitated diffusion using nonspecific hexose transporters of the *HXT* gene family (mainly Hxt5p and Hxt7p), which are the only

Table 11.1. Batch fermentation performance of recombinant yeast strains in media containing xylose as a sole carbon source.

Strain	Modified Genes	Condition	Y_E (g/g)	Y_X (g/g)
S. diastaticus/S. uvarum	*XYL1/XYL2/XKS1*	Fermentative	0.30	0.08
S. cerevisiae	*XYL1/XYL2/XKS1*	Aerobic	0.06	0.06
S. cerevisiae	*XYL1/XYL2/XKS1*	Microaerobic	0.12	0.42
S. cerevisiae	*XYL1/XYL2/XKS1*	Anaerobic	0.09	0.41
S. cerevisiae	*XYL1/XYL2/XKS1*	Anaerobic	0.31	0.29
S. cerevisiae	*XYL1/XYL2/XKS1/Δgnd1*	Anaerobic	0.38	0.13
S. cerevisiae	*XYL1/XYL2/XKS1/Δzwf1*	Anaerobic	0.41	0.05
S. cerevisiae	*XYL1/XYL2/XKS1*	Anaerobic	0.28	0.34
S. cerevisiae	*XYL1/XYL2/XKS1/CTH*	Anaerobic	0.28	0.30
S. cerevisiae	*XYL1/XYL2/XYL3*	Oxygen limited	0.09	0.33
S. cerevisiae	*XYL1/XYL2/XYL3*	Oxygen limited	0.12	0.27
S. cerevisiae	*XYL1/XYL2/XKS1*	Anaerobic	0.18	0.53
S. cerevisiae	*XYL1/XYL2/XKS1/GDP1*	Anaerobic	0.23	0.48
S. cerevisiae	*XYL1/XYL2/XKS1/Δzwf1*	Anaerobic	0.24	0.28
S. cerevisiae	*XYL1/XYL2/XKS1/GDP1/Δzwf1*	Anaerobic	0.41	0.34
S. cerevisiae	*XYL1/XYL2/XKS1*, xylose adaptation	Anaerobic	0.24	0.32
Saccharomyces USM21	*XYL1/XYL2/XKS1*	Anaerobic	0.05	0.59
S. cerevisiae	*XYL1/XYL2/XKS1*, random mutagenesis	Anaerobic	0.18	0.25
S. cerevisiae	$XYLA^P$, xylose adaptation	Anaerobic	0.42	0.021
S. cerevisiae	*XYL1/XYL2/XKS1/ΔGRE3*	Oxygen limited	0.46	0.09
S. cerevisiae	*XYL1/XYL2/XKS1/GRE3*	Oxygen limited	0.46	0.18
S. cerevisiae	$XYLA^T$ */XKS1/TAL1/TKL1/RKI1/RPE1/ΔGRE3*, xylose adaptation	Oxygen limited	0.29	0.23
S. cerevisiae	$XYLA^P$ */XKS1/TAL1/TKL1/RKI1/RPE1/ΔGRE3*	Anaerobic	0.43	0.003
S. cerevisiae	$XYLA^P$ */XKS1/TAL1/TKL1/RKI1/RPE1/ΔGRE3*, glucose and xylose adaptation	Anaerobic	0.41	0.001
S. cerevisiae	*XYL1(K270M one copy)/XYL2/XKS1*	Anaerobic	0.36	0.17
S. cerevisiae	*XYL1(K270M two copies)/XYL2/XKS1*	Anaerobic	0.31	0.09
S. cerevisiae	*XYL1/XYL2/XKS1/TAL1/TKL1/RKI1/RPE1/ΔGRE3*	Anaerobic	0.33	0.22
S. cerevisiae	$XYLA^P$ */XKS1/TAL1/TKL1/RKI1/RPE1/ΔGRE3*	Anaerobic	0.43	0.04
S. cerevisiae	*XYL1/XYL2/XKS1*, random mutagenesis	Anaerobic	0.34	0.29

Table 11.1. *(Continued)*

Strain	Modified Genes	Condition	Y_E (g/g)	Y_X (g/g)
S. cerevisiae	*XYL1/XYL2/XYL3*	Aerobic	0.13	0.00
S. cerevisiae	*XYL1/XYL2/XYL3/Δpho13*	Aerobic	0.25	0.04
S. cerevisiae	*XYL1/XYL2/XYL3*	Low oxygen	0.066	0.007
S. cerevisiae	*XYL1/XYL2/XYL3/Δpho13*	Low oxygen	0.315	0.006
S. cerevisiae	*CtXR(K274-N276D)/GmXDH/XKS1*	Oxygen limited	0.34	0.17
S. cerevisiae	*CtXR(K274/N276D)/GmXDH/XKS1*	Anaerobic	0.34	0.19
S. cerevisiae	*XYL1/XYL2(D207A/I208R/F209S/N211R)/XKS1*	Anaerobic	0.36	0.06
S. cerevisiae	$XYLA^O$ */XKS1/SUT1*	Anaerobic	0.39	0.08
S. cerevisiae	$XYLA^O$ */XKS1/SUT1*, xylose adaptation	Anaerobic	0.43	0.08
S. cerevisiae	$XYLA^O$ */XKS1/SUT1*, xylose adaptation	Anaerobic	0.35	0.16
S. cerevisiae	*XYL1/XYL2/XKS1*	Anaerobic	0.35	0.048
S. cerevisiae	*XYL1/XYL2(D207A/I208R/F209S/N211R)/XKS1*	Anaerobic	0.37	0.038

Modified from Matsushika et al., 2009; Y_E ethanol yield, Y_X xylitol yield.
XYL1, xylose reductase from *P. stipitis; CtXR*, xylose reductase from *C. tenuis; XYL2*, xylitol dehydrogenase from *P. stipitis; GmXDH*, xylose reductase from *Galactocandida mastotermitis;* $XYLA^T$, xylose isomerase from *Thermus thermophilus;* $XYLA^P$, xylose isomerase from *Piromyces* sp.; $XYLA^O$, xylose isomerase from *Orpinomyces* sp.; *XYL3*, xylulokinase from *P. stipitis; XKS1*, xylulokinase from *S. cerevisiae; GRE3*, aldose reductase from *S. cerevisiae; TAL1*, transaldolase from *S. cerevisiae; TKL1*, transketolase from *S. cerevisiae; RKI1*, ribose-5-phosphate ketol-isomerase from *S. cerevisiae; RPE1*, ribulose-5-phosphate-3-epimerase from *S. cerevisiae; gnd1*, 6-phosphogluconate dehydrogenase from *S. cerevisiae; zwf1*, glucose-6-phosphate dehydrogenase from *S. cerevisiae; CTH*, cytoplasmic transhydrogenase from *Azotobacter vinelandii; GDP1*, glyceraldehyde-3-phosphate dehydrogenase from *Kluyveromyces lactis; pho13*, alkaline phosphatase from *S. cerevisiae; SUT1*, sugar transporter from *P. stipitis.*

HXT transporters expressed on xylose (Sedlak and Ho, 2004). These transporters have a lower affinity for xylose than for glucose, and the transport of xylose is therefore competitively inhibited by glucose. In a lignocelluloses hydrolysate, this implies that xylose is only consumed after depletion of glucose. However, overexpression of *HXT7* or introduction of a xylose transporter from *T. reesei* did not produce a positive result (Matsushika et al., 2009), and thus the role of the xylose transporters—if any—is still unclear.

XK is another very obvious candidate for improving ethanol formation from xylose, because its activity was believed to limit xylulose catabolism and thus xylose fermentation (Ho et al., 1998). Consequently, several groups constructed *S. cerevisiae* strains that simultaneously overexpressed its own XK and the XR and XDH genes of *P. stipitis*. Although this led to an improvement of the ethanol yield from D-xylose, xylitol remained being a major by-product. In addition, the optimal XK expression level that results in effective utilization of xylose is still unclear and it seems that high levels of XK result in accumulation of X5P and depletion of ATP, both of which impair metabolism (Jin et al., 2003). Thus, only a fine-tuned overexpression of

XK in *S. cerevisiae* will lead to improved fermentation of xylose to ethanol. Ni et al. (2007) reported that spontaneous or chemically induced mutants of engineered *S. cerevisiae* strains that overcame the growth inhibition caused by overexpression of XK exhibit an increased expression of the transaldolase gene *TAL1* and a deletion of the *PHO13* gene encoding an alkaline phosphatase. Indeed, deletion of *PHO13* in a xylose-utilizing, XK-expressing *S. cerevisiae* strain also improved growth on xylose and increases ethanol production from xylose (Ni et al., 2007; Van Vleet and Jeffries, 2009). The role of Pho13p is unknown, but it may be involved in the dephosphorylation of intermediates in the fungal pentose catabolic pathway.

To counteract the impairment of the intracellular redox balance caused by the different coenzyme specificity of XR and XDH, Jeppsson et al. (2002) constructed recombinant *S. cerevisiae* strains defective in 6-phosphogluconate dehydrogenase and glucose-6-phosphate dehydrogenase, respectively. While this resulted in increased ethanol yields, the rates of xylose consumption were strongly lowered, consistent with the requirement of XR for NADPH. To enable an exchange between NADPH and NADH and compensate for the lack of a transhydrogenase in *S. cerevisiae* (see above), a bacterial transhydrogenase was expressed in *S. cerevisiae*. While this indeed reduced xylitol production, it increases the by-production of glycerol (Hou et al., 2009), indicating a more general problem with the redox balance during xylose consumption. The significance of redox reactions at the level of glycerol-3-phosphate was also demonstrated by several other authors (reviewed by Matsushika et al., 2009): redirection of the flux toward ethanol was achieved by modifying the ammonia assimilation pathway or exchanging the NADPH-dependent glutamate dehydrogenase gene *GDH1* against the NADH-dependent isoenzyme gene *GDH2*. Thus, a subtle manipulation of the intracellular NADPH/NAD ratio is necessary to redirect the flux toward ethanol formation.

The nonoxidative part of the pentose phosphate pathway is essential for the introduction of xylulose into glycolysis and thus ethanol formation, but the flux through it in *S. cerevisiae* is slow compared to other yeast species (Fiaux et al., 2003). Consequently, enhancement of this flux by the overexpression of its enzymes (e.g., transaldolase, EC 2.2.1.2; transketolase, EC 2.2.1.1; ribulose-5-phosphate 3-epimerase, EC 5.1.3.1; and ribose-5-phosphate ketol-isomerase, EC 5.3.1.6) has been attempted in this direction. Of these, only overexpression of the transaldolase gene *TAL1* improved growth on xylose (Jin et al., 2005). Interestingly, the overexpression of the all four nonoxidative PPP genes in a recombinant *S. cerevisiae* strain that also expressed XR and XDH from *P. stipitis* and exhibited enhanced expression of XK did not improve xylose consumption. In contrast, deletion of *GRE3* in a strain overexpressing all four pentose phosphate pathway enzymes and a *Piromyces* XI gene improved growth on xylose (Matsushika et al., 2009). Similar results were also obtained in a strain with the same modifications but exhibiting high activity of both XR and XDH instead of XI. This supports the assumption that a simultaneous increase in the activity converting xylose to xylulose and the PPP activity required for xylulose metabolism is essential for efficient xylose utilization in *S. cerevisiae* (Matsushika et al., 2009).

Additional improvements of engineered strains could be achieved by evolutionary adaptation. This is an alternative technique to obtain improved producer strains, in which strains are grown for a prolonged period in chemostat cultures at low growth rates, which allows spontaneous mutations to be observed. For example, Sonderegger and Sauer (2003) obtained strains of *S. cerevisiae* that were able to anaerobically grow on xylose as the sole carbon source and also exhibited increased xylose consumption rates. This technique has also been used to select for improved growth rates on xylose and xylose fermentation of XI-carrying strains.

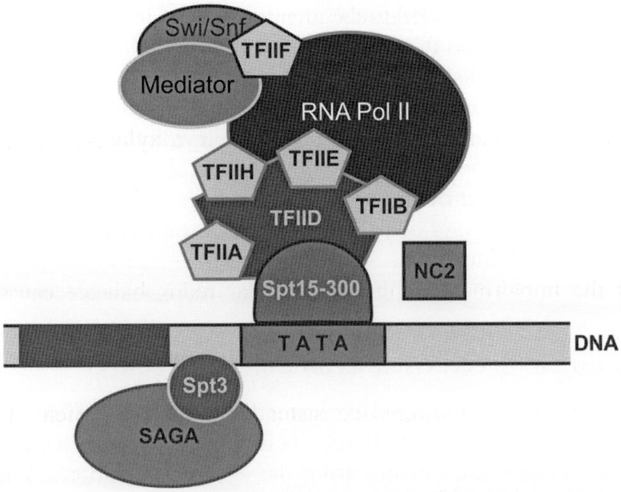

Figure 11.3. A schematic representation of the global transcriptional machinery of yeast.

Engineering Transcriptional Factors

A completely different approach toward improved ethanol formation from xylose is the engineering of transcriptional factors that creates a more global change in metabolism, and this strategy has been applied successfully in both *Escherichia coli* and *S. cerevisiae* for improving biofuels production. The best known example is the process known as global transcriptional machinery engineering originally developed by Stephanopoulos and colleagues (Alper et al., 2006). Using error-prone PCR, they focused on the transcription factor TFIID, which consists of TATA-binding protein and 14 other associated proteins that form the transcription initiation complex (Figure 11.3). Among these, Spt15 is a TATA-binding protein that is involved in the transcription of most of the mRNA gene in the genome. A combined effect of three separate mutations in Spt15 (F177S, Y195H, and K218R) indeed lead to improved tolerance of *S. cerevisiae* to ethanol (Alper et al., 2006). This work provided an alternative to adaptive evolution and demonstrated that complex phenotypic improvements were achievable by concentrating on a single cellular protein.

Engineering Inhibitor Tolerance

The presence of inhibitors in the lignocelluloses hydrolysates poses another problem. Methods for the detoxification by chemical or physical methods before the fermentation have been developed. The most successful one is treatment with $Ca(OH)_2$ or ion exchangers, but this results in increased process cost and also losses of fermentable sugar (Nilvebrant et al., 2003). Considerable work has therefore been dedicated to the mechanism of how *S. cerevisiae* reacts to the presence of the inhibitory components and how it could be selected or engineered to tolerate or detoxify them (reviewed in detail by Liu, 2011). In fact, strains that are capable of converting furfural and 5-(hydroxymethyl)-2-furaldehyde (HMF) into the corresponding alcohols while not altering the yields of ethanol produced were found. This finding can be used by selecting an appropriate feed rate of hydrolysate so that the capacity of the enzymes performing conversion of the inhibitors is not overloaded (Taherzadeh et al., 2000). However, the discovery of this principle also led to the identification of numerous enzymes that exhibit aldehyde reductase activities and thus contribute to the detoxification of furfural, HMF, and also other inhibitory components such as cinnamaldehyde and vanillins (that are derived from the lignin). Heer et al. (2009) used a system's biological approach to demonstrate that *ADH7*, ORF *YKL071W*, and four other reductases are associated with the yeast resistance to furfural challenge. The expression of these genes is under the control of the *YAP1* transcription factor, which acts as a general activator of stress response (Rodrigues-Pousada et al., 2004). Genetic engineering of *S. cerevisiae* toward detoxification of these inhibitors is therefore possible and subject to several current attempts.

11.1.2 Alternatives to S. cerevisiae *as an Ethanol Producer*

While ethanol production by *S. cerevisiae* has a long and successful history with starch hydrolysates or molasses as a carbon source, its disadvantages in ethanol production from lignocelluloses hydrolysates are obvious from the above descriptions. While considerable efforts in genetic engineering *S. cerevisiae* have been made (see above), several laboratories and industries have looked for other organisms that would be superior to *S. cerevisiae* in the one or the other limiting aspect and eventually be easier to improve by genetic engineering. Obviously, no known organism that combines all the properties necessary to be a sustainable bioalcohol producer exist, and thus breeding technologies, genetic engineering, and the search for undiscovered species are thus performed to identify a microorganism exhibiting high alcohol productivities and yields, converting all lignocellulosic sugars or are even able to use carbon dioxide or monoxide, being highly resistant to inhibitors and fermentation products, and easy to cultivate in huge bioreactors (Warnick et al., 2002).

P. stipitis is one of the few naturally xylose-fermenting yeasts and thus considered as a promising candidate for ethanol production from lignocellulosic materials. Its advantages lie in the facts that it is able to ferment glucose, galactose, and cellobiose, and also possesses various cellulase and hemicellulase genes (Warnick et al., 2002). Further, its aldose reductase can use both NADPH as well as NADH as coenzymes, and it also contains an alternative (SHAM-sensitive) respiratory bypass to the respiratory chain, which can resolve cofactor imbalances (Jeppsson et al., 1995). Consequently, 61 g/L ethanol could be obtained from 150 g/L xylose (Slininger et al., 2006). However, *P. stipitis* also has a number of significant drawbacks: it does not grow at temperatures >30°C, and its sugar consumption rates are low, probably due to the fact that they are mainly proton symporters. This permanent need for proton formation by the plasma membrane-bound ATPase could pose energetic constraints to this yeast (Warnick et al., 2002). In addition, high ethanol yields are only obtained when microaerophilic conditions and complex nitrogen sources are used (Agbogbo and Coward-Kelly, 2008).

There are two other yeasts that are capable of fermenting at elevated temperatures and would thus also be compatible with the requirements for the simultaneous saccharification and fermentation (SSF) process: *Kluyveromyces marxianus* and *Hansenula polymorpha*. *K. marxianus* has the attractive property of being able to grow and ferment at up to 52°C. Currently, many isolates are investigated for their ability to grow on and ferment different sugars or substrates, but fermentation of xylose to ethanol by the available isolates is still very poor. This is due to the absence of the Crabtree effect from this species, thus making oxygen limitation essential for fermentation to occur (Bellaver et al., 2004). Ethanol production by *H. polymorpha* suffers from the same fact and is even lower.

Zymomonas mobilis is a gram-negative bacterium, originally occurring in agaves, where it was observed to ferment sugars to ethanol (the biological basis for tequila production). This bacterium has a high potential for ethanol production and also a very high ethanol tolerance (12%, w/v) and has therefore attracted attention already during the development of ethanol fuel technology from molasses already 30 years ago. Moreover, its specific ethanol productivity is 2.5-fold higher than that of *Saccharomyces* sp., and its ethanol yield approaches 97% of the theoretical maximum at 30°C (Rogers et al., 1982). This is due to the fact that *Z. mobilis*—like other *Pseudomonas* spp.—is unable to perform glycolysis and therefore metabolizes glucose via the Entner-Doudoroff pathway, which produces only 1 mol of ATP per mol of hexose (Kalnenieks, 2006). However, this advantage of *Z. mobilis* is only restricted to the fermentation of glucose, fructose, and sucrose because it is unable to utilize or ferment pentoses. Consequently, the necessary catabolic genes (XI, XK, transaldolase and transketolase for xylose; arabinose isomerase, ribulokinase, ribulose-5-phosphate-4-epimerase, transaldolase, and transketolase for arabinose) were transformed into *Z. mobilis* (Zhang et al., 1995; Deanda et al., 1996). Although the utilization of the different sugars was observed to occur in sequential phases, the strains produced reasonable amounts of ethanol.

E. coli is already used in several industrial processes, mainly for production of recombinant proteins, and the experience in its handling in industrial fermentations recommended it as a potential recombinant producer of biofuel. In fact, introduction of the pyruvate decarboxylase and alcohol dehydrogenase genes from *Z. mobilis* enabled it to produce ethanol (Ingram et al., 1998). However, the drawback of using *E. coli* is its only low tolerance for ethanol, and the stringent operation of carbon catabolite repression that prevents the simultaneous consumption of all of the lignocellulose-derived sugars (Jojima et al., 2010). Attempts to bypass this circuit by introducing heterologous permeases have been made, but xylose consumption still lagged behind that of glucose (Ren et al., 2009). One possible reason for this was identified to be the methylglyoxal pathway, which removes carbon from the routes of carbohydrate catabolism at the stage of dihydroxyacetone phosphate. Methylglyoxal, the product of the first reaction, is believed to function as a general inhibitor of sugar metabolism. In fact, deletion of the methylglyoxal synthase gene (*mgsA*) enabled *E. coli* to utilize a mixture of mannose, glucose, arabinose, xylose, and galactose (2% each) simultaneously (Yomano et al., 2009). A constitutively active mutant of the cyclic AMP receptor protein has also been used to engineer *E. coli* for simultaneous utilization of glucose and D-xylose (Cirino et al., 2006).

Clostridium phytofermentans is a gram-positive, anaerobic, and spore-forming prokaryote isolated from forest soil (Warnick et al., 2002). Like many other *Clostridia*, it is capable of growing on cellulose, pectin, and xylan by producing ethanol and minor amounts of acetate, formate, lactate, and H_2 as products of anaerobic fermentation. What makes this organism interesting is its huge number of cellulose an hemicellulase hydrolyzing enzymes, which outnumbers all other clostridial genomes sequenced to date (http://genome.jgi-psf.org/finished_microbes/clopi/clopi.home.html). It has therefore been considered a model organism for consolidated bioprocessing (CBP) processes.

Also *Corynebacterium glutamicum*, a well-known industrial producer of glutamic acid and other amino acids (Hermann, 2003), has been investigated for its potential suitability for ethanol production. Its attraction would be the high resistance to phenols and furans present in pretreated lignocellulosic hydrolysates. Using the same strategy as for *E. coli* (Ingram et al., 1998), Inui et al. (2004) introduced the *Z. mobilis* genes encoding pyruvate decarboxylase and alcohol dehydrogenase in *C. glutamicum.* The resulting strains produced ethanol at a 20% conversion yield, but with excessive accumulation of additional by-products (acetate, lactate, and succinate).

Summarizing, the current stage of knowledge seems to favor the use of *S. cerevisiae* and *Z. mobilis* for industrial lignocellulosic bioalcohol production. The yeast process is well established and simple, the organism is robust and well known, and most parts of the technology of current ethanol production plants can also be used for the next-generation technologies. To establish bioalcohol production technologies for new kinds of species seems to be difficult in the short term due to the huge dimensions of the production plants (Warnick et al., 2002).

11.1.3 Process Improvement

There is still another aspect of the lignocellulose to ethanol fermentation that is subject of considerable attempts for improvement: this is the establishment of a SSF process. Existence of such a process would considerably reduce the capital costs, because it would eliminate the end-product inhibition (and thus prolonged time) of cellulose hydrolysis because of the simultaneous uptake of the formed sugars by the yeast. Also, components that inhibit the enzymes may be detoxified by the yeast (see Chapter 10). Examples show that the ethanol concentration and the final ethanol yield are in fact increased in SSF (Galbe et al., 2007). However, the major drawback of such a process is that hydrolysis and fermentation cannot occur at their optimum conditions.

The SSF technology has further proven advantageous for the simultaneous fermentation of both the hexose and pentose sugars (simultaneous saccharification and cofermentation). The benefit comes from the fact that SSF increases the rate of glycolysis, which also results in faster fermentation of xylose and with higher yield (Ohgren et al., 2006). So far, the enzymes still need to be added to the fermentation. Quite some attempts have therefore been dedicated to create microorganisms that can form sufficiently active cellulolytic and hemicellulolytic enzymes, and are also able to form ethanol at high yield and efficacy ("CBP"; van Zyl et al., 2007). This has been done by genetically modifying cellulose-degrading bacteria such as *Clostridium* spp. toward ethanol production or by introduction of cellulolytic enzymes into ethanol-producing microbes such as yeast However, none of these strategies has so far yielded a breakthrough (van Zyl et al., 2007; Elkins et al., 2010).

11.2 *n*-Butanol

n-Butanol is a promising alternative to ethanol because it has an energy density of 29.2 MJ/L (comparable to that of gasoline, 32 MJ/L), an octane number of 87, and it can be mixed with gasoline at any percentage or even completely replace it. Further, butanol's high hydrophobicity may allow the use of existing fuel transportation and storage infrastructure without major modifications (Warnick et al., 2002).

Clostridium acetobutylicum naturally produces acetone, butanol, and ethanol with typical ratios of 3:6:1 (Dürre, 2008). The process is known since the mid of the last century as acetone-butanol-ethanol (ABE) fermentation. The production of butanol by this process declined after 1950s, and almost all butanol is now produced through petrochemical routes. Presently, the worldwide market for butanol is 370 million gallons per year and is expected to rise dramatically if butanol can be produced economically from lignocelluloses by fermentation (Adsul et al., 2011).

The ABE fermentation usually consists first of an acidogenic phase during which acetate and butyrate are produced, which are then converted to the corresponding butanol and acetone in the solventogenic phase. Several strategies have attempted to selectively improve the production of butanol, which are all hampered by the fact that efficient methods for gene manipulation are still in the infancy in the *Clostridia*. Yet Tummala et al. (2003) used an antisense RNA strategy for inactivating the expression of CoA transferase (*ctfB*) and overexpressed the alcohol-aldehyde-dehydrogenase (*aad*), which increased the *n*-butanol/acetone ratio by more than twofold. However, *Clostridia* have several important disadvantages as industrial producers that have still not been satisfactorily been solved (e.g., being strict ananerobes, slow growth rate, sensitivity to bacteriophage infection, sensitivity to *n*-butanol). While bioprocess optimizations have been shown to overcome the latter (Ezeji et al., 2007), the other disadvantages led to attempts to introduce the clostridial butanol biosynthetic pathway into yeast or yeast and *E. coli* (Steen et al., 2010).

11.3 Advanced Biofuel Alcohols

While ethanol likely can be produced on an industrial scale and with the desired economy soon, it nevertheless has some disadvantages such as incompatibility with the existing fuel infrastructure, a (relatively) low molar energy, a high vapor pressure (these both strongly influencing biofuel transport costs), and corrosivity. In view of the above-described difficulties with the production of the theoretical alternative *n*-butanol, industry is therefore developing processes for the production of so-called "advanced biofuels." This term encompasses components that would bypass at least some of the disadvantageous properties of bioethanol. The major focus thereby is on higher alcohols that are superior to ethanol in many of the above-named properties (Warnick et al., 2002). Isobutanol is currently most preferred, but 1-butanol, 2-methyl-1-butanol, 3-methyl-1-butanol, and isopentenol are also on the list of potentially interesting compounds (Connor and Liao, 2009). In addition, terpenoids, fatty-acid esters, or long-chain alkanes have been suggested to be produced by microorganisms as sustainable replacements for diesel and jetfuels (Peralta-Yahya and Keasling, 2010; Steen et al., 2010).

11.3.1 Isobutanol

Isobutanol is known to be produced by *S. cerevisiae* during catabolism of valine (Hazelwood et al., 2008), but only at low concentrations. Metabolic engineering strategies to increase isobutanol production included the simultaneous overexpression of genes encoding enzymes of the ketoisovalerate pathway: ketoacid decarboxylase and alcohol dehydrogenase (Figure 11.4). Engineering of cofactor specificities of the involved oxidoreductases was also proposed in order to separate valine degradation from fermentative conversion to isobutanol. Also, in order to avoid by-production of ethanol, pyruvate decarboxylase needs to be deleted. Finally, to reach a sufficiently high cytosolic acetyl-CoA concentrations, a bacterial pyruvate formate lyase together and a formate dehydrogenase were introduced and expressed (patent application PCT/EP2009/000181; cited in Warnick et al., 2002). Chen et al. (2011) overexpressed the *ILV2, ILV3, ILV5*, and *BAT2* genes (encoding enzymes of valine metabolism) and achieved an increase in isobutanol production in *S. cerevisiae*. Interestingly, aerobic cultivations of the *ILV2 ILV3 ILV5* overexpression strain and the reference strain showed that supplying amino acids in cultivation media gave a substantial improvement in isobutanol production for the reference strain, but not for the *ILV2 ILV3 ILV5* overexpression strain, thus implying that other constraints besides the enzyme activities for the supply of 2-ketoisovalerate may become

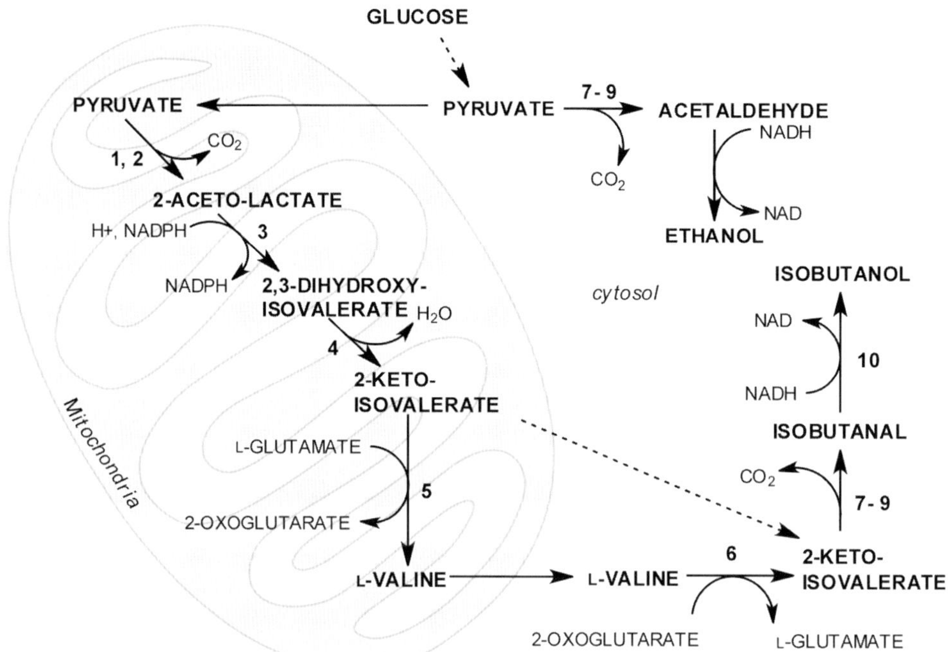

Figure 11.4. Metabolic pathway of isobutanol formation by *S. cerevisiae.* Enzymes are given by numbers: 1 and 2, acetolactate synthase (regulatory and catalytic subunit, respectively); 3, acetohydroxyacid reductoisomerase; 4, dihydroxyacid dehydratase; 5 and 6, branched-chain amino acid aminotransferase; 7–9 pyruvate decarboxylases; 10, alcohol dehydrogenase.

bottlenecks for isobutanol production after *ILV2, ILV3*, and *ILV5* have been overexpressed, most likely involving valine feedback inhibition (Chen et al., 2011).

Isobutanol—as well as some other lower alcohols—can also be produced by recombinant *E. coli* strains (Hanai et al., 2007; Atsumi et al., 2009).

11.3.2 Syngas

In contrast to the biochemical conversion of biomass into bioalcohols by enzymatical hydrolysis and fermentation, alcohols can also be produced by a combination of thermochemical and fermentative pathways. Thereby, the lignocellulose biomass is thermically converted into a mixture of carbon monoxide and hydrogen (the so-called "syngas") (Figure 11.5). Obviously, this combustion is not restricted to the cellulose and hemicelluloses part of the plant biomass but also acts on the lignin, thus having the significant advantage of converting all parts of biomass. The syngas is then converted into ethanol by either catalytic conversion or bacterial fermentation with *Clostridium ljungdahlii* and *Clostridium carboxidivorans* (Munasinghe and Khanal, 2010).

The biggest drawback for syngas to biofuel fermentations is its poor solubility in water. While laboratory fermentations can solve these problems (e.g., by growing the bacteria as a biofilm on porous substrates), the scaling up of such systems to industrial fermentations is still a challenge (Warnick et al., 2002).

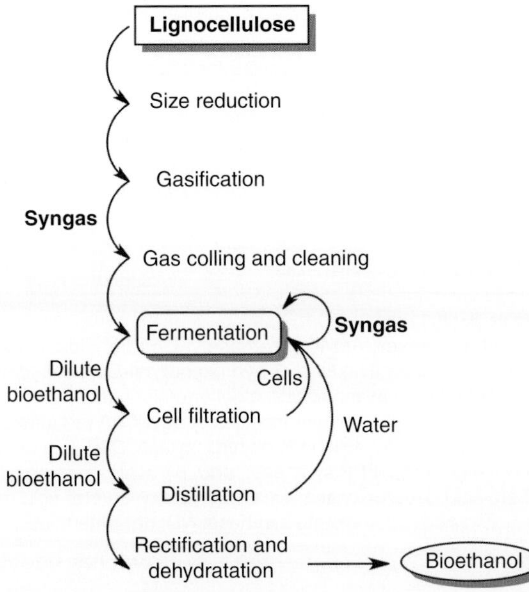

Figure 11.5. Simplified process scheme for ethanol production via syngas fermentation.

11.3.3 Fuels from Isoprenoid and Fatty Acid Pathways

Several lipophilic compounds such as the terpenoids, farnesol, and farnesene; cyclized monoterpene olefins such as limonene, pinene, sabinene, and terpinene; and fatty acid esters have been identified as biofuel targets because of their low water solubility, high energy density, and low toxicity. Keasling et al. (reviewed by Peralta-Yahya and Keasling, 2010; Zhang et al., 2011a) have undertaken extensive metabolic engineering of *E. coli* toward overproduction of these compounds. While the data are impressing, the yields are still low (Table 11.2), and production from lignocelluloses hydrolyzates has not been demonstrated yet.

11.4 Lactic Acid

Naturally occurring lactic acid bacteria, which are obligatory dependent on an anaerobic metabolism, convert hexoses to lactic acid in order to reoxidize NADH generated during glycolysis, using pyruvate as the electron acceptor. These organisms have been used for more than half of a century for the commercial production of lactic acid for different purposes, such as food preservation, food packaging, drug delivery compounds, medical implants, cosmetic ingredients, and biodegradable plastic (for review, read Doran-Peterson et al., 2008). At the end of 2011, the lactic acid global demand was estimated to be around 200,000 metric tons per year, mainly because of the growing market for poly(lactic acid) (PLA). PLA can replace various petrochemical industry-based polymers in applications ranging from packaging to fibers (Wee et al., 2006). Also, it is biocompatible and thus attractive for applications in the medical field (John et al., 2007).

Table 11.2. Advanced fuel production by engineered microbial pathway mutants.

Target Fuel	Strain	Key Enzymes Expressed*	Titer (g/L)	Yield (%)
Isopropanol	*E. coli*	*Thl, AtoAD, Adc, Adh*	143	67
Butanol	*E. coli*	*AtoB, Hbd, Crt, Ter, AdhE2*	30	70–88
Isobutanol	*E. coli*	*AlsS, IlvCD, KDC, Adh*	22	86
Farnesol	*E. coli*	*Mevalonate pathway, IspA*	0.135	ND
Fatty acid ethyl ester	*E. coli*	*TesA, FadD, AtfA, Pdc, AdhB*	0.43	9.4
Fatty alcohols	*E. coli*	*TesA, FadD, Acr1*	0.06	0.7
Alkanes	*E. coli*	*AAR, ADC*	0.3	3.5

Data modified from Zhang et al. (2011a).
**Thl/PhaA*, acetyl-CoA acetyltransferases; *AtoAD*, acetoacetyl-CoA transferases; *Adc*, acetoacetate decarboxylase; *Adh*, alcohol dehydrogenase; *Hbd*, 3-hydroxybutyryl-CoA dehydrogenase; *Crt*, crotonase; *Ter*, butyryl-CoA dehydrogenases; *AdhE2*, aldehyde/alcohol dehydrogenase; *AtoB*, thiolase; *TesA*, acyl-ACP thioesterase; *FadD*, acyl-CoA synthase; *Acr1*, acyl-CoA reductase; *AtfA*, wax-ester synthase; *AAR*, acyl-ACP reductase; *ADC*, aldehyde decarboxylase; $OleT_{JE}$, a cytochrome P450 enzyme that reduces fatty acids to alkenes; *ispA* (FPP synthase); *pdc*, pyruvate decarboxylase; *adhB*, alcohol dehydrogenase; *2-KDC*, keto acid decarboxylase; *ilvCD*, 2-hydroxy-3-ketol-acid reductoisomerase and 2,3-dihydroxy-acid hydro-lyase; *alsS*, acetolyctate synthase; ND, not determined.

Lactic acid exists in two enantiomeric forms, L-(+) and D-(−), and the ratio of the two isomeric forms is important for some of its uses, for example, polylactide polymer production (Pang et al., 2010). Lactic acid bacteria usually possess two lactate dehydrogenase-encoding genes, which are responsible for formation of either L-(+) or D-(−) lactic acid (Makarova et al., 2006), and the ratio production of these two enantiomeric forms thus depends on the level of expression of these two genes. However, it must be noted that D-lactic acid is not metabolized by animals and causes acidosis and decalcification (Wee et al., 2006), and production of the L-form is thus essential if the resulting lactic acid is to be used in any way for human or animal consumption.

One disadvantage of the use of *Lactobacillus* spp. for the production of lactic acid from lignocelluloses hydrolysates is their limited ability to ferment pentoses. Today, mainly four species (*Lactobacillus lactis*, *L. casei*, *L. plantarum*, and *L. buchneri*) are known that are capable of fermenting pentoses. However, these species have two major shortcomings: *L. lactis* utilizes a heterofermentative metabolism when catabolizing pentoses (Figure 11.6), which results in the production of an equal mixture of lactate and acetate. Some lignocellulose pretreatments (such as that with dilute acid) also result in the formation of methylglucuronoxylose from α-1,2-methylglucuronate-substituted xylan, which cannot be fermented by the *Lactobacilli*. In addition, they are usually sensitive to the presence of inhibitors such as furan, phenol compounds, and weak acids that occur in the lignocelluloses hydrolysates.

To overcome this bottleneck, Guo et al. (2010) isolated a strain of *L. brevis* and one of *L. plantarum* that exhibited the ability to utilize the various sugars present in dilute-acid hydrolysate of biomass and also showed strong resistance to inhibitors like ferulic acid and furfural. However, lactic acid production was only 39.1 g/L from an initial total sugar concentration of 56.9 g/L (xylose, 46.4 g/L; glucose, 4.0 g/L; and arabinose, 6.5 g/L).

The above-described shortcomings and the requirement of lactic acid bacteria for complex media have so far hampered the use of PLA as a renewable plastic. Consequently, other microorganisms that would not display these disadvantages were sought. Ou et al. (2011) reported on a *Bacillus coagulans* strain, a thermotolerant bacterium, that can grow and efficiently

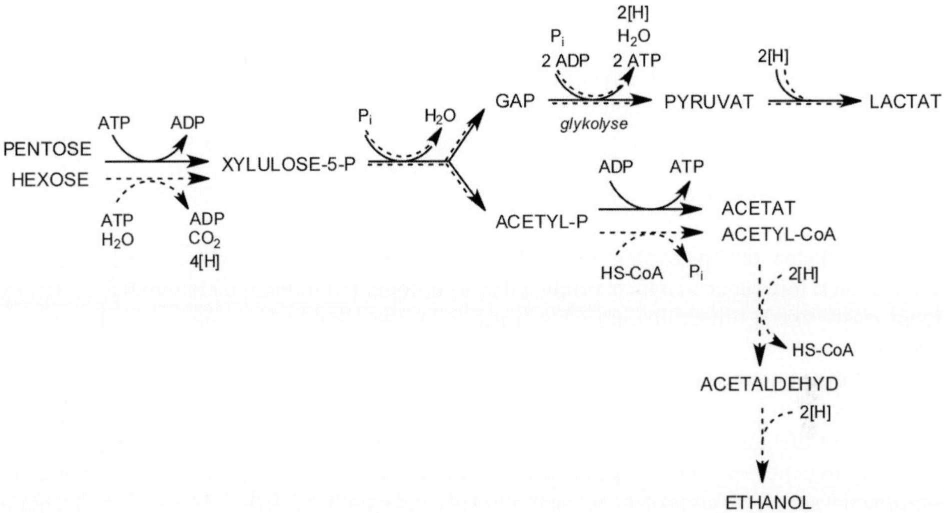

Figure 11.6. Metabolic pathway from xylose to ethanol and lactic acid by heterofermentative lactic acid bacteria.

ferment pentoses and all other sugar constituents of lignocellulosic biomass at temperature and pH conditions that also favor simultaneous enzymatic saccharification and fermentation of cellulose. The strain produced 150–180 g L^{-1} of lactic acid from xylose and glucose with minimal by-products.

Other microorganisms that have been tested for their appropriateness to accumulate high amounts of lactate include *Kluyveromyces, Saccharomyces, Pichia*, and *Rhizopus* (for review, see Doran-Peterson et al., 2008). However, most promising results have been obtained with *E. coli*, which uses the glycolytic pathway to convert hexose sugars into a mixture of acids (lactic, acetic, formic, and succinic) and ethanol (Grabar et al., 2006). In addition, it ferments the pentoses through the pentose phosphate pathway, which feeds intermediates into the EMP pathway toward homolactate production. *E. coli* has thus been engineered for production of optically pure lactic acid from lignocellulose hydrolysates by deleting competing pathways (Yu et al., 2011). *E. coli* produces only the D(−)-lactate dehydrogenase enzyme, and the resulting product is 100% D(−)-lactic acid. Zhou et al. (2003) showed that replacing the native D-*ldh* gene by a *Pediococcus* L-*ldh* gene resulted in the production of L(+)-lactic acid. Further metabolic engineering and directed evolution led to *E. coli* strains that achieved yields of lactic acid equal to or higher than the best values reported for lactic acid bacteria (Grabar et al., 2006). Similarly, Bi et al. (2009) isolated an *Enterobacter asburiae* strain that was able to efficiently ferment both methylglucuronoxylose and xylose in acid hydrolysates of sweetgum xylan, and—after deletion of the pyruvate formate lyase and the acetolactate synthase genes—produced lactate in almost 100% of the theoretical maximum yield.

11.5 Succinic Acid

The dicarboxylic acid succinate is currently used as a surfactant, an ion chelator, and as an additive in the pharmaceutical and food industries, which, however, has only a limited market

(15,000 t/year worldwide; McKinlay et al., 2007). It is currently produced chemically, but its production by fermentation is possible. The research and development of bio-based succinate production is targeting a much larger commodity chemical market (i.e., 270,000 t/year; Willke and Vorlop, 2004), because it could serve as the starting material for producing bulk chemicals such as 1,4-butanediol (a precursor to "stronger-than-steel" and biodegradable plastics), ethylene diamine disuccinate (a biodegradable chelator), diethyl succinate (a "green" solvent replacement for methylene chloride), and adipic acid (a nylon precursor; McKinlay et al., 2007). More importantly, the biochemistry of fermentative succinate production involves a CO_2 fixation step, thus providing the benefit of using a greenhouse gas as a substrate. Succinate was therefore suggested to be produced as a coproduct of industrial ethanol fermentations (Lynd et al., 2002), thus allowing CO_2 waste from the ethanol fermentation to be used as substrate.

One of the biggest obstacles to economic succinate production, however, is the significant costs for its purification, which generally involves purification of the free acid. This currently makes use of precipitation procedures, electrodialysis, and ion exchange chromatography, all of which are expensive. Less expensive methods that have been proposed would involve the addition of $Ca(OH)_2$ to control the fermentation pH and simultaneously precipitating succinate. Succinic acid would then be released from the precipitate by the addition of sulfuric acid and purified by ion exchange chromatographies. However, this would create a large amount of contaminated gypsum and thus present a problem for disposal. Another method proposed used NH_4OH for pH control, leading to diammonium succinate formation. This is then reacted with ammonium bisulfate to form ammonium sulfate and succinic acid, and the latter can be crystallized from the broth by lowering the pH below 2 (McKinlay et al., 2007).

Several microorganisms have been investigated as potential succinate producers, but best results are obtained today by engineered strains of *E. coli*. Their construction provides an impressive example of drastic manipulation of bacterial metabolism based on the knowledge of complementary pathways: today, engineered strains produce succinate at concentrations comparable to those attained by natural succinate producers and with high succinate product ratios (McKinlay et al., 2007).

E. coli naturally produces succinate by the reductive branch of the tricarboxylic acid (TCA) cycle but only as a minor product of a mixed acid fermentation (Figure 11.7). Succinate production was originally enhanced by overexpressing the genes encoding phosphoenol-pyruvate kinase (PEPCK) and pyruvate carboxylase (PYC). Subsequent attempts to overproduce pantothenate kinase and exogenously supplying pantothenic acid led to an increase in intracellular acetyl-CoA concentrations and (because of the activation of PEPCK and PYC by acetyl-CoA) also an increase in succinate production. A further significant breakthrough in succinate production by *E. coli* was the isolation of a strain with a spontaneous mutation in a gene that encodes a component of the glucose-specific phosphotransferase system that is necessary for fermentative growth on glucose. The resulting mutant produced succinate as the main fermentation product and consumed glucose and other sugars simultaneously (Chatterjee et al., 2001).

The yield of succinate is dependent on the stochiometry of NADH formation: if all the NADH for formation of succinate via the reductive TCA branch comes from glycolysis, a total of 1 mol succinate per mol glucose can be achieved. Vemuri et al. (2002), therefore, investigated a new route for succinate production involving the glyoxylate shunt, which is activated during aerobic growth. The glyoxylate shunt uses less NADH than the TCA cycle and thus safes reducing equivalents that could theoretically allow for a higher succinate yield (Figure 11.7). Vemuri et al. (2002) started with an aerobic growth phase and then shifted to an

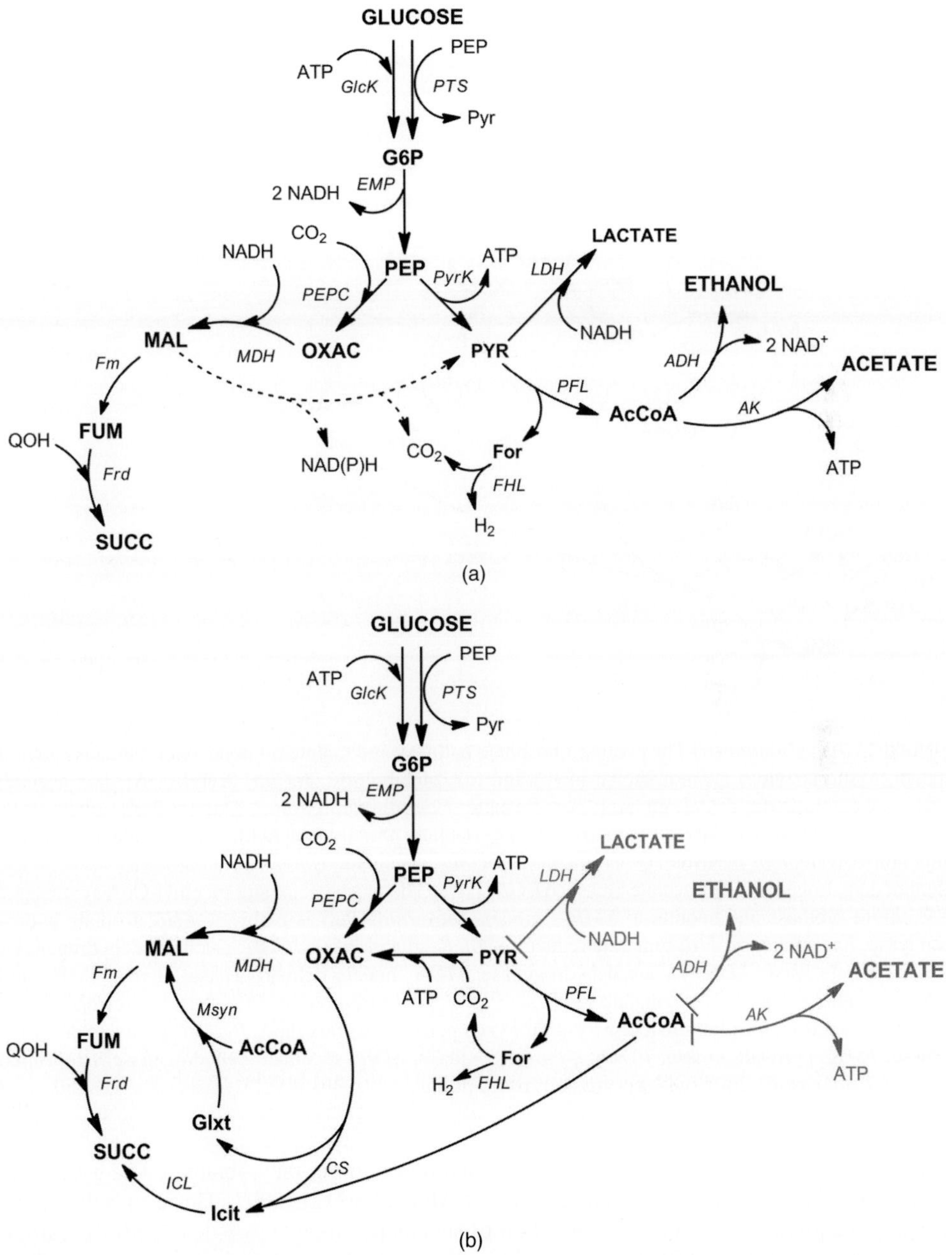

Figure 11.7. (a) Simplified metabolic map of wild-type *E. coli* mixed acid fermentation, (b) simplified metabolic map of engineered *E. coli* anaerobic succinate-producing metabolism, and (c) simplified metabolic map of engineered *E. coli* aerobic succinate-producing metabolism. Crossed-out gray arrows indicate missing enzyme activities due to gene deletions. The double-lined arrow indicates introduced or enhanced activity. Dotted lines indicate enzyme activities with unknown *in vivo* contributions.

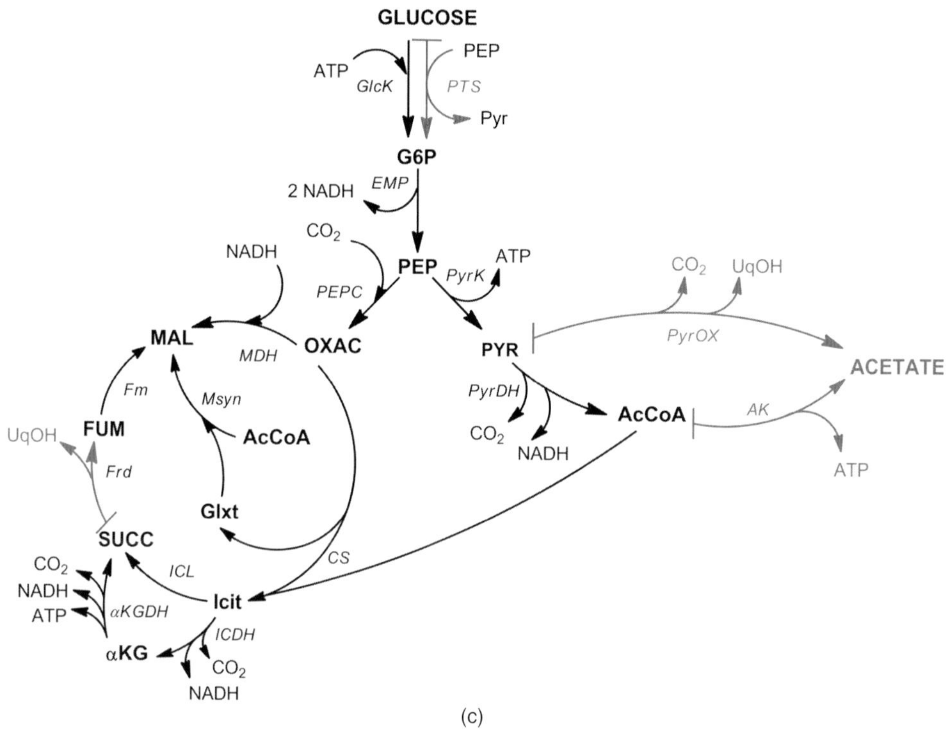

Figure 11.7. (*Continued*) The pentose phosphate pathway and malate:quinone oxidoreductase are not shown. Malic enzymes are not shown in (b) and (c). Metabolites: *AcCoA*, acetyl-CoA; *Ace*, acetate; *αKG*, α-ketoglutarate; *EtOH*, ethanol; *For*, formate; *Fum*, fumarate; *G6P*, glucose-6-phosphate; *Glc*, glucose; *Glxt*, glyoxylate; *Icit*, citrate/isocitrate; *Lac*, lactate; *Mal*, malate; *OAA*, oxaloacetate; *PEP*, phosphoenolpyruvate; *Pyr*, pyruvate; *QOH*, menaquinol; *Suc*, succinate; *UqOH*, ubiquinol. Reactions: *ADH*, alcohol dehydrogenase; *AK*, acetate kinase; *αKGDH*, αKG dehydrogenase and succinyl-CoA synthetase; *CS*, citrate synthase and aconitase; *EMP*, Embden–Meyerhoff–Parnas pathway; *FHL*, formate hydrogen lyase; *Fm*, fumarase; *Frd*, fumarate reductase; *GlcK*, glucokinase; *ICDH*, isocitrate dehydrogenase; *ICL*, isocitrate lyase; *LDH*, lactate dehydrogenase; *MDH*, malate dehydrogenase; *Menz*, NAD^+- and $NADP^+$-malic enzymes; *Msyn*, malate synthase; *PEPC*, PEP carboxylase; *PFL*, pyruvate formate-lyase; *PTS*, PEP:glucose phosphotransferase system; *PyrC*, pyruvate carboxylase; *PyrDH*, pyruvate dehydrogenase; *PyrK*, pyruvate kinase; *PyrOx*, pyruvate oxidase; *SDH*, succinate dehydrogenase. (Reprinted from McKinlay et al., 2007, with permission from Springer.)

anaerobic fed-batch fermentation. With this operation regime, they obtained 99.2 g succinic acid/L, one of the higher succinate titers reported (McKinlay et al., 2007). However, a drawback of using the glyoxylate shunt for production of succinate is that CO_2 is lost from the carbon balance, because 3 mols of PEP are required to generate the 1 mol of oxaloacetate and 2 mols of acetyl-CoA that form 2 mols succinate via the glyoxylate shunt (Figure 11.8). In comparison, the reductive TCA branch produces 1 mol succinate from 1 mol PEP, as long as the reductant and CO_2 are not limiting (McKinlay et al., 2007).

E. coli was also engineered for aerobic succinate production by deleting the succinate dehydrogenase gene, *sdhA*, inactivating the glyoxylate shunt (*iclR*), and prohibiting acetate production (*poxB* and *ackA-pta* deletion) (Lin et al., 2005). Further inactivation of *ptsG* and

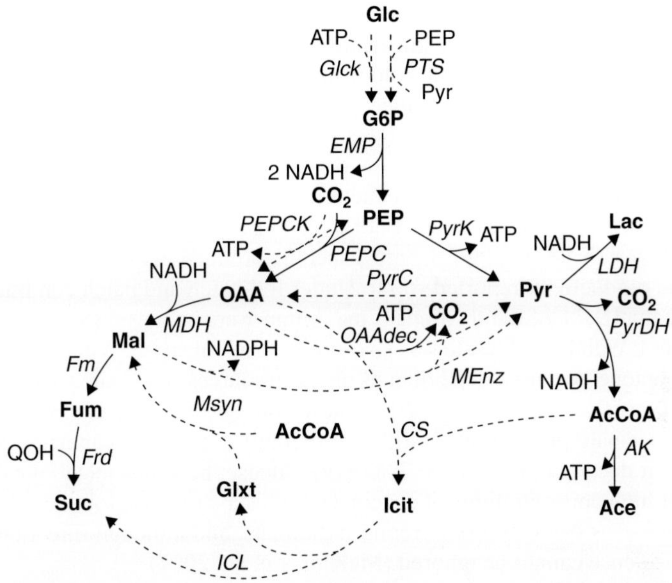

Figure 11.8. Simplified map of wild-type *C. glutamicum* succinate-producing metabolism. The pentose phosphate pathway, malate:quinone oxidoreductase, and pyruvate:quinone oxidoreductase are not shown. *OAAdec*, OAA decarboxylase; *PEPCK*, PEP carboxykinase. (Reprinted from McKinlay et al., 2007, with permission from Springer.)

overproduction of *S. vulgare* PEPC resulted in the production of 1 mol succinate per mol glucose, which is the theoretical maximum (Cox et al., 2006). In fed-batch aerobic cultures, the strain produced 58.3 g succinate/L at a rate of 1.1 g/L/h. Yet this yield of 1 mol of succinate per glucose is less than the yield in the anaerobic process. And maximum succinate yields under aerobic conditions result in a net CO_2 production rather than fixation (Cox et al., 2006), thus omitting the environmental benefits of succinate production.

Despite these obvious stories of success, there are other disadvantages in *E. coli*-based succinate production (mainly the costs associated with dual-phase fermentations and medium aeration, and the high costs for nutrients to cultivate the engineered strains and induce the manipulated genes) and they have so far been performed under laboratory conditions (expensive nutrients, induction of gene expression by artificial inducers), which prompted to search for other potential producers. A well-studied alternative to succinate production by *E. coli* is *C. glutamicum.* This bacterium is well known for its industrial production of L-glutamic acid and other amino acids. Succinate by-production by *C. glutamicum* under less well-aerated conditions was known. Under anoxic conditions and in the presence of bicarbonate, *C. glutamicum* converts glucose to succinate and lactate with acetate as a minor product (Inui et al., 2004). The *C. glutamicum* succinate-producing pathway involves formation of oxaloacetate mainly by the phosphoenolpyruvate carboxykinase and subsequent conversion to succinate by the reductive TCA cycle branch (Figure 11.8; Inui et al., 2004). To increase the range of lignocellulose-based sugars to be used as feedstock, *C. glutamicum* was engineered to consume xylose. Completing the partial xylose-catabolizing pathway with *E. coli* XI and overexpressing *E. coli* XK in *C. glutamicum* allowed it to convert glucose and xylose simultaneously to succinate and lactate (Kawaguchi et al., 2007). Unfortunately, strains that do not accumulate lactate as a by-product have not yet been reported. Thus the use of *C. glutamicum* as a producer of succinate would require expensive purification procedures.

A. succiniciproducens is a natural colonizer of human and animal feces that produces succinate as a major fermentation product. At high CO_2 concentrations and a pH near 6, it forms succinate and acetate at a ratio of 1.8:1 (McKinlay et al., 2007). It can also use H_2 as an electron source, and supplying H_2 favors succinate production (Zeikus et al., 1999). *A. succiniciproducens*'s fermentation characteristics were analyzed in industrially relevant media: in media with light corn steep water as a source of nitrogen, *A. succiniciproducens* produced 32 g/L succinate at 1.2 g/L/h with a specific productivity of 1,650 mg/g-DCW/h, the highest specific productivity reported so far. Under fed-batch and batch conditions, succinate yields of more than 91% of the theoretical maximum were achieved (Samuelov et al., 1999; Lee et al., 2000a). Wood hydrolysate was also explored as substrate. When supplemented with yeast extract, peptone, or corn-steep liquor as nitrogen sources, succinate yields near 90% were achieved in all cases (Lee et al., 2003a). While these findings recommend *A. succiniciproducens* for industrial succinate production, it still also exhibits some disadvantages that need to be overcome first: it does not grow above 7% (w/v) glucose, becomes metabolically inactive in the presence of high concentrations of succinate, and is a strict anaerobe. Even more critical, *A. succiniciproducens* can cause bacteremia in humans, and although infections are rare, its potential for virulence cannot be ignored (McKinlay et al., 2007).

11.6 Xylitol

Xylitol has attracted global attention because of its sweetening power similar to that of sucrose. Since xylitol is metabolized in humans through insulin-independent pathways, it can be used as sugar substitute for diabetics. In addition, xylitol has anticarcinogenic properties that can aid in oral health and prevent caries (Prakasham et al., 2009). The present industrial production of xylitol occurs by chemical hydrogenation of the xylose present in hemicellulosic hydrolysates in the presence of a metal catalyst under high temperature and pressure, and thus by an expensive and energy-intensive process.

From a biochemical perspective, the conversion of D-xylose to xylitol is a reaction that can be catalyzed by the aldose reductase protein family, which is widespread in eukaryotes. Microorganisms that express this protein at high specific activities are particularly found in the yeast genus *Candida*. Attempts have therefore been made to use species of *Candida* for fermentative production of xylitol (Prakasham et al., 2009). However, at the time of this writing, yields are in the range of a few grams per liter, thus far from commercialization.

11.7 1,3-Propanediol

1,3-Propanediol can be produced by fermentation with *Klebsiella pneumoniae* from glycerol as a carbon source (Nakamura and Whited, 2003; Liu et al., 2010a). It is being exploited to supplement the use of petrochemically derived 1,3-propanediol for the manufacturing of poly(trimethylene terephthalate), a polymer fiber with properties related to nylon and polyethylene terephthalate and use in the apparel and carpet industries.

1,3-Propanediol production from glycerol needs two enzymes: glycerol dehydratase (encoded by the gene *dhaB*1-3) and 1,3-propanediol oxidoreductase (encoded by the gene *dhaT*) (Nakamura and Whited, 2003; Figure 11.9). Interestingly, experiments to overexpress these

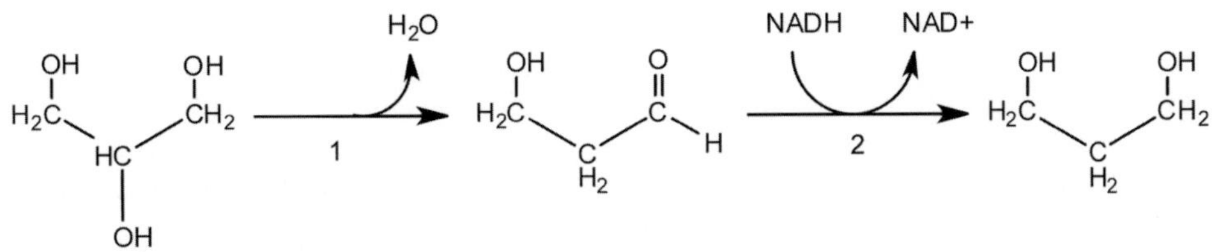

Figure 11.9. Pathway for 1,3-propanediol formation from glycerol. Enzymes involved are given by numbers. 1, glycerol dehydratase; 2, 1,3-propanediol oxidoreductase.

genes in *E. coli* gave rise to only 9 g/L, whereas expressing a nonspecific alcohol dehydratase instead of *dhaT* resulted in a final concentration of 129 g/L (Yu et al., 2010). Dupont and Genencor have commercialized 1,3-propanediol production by engineered *E. coli* using glucose as the feedstock (Nakamura and Whited, 2003).

11.8 Polyhydroxyalkanoate

Some biopolymers have been produced completely within microbial cells, most notably the family of polyhydroxyalkanoates (PHAs). This technology can be incorporated into a biorefinery as a separate processing stream since the bacteria that produce PHAs have been shown to be amenable to feeding from a wide range of carbon sources. PHAs are a family of natural polymers that comprised of over 150 different hydroxyalkanoates, produced by a range of bacteria (at least 75 genera) for carbon and energy storage (Lee et al., 1999). PHAs are formed as intracellular granules that have been reported at upward of 90% of dry cell weight (Lee, 2006). Their composition can be modulated by the addition of different precursor hydroalkanoates to the medium, which exhibit a wide range of properties, enabling them to be viable in a large fraction of the plastics industry (Suriyamongkol et al., 2007). Typically, bacteria synthesize PHA when carbon is abundant in excess, while one essential growth nutrient such as nitrogen or phosphorous is limited. It has also been shown that in some organisms, a carbon-limited feeding strategy can be employed for the production of PHA, with no limiting of essential growth nutrients (FitzPatrick et al., 2010).

Despite these promising facts, microbial production of PHA by bacteria is still uneconomic mainly due to low yields. To overcome this situation, engineering of the PHA synthase toward more effective P(3HA) production has been attempted because of the possibility to manipulate the PHA content and monomeric composition. To this end, extensive attempts at molecular evolution were recently performed for several PHA synthases without any tertiary structure information (reviewed by Nomura and Taguchi, 2007). However so far, no breakthrough has been achieved.

11.9 Other Products

A number of further components have been envisaged as potential products obtainable from lignocellulose hydrolysates. However, all of them still need major developments to actually compete to the existing production processes.

Itaconic acid is industrially produced by fermentation of starch hydrolysates or molasses by *Aspergillus terreus* and *A. itaconicus* in yields >200 g/L (Kubicek et al., 2009). Its polymerized esters (methyl, ethyl, and vinyl) are used in adhesives and coatings. Itaconic acid is also used in emulsion paints to aid in polymer adhesion and as a hardening agent for organosiloxanes, which are used in contact lenses (Okabe et al., 2009). In addition, because of its two reactive carboxyl groups, itaconic acid has the potential to be incorporated into polymers and is being assessed as a biofriendly substitute for acrylic and methacrylic acid in polymers and in styrene–butadiene systems (Lucia et al., 2006). The problem with its production from lignocelluloses, however, is the necessity for a high sugar concentration to achieve reasonable yields, and the fact that these yields have so far not been able to be obtained on pentoses (Kubicek et al., 2009).

Acrylic acid and its amide and ester derivatives are key components in polymer manufacturing, appearing in a wide range of products, including surface coatings, textiles, detergents, and absorbent materials. Xu et al. (2006) reported that some organisms express acrylic acid biosynthetic pathways, thus rendering its production by direct fermentation from biomass possible. Converting the carbohydrates first to lactic acid by fermentation and then dehydrating lactic acid to acrylic acid is hitherto the most effective way for producing acrylic acid from biomass. While the lactic acid fermentation has been commercialized since longer times (see also Section 11.4), the dehydration process of lactic acid is still under development because of its low yield.

Because of the existence of acrylic acid pathways in some microorganisms, strain improvement and metabolic engineering provides an alternative to produce acrylic acid directly from biomass by fermentation. However, most microbial processes resulted in only very low yields (around 1 g/L) of acrylic acid (Straathof et al., 2005). These authors, therefore, compared different metabolic routes that could lead to acrylic acid accumulation and arrived at a modification of the methylcitrate pathway as the most economic alternative (Figure 11.10). However, the process has not yet been realized in practice.

Ethylene, which can be polymerized into widely used polyethylene, is currently exclusively produced by steam cracking in the petrochemical industry. Its formation directly from biomass hydrolysates by bacteria has been shown (Danner and Braun, 1999). However, at the time of this writing, it cannot be predicted whether a biotechnology-based process would be competitive.

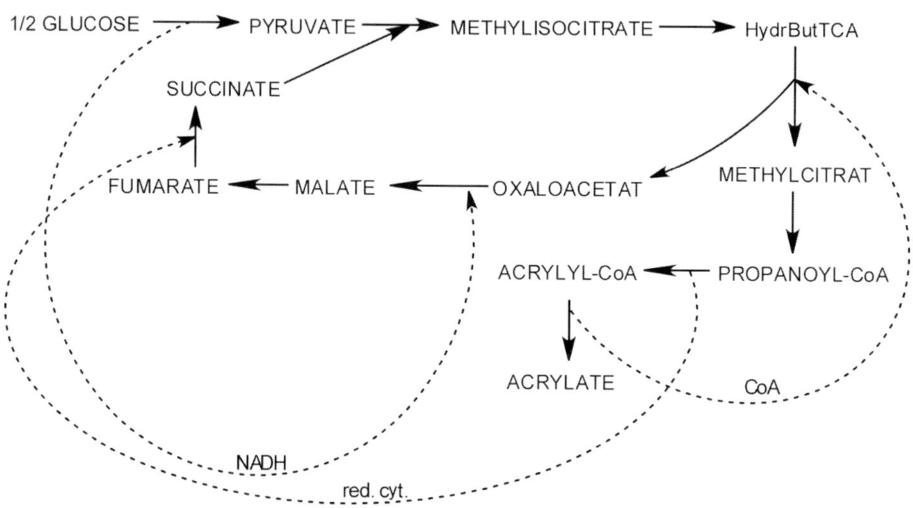

Figure 11.10. Theoretical pathway for conversion of glucose to acrylate via methylcitrate. Dashed arrows indicate cofactor regeneration pathways. HydrButTCA, (2*R*,3*S*)-2-hydroxybutane-1,2,3-tricarboxylate. (Modified from Straathof et al., 2005.)

11.10 Refinement by Chemical Processes

Biological processing is not the only refining approach, however. Although, as explained above, many biological protocols of converting polysaccharides to biorefineries are very well developed processes, chemical technologies are still being pursued and present promising alternatives. Their strategy can be summarized as the attempt to remove the oxygen from carbohydrates and obtain oxygenated hydrocarbons, mostly controlled elimination of water from the sugars. By this means, products like HMF, levulinic acid, and some other organic acids can be obtained.

These chemicals are yet too polar to be directly used as fuels, but reactions are available that can further convert them into alternative fuels. Examples are valerolactone or 2-methylfuran, two nonpolar compounds that are considered as components for novel gasoline blends, which are typically dependent on ~C5–C10 hydrocarbons. While these compounds cannot serve as a diesel supplement directly because the latter typically relies on C12–C20 hydrocarbons, this could be overcome by subsequent dimerization valerolactone or 2-methylfuran to increase the chain length. A tight control of sugar dehydration is critical to these strategies, as overdehydration will lead to polymerization, which will spoil the material.

11.10.1 Furfural

Furfural is commercially produced by acid hydrolysis of pentosans such as xylan and arabinan. Its major application is its use as a feedstock for furfuryl alcohol production, most of which is then used in the production of furan resins, which in turn are used to produce foundry sand binders. It is also used as a solvent in petroleum refining, in the production of phenolic resins, and in the production of tetrahydrofuran. The latter is also used as a commercial solvent, but—more importantly—also for the preparation of nylon (Rüsch genannt Klaas and Schöne, 2009).

11.10.2 Levulinic Acid

Levulinic acid is a short chain fatty acid having a ketone carbonyl group and an acidic carboxyl group. It has numerous potential uses, such as in the production of textile dyes, antifreeze, animal feed, coating material, solvent, food flavoring agent, pharmaceutical compounds, and resin (Chang et al., 2007). It can be produced by heating any hexose with a dilute mineral acid for a longer time period (Harris, 1975). The hexose can still be present in biomass materials such as wood, starch, cane sugar, grain sorghum, and agricultural wastes. At elevated temperatures, the process can result in a variety of soluble and humin-like products, with levulinic acid and formic acid being the final soluble products from hexoses through an intermediary formation and further reaction of HMF (Ragauskas et al., 2006; Chang et al., 2007).

11.10.3 Uses of Lignin

Lignin can be used as fuel to produce heat but more complex applications are also possible (Ragauskas et al., 2006): it can be incorporated in resins to substitute phenols or acting as a cross-linker in epoxy resins. New developments as adhesives are possible. Lignin incorporation in polyolefin increases resistance to UV without major modification in mechanical properties. The utilization of lignin in chemical industries is a new and attractive option because the thermal cracking at temperatures of ~250–600°C produces low molecular weight chemicals that can be further processed (Britt et al., 2000). In such a case study, lignin is first depolymerized by a base-catalyzed treatment, which results in the accumulation of a series of low molecular weight phenolic compounds. This mixture is then subjected to hydroprocessing, yielding a mixture of alkylbenzenes useful as a potential liquid biofuel.

11.10.4 Extraction of Chemicals from Lignocelluloses Biomass

In order to benefit from their production, efficient extraction technologies need to be implemented into biorefineries early on to allow for extraction before further processing. For example, commodity chemicals have been directly extracted from biomass via conventional chemical extraction techniques. Future biorefinery operations will first extract high-value chemicals already present in the biomass, such as fragrances, flavoring agents, food-related products, and high-value nutraceuticals that provide health and medical benefits.

One of the most notable chemicals derived in this manner is ferulic acid, which can be extracted in high concentrations and yields from corn fiber. Ferulic acid is a chemical feedstock for the production of fine chemicals such as vanillin and guaiacol. Vanillin is one of the most important aromatic compounds used in foods, beverages, perfumes, and pharmaceuticals. It is mainly produced from lignin and petrochemicals through chemical synthesis on a scale of 10,000 tons per year (Hua et al., 2007). Most biotechnological production processes of vanillin are based on biocatalysis or fermentation with a variety of microorganisms, applying bacteria, fungi, plant cells, or genetically engineered microorganisms (Priefert et al., 2001; Figure 11.11a), but the vanillin yields from most processes are very low (cf. Barghini et al., 2007). Hua et al. (2007) have described a biotransformation strategy in which vanillin was

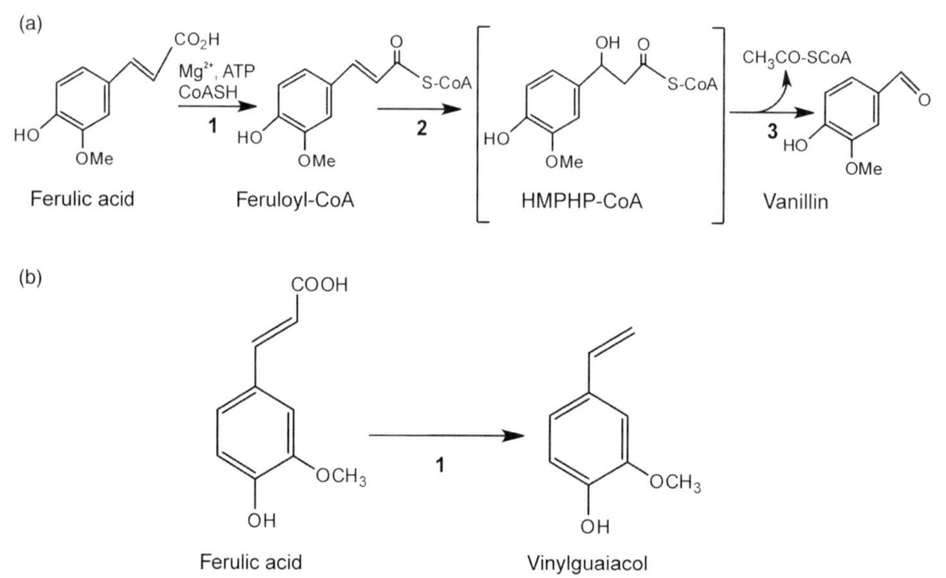

Figure 11.11. Production of vanillin (a) and vinylguaiacol (b) from ferulic acid. Enzymes are specified by numbers. Top scheme: 1, ATP-dependent hydroxycinnamic acid–CoA ligase; 2, benzylic acid hydratase; 3, hydroxycinnamoyl-CoA hydratase-lyase (HCHL). Bottom scheme: 1, ferulic acid decarboxylase. (Adapted from Yang et al., 2009.)

removed from the solution by specific adsorption on a resin, which enabled the conversion of 45 g/L ferulic acid to 19.2 g/L vanillin in a continuous process, using a *Streptomyces* sp. as whole cell biocatalyst.

4-Vinylguaiacol is a valuable product in the brewing industry by being used as flavoring in beers such as Belgian wheat and German Rauch (Mathew et al., 2007), and is also used in the ophthalmic field for preparation of flavoring solutions (Anonymous, 2006). It can be produced by biotransformation with yeasts and *Aspergillus* spp. (Mathew and Abraham, 2006) (Figure 11.11b), using a ferulic acid decarboxylase in a two-phase biotransformation. Yang et al. (2009) used whole cells of *E. coli* that expressed ferulic acid decarboxylase from *Bacillus pumilus* for synthesis of 4-vinylguaiacol from ferulic acid and reached 13.8 g vinyl guaiacol from 25 g of ferulic acid in a 2-L working volume bioreactor with octane as organic phase.

Acknowledgments

I would like to express my gratitude to the editors of this book, Anna Ehler and Justin Jeffryes, for their continuous friendly, helpful, and constructive support in the process of getting this book published. I would also like to thank many of my international and local colleagues for reading drafts of the chapters or discussing topics to be included. I also acknowledge the support by the Austrian Science Foundation, the Austrian Institute of Industrial Biotechnology, and several international companies active in this field for my own scientific projects related to this topic. Finally, I thank my wife Eva for persuading me to write this book, and providing me with the extra nonbusiness time to complete it.

Christian P. Kubicek

References

Aalto MK, Ronne H, Keränen S. Yeast syntaxins Sso1p and Sso2p belong to a family of related membrane proteins that function in vesicular transport. EMBO J, 1993; 12: 4095–4104.

Abbott DW, Boraston AB. The structural basis for exopolygalacturonase activity in a family 28 glycoside hydrolase. J Mol Biol, 2007; 368: 1215–1222.

Abramson J, Iwata S, Kaback HR. Lactose permease as a paradigm for membrane transport proteins. Mol Membr Biol, 2004; 21: 227–236.

Ademark P, de Vries RP, Hägglund P, Stålbrand H, Visser J. Cloning and characterization of *Aspergillus niger* genes encoding an alpha-galactosidase and a beta-mannosidase involved in galactomannan degradation. Eur J Biochem, 2001; 268: 2982–2990.

Ademark P, Lundqvist J, Hägglund P, Tenkanen M, Torto N, Tjerneld F, Stålbrand H. Hydrolytic properties of a beta-mannosidase purified from *Aspergillus niger*. J Biotechnol, 1999; 75: 281–289.

Adsul MG, Singhvi MS, Gaikaiwari SA, Gokhale DV. Development of biocatalysts for production of commodity chemicals from lignocellulosic biomass. Bioresour Technol, 2011; 102: 4304–4312.

Aehle W, Goedegebuur F, Dankmeyer L, Mitchinson C, Neefe P, Kelemen B, Caldwell R, Teunissen P. Novel variant *Hypocrea jecorina cbh2* cellulases. 2006; US Patent: 20060205042.

Agbogbo FK, Coward-Kelly G. Cellulosic ethanol production using the naturally occurring xylose-fermenting yeast, *Pichia stipitis*. Biotechnol Lett, 2008; 30: 1515–1524.

Ahmed S, Riaz S, Jamil A. Molecular cloning of fungal xylanases: an overview. Appl Microbiol Biotechnol, 2009; 84: 19–35.

Akao T, Yamaguchi M, Yahara A, Yoshiuchi K, Fujita H, Yamada O, Akita O, Ohmachi T, Asada Y, Yoshida T. Cloning and expression of 1,2-alpha-mannosidase gene (*fmanIB*) from filamentous fungus *Aspergillus oryzae*: *in vivo* visualization of the FmanIBp-GFP fusion protein. Biosci Biotechnol Biochem, 2006; 70: 471–479.

Akel E, Metz B, Seiboth B, Kubicek CP. Molecular regulation of arabinan and L-arabinose metabolism in *Hypocrea jecorina* (*Trichoderma reesei*). Eukaryot Cell, 2009; 8: 1837–1844.

Albenne C, Canut H, Boudart G, Zhang Y, San Clemente H, Pont-Lezica R, Jamet E. Plant cell wall proteomics: mass spectrometry data, a trove for research on protein structure/function relationships. Mol Plant, 2009; 2: 977–989.

Albersheim P, Darvill A, Roberts K, Staehelin LA, Varner JE. Do the structures of cell wall polysaccharides define their mode of synthesis? Plant Physiol, 1997; 113: 1–3.

Fungi and Lignocellulosic Biomass, First Edition. Christian P. Kubicek.

Alder E. Lignin chemistry—past, present and future. Wood Science Technol, 1977; 11: 169–218.

Alig RJ, Adams DM, McCarl BA, Ince PJ. Economic potential of short-rotation woody crops on agricultural land for pulp fiber production in the United States. Forest Prod J, 2000; 50: 67–74.

Alizadeh H, Teymouri F, Gilbert TI, Dale BE. Pretreatment of switchgrass by ammonia fiber explosion (AFEX). Appl Biochem Biotechnol Part A 2005; 124: 1133–1141.

Allen AL, Roche CD. Effect of strain and fermentation conditions on production of cellulases by *Trichoderma reesei*. Biotechnol Bioengin, 1989; 33: 650–656.

Alper H, Moxley J, Nevoigt E, Fink GR, Stephanopoulos G. Engineering yeast transcription machinery for improved ethanol tolerance and production. Science, 2006; 314: 1565–1568.

Alshahni MM, Makimura K, Yamada T, Takatori K, Sawada T. Nourseothricin acetyltransferase: a new dominant selectable marker for the dermatophyte *Trichophyton mentagrophytes*. Med Mycol, 2010; 48: 665–668.

Alvarez JM, Canessa P, Mancilla RA, Polanco R, Santibáñez PA, Vicuña R. Expression of genes encoding laccase and manganese-dependent peroxidase in the fungus *Ceriporiopsis subvermispora* is mediated by an ACE1-like copper-fist transcription factor. Fungal Genet Biol, 2009; 46: 104–111.

Alvira P, Tomás-Pejó E, Ballesteros M, Negro MJ. Pretreatment technologies for an efficient bioethanol production process based on enzymatic hydrolysis: a review. Bioresour Technol, 2010; 101: 4851–4861.

Amor Y, Haigler CH, Johnson S, Wainscott M, Delmer DP. A membrane-associated form of sucrose synthase and its potential role in synthesis of cellulose and callose in plants. Proc Natl Acad Sci USA, 1995; 92: 9353–9357.

Anderson JP, Gleason CA, Foley RC, Thrall PH, Burdon JB, Singh KB. Plants versus pathogens: an evolutionary arms race. Funct Plant Biol, 2010; 37: 499–512.

Anonymous. Ophthalmic solution with flavoring agents. 2006; US Patent: 20060292188.

Aparicio R, Fischer H, Scott DJ, Verschueren KH, Kulminskaya AA, Eneiskaya EV, Neustroev KN, Craievich AF, Golubev AM, Polikarpov I. Structural insights into the beta-mannosidase from *T. reesei* obtained by synchrotron small-angle X-ray solution scattering enhanced by X-ray crystallography. Biochemistry, 2002; 41: 9370–9375.

Arantes V, Saddler JN. Access to cellulose limits the efficiency of enzymatic hydrolysis: the role of amorphogenesis. Biotechnol Biofuels, 2010a; 3: 4.

Arantes V, Saddler JN. Cellulose accessibility limits the effectiveness of minimum cellulase loading on the efficient hydrolysis of pretreated lignocellulosic substrates. Biotechnol Biofuels, 2011b; 4: 3.

Archer DB. Filamentous fungi as microbial cell factories for food use. Curr Opin Biotechnol, 2000; 11: 478–483.

Aro N, Ilmén M, Saloheimo A, Penttilä M. ACEI of *Trichoderma reesei* is a repressor of cellulase and xylanase expression. Appl Environ Microbiol, 2003; 69: 56–65.

Aro N, Pakula T, Penttilä M. Transcriptional regulation of plant cell wall degradation by filamentous fungi. FEMS Microbiol Rev, 2005; 29: 719–739.

Asada Y, Watanabe A, Irie T, Nakayama T, Kuwahara M. Structures of genomic and complementary DNAs coding for *Pleurotus ostreatus* manganese (II) peroxidase. Biochim Biophys Acta, 1995; 1251: 205–209.

Asgher M, Bhatti HN, Ashraf M, Legge RL. Recent developments in biodegradation of industrial pollutants by white rot fungi and their enzyme system. Biodegradation, 2008; 19: 771–783.

Atanassov II, Pittman JK, Turner SR. Elucidating the mechanisms of assembly and subunit interaction of the cellulose synthase complex of *Arabidopsis* secondary cell walls. J Biol Chem, 2009; 284: 3833–3841.

Atsumi S, Higashide W, Liao JC. Direct photosynthetic recycling of carbon dioxide to isobutyraldehyde. Nat Biotech, 2009,; 27: 1177–1180.

Avalos J, Geever RF, Case ME. Bialaphos resistance as a dominant selectable marker in *Neurospora crassa*. Curr Genet, 1989; 16: 369–372.

Baik SH, Saito K, Yokota A, Asano K, Tomita F. Molecular cloning and high-level expression in *Escherichia coli* of fungal alpha-galactosidase from *Absidia corymbifera* IFO 8084. J Biosci Bioeng, 2000; 90: 168–173.

Bak JS, Kim MD, Choi IG, Kim KH. Biological pretreatment of rice straw by fermenting with *Dichomitus squalens*. N Biotechnol, 2010; 27: 424–434.

Baldrian P, Kolařík M, Stursová M, Kopecký J, Valášková V, Větrovský T, Zifčáková L, Snajdr J, Rídl J, Vlček C, Voříšková J. Active and total microbial communities in forest soil are largely different and highly stratified during decomposition. ISME J, 2011, doi: 10.1038/ismej.2011.95.

Baldrian P, Valásková V. Degradation of cellulose by basidiomycetous fungi. FEMS Microbiol Rev, 2008; 32: 501–521.

Baldrian P. Fungal laccases—occurrence and properties. FEMS Microbiol Rev, 2006; 30: 215–242.

Ballance DJ, Buxton FP, Turner G. Transformation of *Aspergillus nidulans* by the orotidine-5′-phosphate decarboxylase gene of *Neurospora crassa*. Biochem Biophys Res Comm, 1983; 112: 284–289.

Ballesteros I, Negro MJ, Oliva JM, Cabanas A, Manzanares P, Ballesteros M. Ethanol production from steam-explosion pretreated wheat straw. Appl Biochem Biotechnol, 2006; 130: 496–508.

Bals B, Wedding C, Balan V, Sendich E, Dale B. Evaluating the impact of ammonia fiber expansion (AFEX) pretreatment conditions on the cost of ethanol production. Bioresour Technol, 2011; 102: 1277–1283.

Banerjee G, Car S, Scott-Craig JS, Borrusch MS, Bongers M, Walton JD. Synthetic multicomponent enzyme mixtures for deconstruction of lignocellulosic biomass. Bioresour Technol, 2010a; 101: 9097–9105.

Banerjee G, Car S, Scott-Craig JS, Borrusch MS, Aslam N, Walton JD. Synthetic enzyme mixtures for biomass deconstruction: production and optimization of a core set. Biotechnol Bioeng, 2010b; 106: 707–720.

Bar-Peled M, O'Neill MA. Plant nucleotide sugar formation, interconversion, and salvage by sugar recycling. Annu Rev Plant Biol, 2011; 62: 127–155.

Bardiya N, Shiu PK. Cyclosporin A-resistance based gene placement system for *Neurospora crassa*. Fungal Genet Biol, 2007; 44: 307–314.

Barghini P, Di Gioia D, Fava F, Ruzzi M. Vanillin production using metabolically engineered *Escherichia coli* under non-growing conditions. Microbial Cell Fact, 2007; 6: 13.

Battaglia E, Hansen SF, Leendertse A, Madrid S, Mulder H, Nikolaev I, de Vries RP. Regulation of pentose utilisation by AraR, but not XlnR, differs in *Aspergillus nidulans* and *Aspergillus niger*. Appl Microbiol Biotechnol, 2011; 91: 387–397.

Bauchop T. The rumen anaerobic fungi: colonizers of plant fibre. Ann Rech Vet, 1979; 10: 246–248.

Bauchop T. Biology of gut anaerobic fungi. Biosystems, 1989; 23: 53–64.

Bauer S, Vasu P, Mort AJ, Somerville CR. Cloning, expression, and characterization of an oligoxyloglucan reducing end-specific xyloglucanobiohydrolase from *Aspergillus nidulans*. Carbohydr Res, 2005; 340: 2590–2597.

Bauer S, Vasu P, Persson S, Mort AJ, Somerville CR. Development and application of a suite of polysaccharide-degrading enzymes for analyzing plant cell walls. Proc Natl Acad Sci USA, 2006; 103: 11417–11422.

Bayer EA (ed). The life and lifework of mary mandels—first lady of cellulase research. Biotechnol Biofuels, 2009; 2(1): 23.

Bayer EA, Lamed R, White BA, Flint HJ. From cellulosomes to cellulosomics. Chem Rec, 2008; 8: 364–377.

Beadle GW. Recollections. Annu Rev Biochem, 1974; 43: 1–14.

Becherer KA, Rieder SE, Emr SD, Jones EW. Novel syntaxin homologue, Pep12p, required for the sorting of lumenal hydrolases to the lysosome-like vacuole in yeast. Mol Biol Cell, 1996; 7: 579–594.

Beckham GT, Bomble YJ, Matthews JF, Taylor CB, Resch MG, Yarbrough JM, Decker SR, Bu L, Zhao X, McCabe C, Wohlert J, Bergenstråhle M, Brady JW, Adney WS, Himmel ME, Crowley MF. The *O*-glycosylated linker from the *Trichoderma reesei* family 7 cellulase is a flexible, disordered protein. Biophys J, 2010; 99: 3773–3781.

Belinky PA, Flikshtein N, Lechenko S, Gepstein S, Dosoretz CG. Reactive oxygen species and induction of lignin peroxidase in *Phanerochaete chrysosporium*. Appl Environ Microbiol, 2003; 69: 6500–6506.

Bellaver LH, de Carvalho NM, Abrahao-Neto J, Gombert AK. Ethanol formation and enzyme activities around glucose-6-phosphate in *Kluyveromyces marxianus* CBS 6556 exposed to glucose or lactose excess. FEMS Yeast Res, 2004; 4: 691–698.

Benech RO, Li X, Patton D, Powlowski J, Storms R, Bourbonnais R, Paice M, Tsang A. Recombinant expression, characterization, and pulp prebleaching property of a *Phanerochaete chrysosporium* endo-β-1,4-mannanase. Enzyme Microb Technol, 2008,; 41: 740–747.

Benen JA, Kester HC, Parenicová L, Visser J. Characterization of *Aspergillus niger* pectate lyase A. Biochemistry, 2000; 39: 15563–15569.

Berg C. World Ethanol Production 2001. 2002. Available at http://www.distill.com/world_ethanol_production.htm.

Berghäll S, Hilditch S, Penttilä M, Richard P. Identification in the mould *Hypocrea jecorina* of a gene encoding an NADP(+):D-xylose dehydrogenase. FEMS Microbiol Lett, 2007; 277: 249–253.

Berkeley MJ, Broome CE. Enumeration of the fungi of Ceylon. Part II. Containing the remainder of the Hymenomycetes, with the remaining established tribes of fungi. J Linn Soc London Botany, 1875; 14: 29–140.

Berney C, Pawlowski J. A molecular time-scale for eukaryote evolution recalibrated with the continuous microfossil record. Proc Biol Sci, 2006; 273: 1867–1872.

Bessueille L, Sindt N, Guichardant M, Djerbi S, Teeri TT, Bulone V. Plasma membrane microdomains from hybrid aspen cells are involved in cell wall polysaccharide biosynthesis. Biochem J, 2009; 420: 93–103.

Bhiri F, Gargouri A, Ali MB, Belghith H, Blibech M, Ellouz-Chaabouni S. Molecular cloning, gene expression analysis and structural modelling of the cellobiohydrolase I from *Penicillium occitanis*. Enzyme Microb Technol, 2010; 46: 74–81.

Bi C, Zhang X, Ingram LO, Preston JF. Genetic engineering of *Enterobacter asburiae* strain JDR-1 for efficient production of ethanol from hemicelluloses hydrolysates. Appl Environ Microbiol, 2009; 75: 5743–5749.

Biely P, de Vries RP, Vrsanská M, Visser J. Inverting character of alpha-glucuronidase A from *Aspergillus tubingensis*. Biochim Biophys Acta, 2000; 1474: 360–364.

Biely P, Mislovicová D, Toman R. Soluble chromogenic substrates for the assay of endo-1,4-beta-xylanases and endo-1,4-beta-glucanases. Anal Biochem, 1985; 144: 142–146.

Biely P, Vrsanská M, Tenkanen M, Kluepfel D. Endo-beta-1,4-xylanase families: differences in catalytic properties. J Biotechnol, 1997; 57: 151–166.

Binder JB, Raines RT. Fermentable sugars by chemical hydrolysis of biomass. Proc Natl Acad Sci USA, 2010; 107: 4516–4521.

Birsan C, Johnson P, Joshi M, MacLeod A, McIntosh L, Monem V, Nitz M, Rose DR, Tull D, Wakarchuck WW, Wang Q, Warren RA, White A, Withers SG. Mechanisms of cellulases and xylanases. Biochem Soc Trans, 1998; 26: 156–160.

Bisby GR. *Trichoderma viride* Pers. ex Fries, and notes on Hypocrea. Trans Brit Mycol Soc, 1939; 23: 149–168.

Bissett J. A revision of the genus *Trichoderma*. I. Section *Longibrachiatum* sec. nov. Can J Bot, 1984; 62: 924–931.

Boase NA, Kelly JM. A role for creD, a carbon catabolite repression gene from *Aspergillus nidulans*, in ubiquitination. Mol Microbiol, 2004; 53: 929–940.

Boerjan W, Ralph J, Baucher M. Lignin biosynthesis. Annu Rev Plant Biol, 2003; 54: 519–546.

Bootten TJ, Harris PJ, Melton LD, Newman RH. Solid-state 13C NMR study of a composite of tobacco xyloglucan and *Gluconacetobacter xylinus* cellulose:molecular interactions between the component polysaccharides. Biomacromolecules, 2009; 10: 2961–2967.

Boraston AB, Bolam DN, Gilbert HJ, Davies GJ. Carbohydrate-binding modules: fine-tuning polysaccharide recognition. Biochem J, 2004; 382: 769–781.

Boraston AB, McLean BW, Kormos JM, Alam M, Gilkes NR, Haynes CA, Tomme P, Kilburn DG, Warren RA. Carbohydrate-binding modules: diversity of structure and function. In: Gilbert HJ, Davies GJ, Henrissat B, Svensson B (eds) *Recent Advances in Carbohydrate Bioengineering*, 1999, pp. 202–211. Royal Society of Chemistry, Cambridge.

Bourgois TM, Van Craeyveld V, Van Campenhout S, Courtin CM, Delcour JA, Robben J, Volckaert G. Recombinant expression and characterization of XynD from *Bacillus subtilis* subsp. *subtilis* ATCC 6051: a GH 43 arabinoxylan arabinofuranohydrolase. Appl Microbiol Biotechnol, 2007; 75: 1309–1317.

Brijwani K, Vadlani PV. Cellulolytic enzymes production via solid-state fermentation: effect of pretreatment methods on physicochemical characteristics of substrate. Enzyme Res, 2011; 2011: 860134.

Britt PF, Buchanan AC, Cooney MJ, Martineau DR. Flash vacuum pyrolysis of methoxy-substituted lignin model compounds. J Org Chem, 2000; 65: 1376–1389.

Brodeur G, Yau E, Badal K, Collier J, Ramachandran KB, Ramakrishnan S. Chemical and physicochemical pretreatment of lignocellulosic biomass: a review. Enzyme Res, 2011; 787532. Epub 2011 May 24.

Brotman Y, Briff E, Viterbo A, Chet I. Role of swollenin, an expansin-like protein from *Trichoderma*, in plant root colonization. Plant Physiol, 2008; 147: 779–789.

Brown Jr RM. Biogenesis of natural polymer systems, with special reference to cellulose assembly and deposition. In: Proceedings of the Third Phillip Morris U.S.A. Operations Center, Richmond, Virginia, 1979, pp. 50–123.

Bruinenberg PM, de Bot PHM, van Dijken JP, Scheffers WA. NADH-linked aldose reductase: the key to anaerobic fermentation of xylose by yeasts. Appl Microbiol Biotechnol, 1984; 19: 256–260.

Brunow G. Methods to reveal the structure of lignin. In: M. Hofrichter, A. Steinbuchel (eds), Lignin, Humic Substances and Coal In: *Biopolymers*, vol. 1, 2000. Weinheim, Germany: Wiley-VCH.

Brüx C, Ben-David A, Shallom-Shezifi D, Leon M, Niefind K, Shoham G, Shoham Y, Schomburg D. The structure of an inverting GH43 beta-xylosidase from *Geobacillus stearothermophilus* with its substrate reveals the role of the three catalytic residues. J Mol Biol, 2006; 359: 97–109.

Bu L, Beckham GT, Crowley MF, Chang CH, Matthews JF, Bomble YJ, Adney WS, Himmel ME, Nimlos MR. The energy landscape for the interaction of the family 1 carbohydrate-binding module and the cellulose surface is altered by hydrolyzed glycosidic bonds. J Phys Chem B, 2009; 113: 10994–11002.

Buée M, Reich M, Murat C, Morin E, Nilsson RH, Uroz S, Martin F. 454 Pyrosequencing analyses of forest soils reveal an unexpectedly high fungal diversity. New Phytol, 2009; 184: 449–456.

Burris JN, Mann DJG, Joyce BL, Stewart CN Jr. An improved tissue culture system for embyrogenic callus production and plant regeneration in switchgrass (*Panicum virgatum* L.). BioEnergy Research, 2009; 2: 267–274.

Burton RA, Jobling SA, Harvey AJ, Shirley NJ, Mather DE, Bacic A, Fincher GB. The genetics and transcriptional profiles of the cellulose synthase-like HvCslF gene family in barley. Plant Physiol, 2008; 146: 1821–1833.

Buxton FP, Gwynne DI, Davies RW. Transformation of *Aspergillus niger* using the *argB* gene of *Aspergillus nidulans*. Gene, 1985; 37: 207–214.

Caffall KH, Mohnen D. The structure, function, and biosynthesis of plant cell wall pectic polysaccharides. Carbohydr Res, 2009; 344: 1879–1900.

Calmels TP, Martin F, Durand H, Tiraby G. Proteolytic events in the processing of secreted proteins in fungi. J Biotechnol, 1991; 17: 51–66.

Calvo-Flores FG, Dobado JA. Lignin as renewable raw material. ChemSusChem, 2010; 3: 1227–1235.

Cantarel BL, Coutinho PM, Rancurel C, Bernard T, Lombard V, Henrissat B. The Carbohydrate-Active EnZymes database (CAZy): an expert resource for glycogenomics. Nucleic Acids Res, 2009; 37: D233–D238.

Cao Y, Wang Y, Luo H, Shi P, Meng K, Zhou Z, Zhang Z, Yao B. Molecular cloning and expression of a novel protease-resistant GH-36 alpha-galactosidase from Rhizopus sp. F78 ACCC 30795. J Microbiol Biotechnol, 2009; 19: 1295–1300.

Cao Y, Wang Y, Meng K, Bai Y, Shi P, Luo H, Yang P, Zhou Z, Zhang Z, Yao B. A novel protease-resistant alpha-galactosidase with high hydrolytic activity from *Gibberella* sp. F75: gene cloning, expression, and enzymatic characterization. Appl Microbiol Biotechnol, 2009; 83: 875–884.

Carpita NC. Progress in the biological synthesis of the plant cell wall: new ideas for improving biomass for bioenergy. Curr Opin Biotechnol, 2011. [Epub ahead of print]

Carapito R, Imberty A, Jeltsch JM, Byrns SC, Tam PH, Lowary TL, Varrot A, Phalip V. Molecular basis of arabinobio-hydrolase activity in phytopathogenic fungi: crystal structure and catalytic mechanism of *Fusarium graminearum* GH93 exo-alpha-L-arabinanase. J Biol Chem, 2009; 284: 12285–12296.

Carder JH. Detection and quantitation of cellulase by Congo red staining of substrates in a cup-plate diffusion assay. Anal Biochem, 1986; 153: 75–79.

Carle-Urioste JC, Escobar-Vera J, El-Gogary S, Henrique-Silva F, Torigoi E, Crivellaro O, Herrera-Estrella A, El-Dorry H. Cellulase induction in *Trichoderma reesei* by cellulose requires its own basal expression. J Biol Chem, 1997; 272: 10169–10174.

Caufrier F, Martinou A, Dupont C, Bouriotis V. Carbohydrate esterase family 4 enzymes: substrate specificity. Carbohydr Res., 2003,; 338: 687–692.

Chacon Martinez CA, Junca H, Hernandez JF, Rojas A, Anzola JM, Ocampo W, Del Portillo P. Identification and characterization of the alpha-L-arabinofuranosidase B of *Fusarium oxysporum* f. sp. *dianthi*. Physiol Mol Plant Pathol, 2004; 64: 201–208.

Chandel AK, Singh OV. Weedy lignocellulosic feedstock and microbial metabolic engineering: advancing the generation of "Biofuel." Appl Microbiol Biotechnol, 2011; 89: 1289–1303.

Chatterjee R, Millard CS, Champion K, Clark DP, Donnelly MI. Mutation of the *ptsG* gene results in increased production of succinate in fermentation of glucose by *Escherichia coli*. Appl Environ Microbiol, 2001; 67: 148–154.

Chang C, Cen P, Ma X. Levulinic acid production from wheat straw. Bioresource Technol, 2007; 98: 1448–1453.

Chen CH, Dunlap JC, Loros JJ. *Neurospora* illuminates fungal photoreception. Fungal Genet Biol, 2010a; 47: 922–929.

Chen CH, DeMay BS, Gladfelter AS, Dunlap JC, Loros JJ. Physical interaction between VIVID and white collar complex regulates photoadaptation in *Neurospora*. Proc Natl Acad Sci USA, 2010b; 107: 16715–16720.

Chen X, Nielsen KF, Borodina I, Kielland-Brandt MC, Karhumaa K. Increased isobutanol production in *Saccharomyces cerevisiae* by overexpression of genes in valine metabolism. Biotechnol Biofuels, 2011; 4: 21.

Chilton IJ, Delaney CE, Barham-Morris J, Fincham DA, Hooley P, Whitehead MP. The *Aspergillus nidulans* stress response transcription factor StzA is ascomycete-specific and shows species-specific polymorphisms in the C-terminal region. Mycol Res, 2008; 112: 1435–1446.

Choy V, Patel N, Thibault J. Application of image analysis in the fungal fermentation of *Trichoderma reesei* RUT-C30. Biotechnol Prog, 2011; doi:10.1002/btpr.667.

Chundawat SPS, Venkatesh B, Dale BE. Effect of particle size-based separation of milled corn stover on AFEX pretreatment and enzymatic digestibility. Biotechnol Bioengin, 2007; 96: 219–231.

Cirino PC, Chin JW, Ingram LO. Engineering *Escherichia coli* for xylitol production from glucose–xylose mixtures. Biotechnol Bioeng, 2006; 95: 1167–1176.

Cole L, Davies D, Hyde GJ, Ashford AE. Brefeldin A affects growth, endoplasmic reticulum, Golgi bodies, tubular vacuole system, and secretory pathway in *Pisolithus tinctorius*. Fungal Genet Biol, 2000; 29: 95–106.

Coleman HD, Yan J, Mansfield SD. Sucrose synthase affects carbon partitioning to increase cellulose production and altered cell wall ultrastructure. Proc Natl Acad Sci USA, 2009; 106: 13118–13123.

Collins T, Gerday C, Feller G. Xylanases, xylanase families and extremophilic xylanases. FEMS Microbiol Rev, 2005; 29: 3–23.

Comfort DA, Bobrov KS, Ivanen DR, Shabalin KA, Harris JM, Kulminskaya AA, Brumer H, Kelly RM. Biochemical analysis of *Thermotoga maritima* GH36 alpha-galactosidase (TmGalA) confirms the mechanistic commonality of clan GH-D glycoside hydrolases. Biochemistry, 2007; 46: 3319–3330.

Conesa A, Jeenes D, Archer DB, van den Hondel CA, Punt PJ. Calnexin overexpression increases manganese peroxidase production in *Aspergillus niger*. Appl Environ Microbiol, 2002; 68: 846–851.

Conibear E, Stevens TH. Multiple sorting pathways between the late Golgi and the vacuole in yeast. Biochim Biophys Acta, 1998; 1404: 211–230.

Connor MR, Liao JC. Microbial production of advanced transportation fuels in non-natural hosts. Curr Opin Biotechnol, 2009; 20: 307–315.

Corrochano LM. Fungal photoreceptors: sensory molecules for fungal development and behaviour. Photochem Photobiol Sci, 2007; 6: 725–736.

Coughlan CM, Walker JL, Cochran JC, Wittrup KD, Brodsky JL. Degradation of mutated bovine pancreatic trypsin inhibitor in the yeast vacuole suggests post-endoplasmic reticulum protein quality control. J Biol Chem, 2004; 279: 15289–15297.

Coutinho PM, Andersen MR, Kolenova K, vanKuyk PA, Benoit I, Gruben BS, Trejo-Aguilar B, Visser H, van Solingen P, Pakula T, Seiboth B, Battaglia E, Aguilar-Osorio G, de Jong JF, Ohm RA, Aguilar M, Henrissat B, Nielsen J, Stalbrand H, de Vries RP. Post-genomic insights into the plant polysaccharide degradation potential of *Aspergillus nidulans* and comparison to *Aspergillus niger* and *Aspergillus oryzae*. Fungal Genet Biol, 2009; 46: S161–S169.

Cowan D, Meyer Q, Stafford W, Muyanga S, Cameron R, Wittwer P. Metagenomic gene discovery: past, present and future. Trends Biotechnol, 2005; 23: 321–329.

Cox SJ, Shalel Levanon S, Sanchez AM, Lin H, Peercy B, Bennett GN, San KY. Development of a metabolic network design and optimization framework incorporating implementation constraints: a succinate production case study. Metab Eng, 2006; 8: 46–57.

Crepin VF, Faulds CB, Connerton IF. Functional classification of the microbial feruloyl esterases. Appl Microbiol Biotechnol, 2004; 63: 647–652.

Crowell EF, Gonneau M, Stierhof YD, Höfte H, Vernhettes S. Regulated trafficking of cellulose synthases. Curr Opin Plant Biol, 2010; 13: 700–705.

Cullen D, Gray GL, Berka RM. Molecular cloning vectors for *Aspergillus* and *Neurospora*. Bio/Technology, 1988; 10: 419–433.

Cullen D, Leong SA, Wilson LJ, Henner DJ. Transformation of *Aspergillus nidulans* with the hygromycin-resistance gene, *hph*. Gene, 1987; 57: 21–26.

Cuomo CA, Güldener U, Xu JR, Trail F, Turgeon BG, Di Pietro A, Walton JD, Ma LJ, Baker SE, Rep M, Adam G, Antoniw J, Baldwin T, Calvo S, Chang YL, Decaprio D, Gale LR, Gnerre S, Goswami RS, Hammond-Kosack K, Harris LJ, Hilburn K, Kennell JC, Kroken S, Magnuson JK, Mannhaupt G, Mauceli E, Mewes HW, Mitterbauer R, Muehlbauer G, Münsterkötter M, Nelson D, O'donnell K, Ouellet T, Qi W, Quesneville H, Roncero MI, Seong KY, Tetko IV, Urban M, Waalwijk C, Ward TJ, Yao J, Birren BW, Kistler HC. The *Fusarium graminearum* genome reveals a link between localized polymorphism and pathogen specialization. Science, 2007; 317: 1400–1402.

Curtis TP, Sloan WT, Scannell JW. Estimating prokaryotic diversity and its limits. Proc Natl Acad Sci USA, 2002; 99: 10494–10499.

Cziferszky A, Mach RL, Kubicek CP. Phosphorylation positively regulates DNA binding of the carbon catabolite repressor Cre1 of *Hypocrea jecorina* (*Trichoderma reesei*). J Biol Chem, 2002; 277: 14688–14694.

Czlapinski JL, Bertozzi CR. Synthetic glycobiology: exploits in the Golgi compartment. Curr Opin Chem Biol, 2006; 10: 645–651.

da Costa Sousa L, Chundawat SP, Balan V, Dale BE. "Cradle-to-grave" assessment of existing lignocellulose pretreatment technologies. Curr Opin Biotechnol, 2009; 20: 339–347.

Daboussi MJ, Djeballi A, Gerlinger C, Blaiseau PL, Bouvier I, Cassan M, Lebrun MH, Parisot D, Brygoo Y. Transformation of seven species of filamentous fungi using the nitrate reductase gene of *Aspergillus nidulans*. Curr Genet, 1989; 15: 453–456.

Dacks JB, Dyal PL, Embley TM, van der Giezen M. Hydrogenosomal succinyl-CoA synthetase from the rumen-dwelling fungus *Neocallimastix patriciarum*: an energy-producing enzyme of mitochondrial origin. Gene, 2006; 373: 75–82.

Dalrymple BP, Cybinski DH, Layton I, McSweeney CS, Xue GP, Swadling YJ, Lowry JB. Three *Neocallimastix patriciarum* esterases associated with the degradation of complex polysaccharides are members of a new family of hydrolases. Microbiology UK, 1997; 143: 2605–2614.

Daniel G, Volc J, Kubatova E. Pyranose oxidase, a major source of H_2O_2 during wood degradation by *Phanerochaete chrysosporium*, *Trametes versicolor*, and *Oudemansiella mucida*. Appl Environ Microbiol, 1994; 60: 2524–2532.

Danner H, Braun R. Biotechnology for the production of commodity chemicals from biomass. Chem Soc Rev, 1999; 28: 395–405.

Darrah PR, Tlalka M, Ashford A, Watkinson SC, Fricker MD. The vacuole system is a significant intracellular pathway for longitudinal solute transport in basidiomycete fungi. Eukaryot Cell, 2006; 5: 1111–1125.

Darvill JE, McNeil M, Darvill AG, Albersheim P. Structure of Plant Cell Walls: XI. glucuronoarabinoxylan, a second hemicelluloses in the primary cell walls of suspension-sultured *Sycamore* cells. Plant Physiol, 1980; 66: 1135–1139.

Davies G. Glycoside Hydrolase Family 5 in *CAZypedia*. 2011. Available at: http://www.cazypedia.org/ (accessed October 31, 2011).

Davies GJ, Dodson GG, Hubbard RE, Tolley SP, Dauter Z, Wilson KS, Hjort C, Mikkelsen JM, Rasmussen G, Schülein M. Structure and function of endoglucanase V. Nature, 1993; 365: 362–364.

Davies GJ, Mackenzie L, Varrot A, Dauter M, Brzozowski AM, Schülein M, Withers SG. Snapshots along an enzymatic reaction coordinate: analysis of a retaining beta-glycoside hydrolase. Biochemistry, 1998; 37: 11707–11713.

Davies GJ, Tolley SP, Henrissat B, Hjort C, Schülein M. Structures of oligosaccharide-bound forms of the endoglucanase V from *Humicola insolens* at 1.9 A resolution. Biochemistry, 1995; 34: 16210–16220.

Davin LB, Lewis NG. Lignin primary structures and dirigent sites. Curr Opin Biotechnol, 2005; 16: 407–415.

Davis RH, Perkins DD. Timeline: *Neurospora*: a model of model microbes. Nat Rev Genet, 2002; 3: 397–403.

Day A, Goedegebuur F, Gualfetti P, Mitchinson C, Neefe P, Sandgren M, Shaw S, Stahlberg J. Novel variant hypocrea jercorina cbh1 cellulases. 2007; Patent: 20070173431.

de Groot MJ, Bundock P, Hooykaas PJ, Beijersbergen AG. *Agrobacterium tumefaciens*-mediated transformation of filamentous fungi. Nat Biotechnol, 1998; 16: 839–842.

de Groot MJ, Prathumpai W, Visser J, Ruijter GJ. Metabolic control analysis of *Aspergillus niger* L-arabinose catabolism. Biotechnol Prog, 2005; 21: 1610–1616.

De Matteis MA, Luini A. Exiting the Golgi complex. Nat Rev Mol Cell Biol, 2008; 9: 273–284.

de Vries RP. Regulation of *Aspergillus* genes encoding plant cell wall polysaccharide-degrading enzymes;: relevance for industrial production. Appl Microbiol Biotechnol, 2003; 61: 10–20.

de Vries RP, Poulsen CH, Madrid S, Visser J. *aguA*, the gene encoding an extracellular alpha-glucuronidase from *Aspergillus tubingensis*, is specifically induced on xylose and not on glucuronic acid. J Bacteriol, 1998; 180: 243–249.

de Vries RP, van den Broeck HC, Dekkers E, Manzanares P, de Graaff LH, Visser J. Differential expression of three alpha-galactosidase genes and a single beta-galactosidase gene from *Aspergillus niger*. Appl Environ Microbiol, 1999; 65: 2453–2460.

de Vries RP, Visser J. *Aspergillus* enzymes involved in degradation of plant cell wall polysaccharides. Microbiol Mol Biol Rev, 2001; 65: 497–522.

de Vries RP, Visser J, de Graaff LH. CreA modulates the XlnR-induced expression on xylose of *Aspergillus niger* genes involved in xylan degradation. Res Microbiol, 2000; 150: 281–285.

Deanda K, Zhang M, Eddy C, Picataggio S. Development of an arabinose-fermenting *Zymomonas mobilis* strain by metabolic pathway engineering. Appl Environ Microbiol, 1996; 62: 4465–4470.

del Río JC, Rencoret J, Marques G, Gutiérrez A, Ibarra D, Santos JI, Jiménez-Barbero J, Zhang L, Martínez AT. Highly acylated (acetylated and/or p-coumaroylated) native lignins from diverse herbaceous plants. J Agric Food Chem, 2008; 56: 9525–9534.

Delattre C, Michaud P, Keller C, Elboutachfaiti R, Beven L, Courtois B, Courtois J. Purification and characterization of a novel glucuronan lyase from *Trichoderma* sp. GL2. Appl Microbiol Biotechnol, 2006; 70: 437–443.

Delmer DP. Cellulose biosynthesis: exciting times for a difficult field of study. Annu Rev Plant Physiol Plant Mol Biol, 1999; 50: 245–276.

den Herder IF, Rosell AM, van Zuilen CM, Punt PJ, van den Hondel CA. Cloning and expression of a member of the *Aspergillus niger* gene family encoding alpha-galactosidase. Mol Gen Genet, 1992; 233: 404–410.

Deshpande N, Wilkins MR, Packer N, Nevalainen H. Protein glycosylation pathways in filamentous fungi. Glycobiology, 2008; 18: 626–637.

Desmet T, Cantaert T, Gualfetti P, Nerinckx W, Gross L, Mitchinson C, Piens K. An investigation of the substrate specificity of the xyloglucanase Cel74A from *Hypocrea jecorina*. FEBS J, 2007; 274: 356–363.

Dey PM. Biochemistry of plant galactomannans. Adv Carbohydr Chem Biochem, 1978; 35: 3411–3376.

Dhawale SS. Is an activator protein-2-like transcription factor involved in regulating gene expression during nitrogen limitation in fungi? Appl Environ Microbiol, 1993; 59: 2335–2338.

Dijkerman R, Op den Camp HJ, Van der Drift C, Vogels GD. The role of the cellulolytic high molecular mass (HMM) complex of the anaerobic fungus *Piromyces* sp. strain E2 in the hydrolysis of microcrystalline cellulose. Arch Microbiol, 1997; 167: 137–142.

Ding SY, Lamed R, Bayer EA, Himmel ME. The bacterial scaffoldin: structure, function and potential applications in the nanosciences. Genet Eng (NY), 2003; 25: 209–225.

Diotallevi F, Mulder B. The cellulose synthase complex: a polymerization driven supramolecular motor. Biophys J, 2007; 92: 2666–2673.

Divne C, Ståhlberg J, Reinikainen T, Ruohonen L, Pettersson G, Knowles JKC, Teeri TT, Jones TA. The three-dimensional structure of the catalytic core of cellobiohydrolase I from *Trichoderma reesei*. Science, 1994; 265: 524–528.

Doblin MS, Kurek I, Jacob-Wilk D, Delmer DP. Cellulose biosynthesis in plants: from genes to rosettes. Plant Cell Physiol, 2002; 43: 1407–1420.

Dodds DR, Gross D. Chemistry: chemicals from biomass. Science, 2007; 318: 1250–1251.

Dominguez R, Souchon H, Spinelli S, Dauter Z, Wilson KS, Chauvaux S, Béguin P, Alzari PM. A common protein fold and similar active site in two distinct families of beta-glycanases. Nat Struct Biol, 1995; 2: 569–576.

Dooley DM, Rawlings J, Dawson JH, Stephens PJ, Andreasson L-E, Malmstrom BC, Gray HB. Spectroscopic studies of *Rhus vernicifera* and *Polyporus versicolor* laccase. Electronic structures of the copper sites. J Amer Chem Soc, 1979; 101: 5038–5046.

Doran-Peterson J, Cook DM, Brandon SK. Microbial conversion of sugars from plant biomass to lactic acid or ethanol. Plant J, 2008; 54: 582–592.

Dowzer CE, Kelly JM. Analysis of the *creA* gene, a regulator of carbon catabolite repression in *Aspergillus nidulans*. Mol Cell Biol, 1991; 11: 5701–5709.

Doyle WA, Blodig W, Weitch NC, Piontek K, Smith AT. Two substrate interaction sites in lignin peroxidase revealed by site-directed mutangesis. Biochemistry, 1998; 37: 15097–15105.

Drocourt D, Calmels T, Reynes J-P, Baron M, Tiraby G. Cassettes of the *Streptoalloteichus hidustanus ble* gene for transformation of lower and higher eukaryotes to phleomycin resistance. Nucl Acids Res, 1990; 18: 4009.

Druzhinina IS, Komoń-Zelazowska M, Atanasova L, Seidl V, Kubicek CP. Evolution and ecophysiology of the industrial producer *Hypocrea jecorina* (anamorph *Trichoderma reesei*) and a new sympatric agamospecies related to it. PLoS One, 2010; 5: e9191.

Druzhinina IS, Seidl-Seiboth V, Herrera-Estrella A, Horwitz BA, Kenerley CM, Monte E, Mukherjee PK, Zeilinger S, Grigoriev IV, Kubicek CP. *Trichoderma*: the genomics of opportunistic success. Nature Rev Microbiol, 2011; 9: 749–759.

Du J, Li S, Zhao H. Discovery and characterization of novel d-xylose-specific transporters from *Neurospora crassa* and *Pichia stipitis*. Mol Biosyst, 2010; 6: 2150–2156.

Dunlap JC, Borkovich KA, Henn MR, Turner GE, Sachs MS, Glass NL, McCluskey K, Plamann M, Galagan JE, Birren BW, Weiss RL, Townsend JP, Loros JJ, Nelson MA, Lambreghts R, Colot HV, Park G, Collopy P, Ringelberg C, Crew C, Litvinkova L, DeCaprio D, Hood HM, Curilla S, Shi M, Crawford M, Koerhsen M, Montgomery P, Larson L, Pearson M, Kasuga T, Tian C, Baştürkmen M, Altamirano L, Xu J. Enabling a community to dissect an organism: overview of the *Neurospora* functional genomics project. Adv Genet, 2007; 57: 49–96.

Durand H, Clanet M, Tiraby G. Genetic improvement of *Trichoderma reesei* for large scale cellulase production. Enzyme Microbiol Technol, 1988; 10: 341–346.

Duranová M, Spániková S, Wösten HA, Biely P, de Vries RP. Two glucuronoyl esterases of *Phanerochaete chrysosporium*. Arch Microbiol, 2009; 191: 133–140.

Dürre P. Fermentative butanol production: bulk chemical and biofuel. Ann N Y Acad Sci, 2008; 1125: 353–362.

Eastwood DC, Floudas D, Binder M, Majcherczyk A, Schneider P, Aerts A, Asiegbu FO, Baker SE, Barry K, Bendiksby M, Blumentritt M, Coutinho PM, Cullen D, de Vries RP, Gathman A, Goodell B, Henrissat B, Ihrmark K, Kauserud H, Kohler A, LaButti K, Lapidus A, Lavin JL, Lee YH, Lindquist E, Lilly W, Lucas S, Morin E, Murat C, Oguiza JA, Park J, Pisabarro AG, Riley R, Rosling A, Salamov A, Schmidt O, Schmutz J, Skrede I, Stenlid J, Wiebenga A, Xie X, Kües U, Hibbett DS, Hoffmeister D, Högberg N, Martin F, Grigoriev IV, Watkinson SC. The plant cell wall-decomposing machinery underlies the functional diversity of forest fungi. Science, 2011; 333: 762–765.

Eberhardt RY, Gilbert HJ, Hazlewood GP. Primary sequence and enzymic properties of two modular endoglucanases, Cel5A and Cel45A, from the anaerobic fungus *Piromyces equi*. Microbiology UK, 2000; 146: 1999–2008.

Ebringerova A, Heinze T. Xylan and xylan derivatives—biopolymers with valuable properties. 1. Naturally occurring xylan structures, isolation procedures and properties. Macromol Rapid Commun, 2000; 21: 542–556.

Eggert C, Temp U, Eriksson KEL. The ligninolytic system of the white rot fungus *Pycnoporus cinnabarinus*: purification and characterization of the laccase. Appl Environ Microbiol, 1996; 62: 1151–1158.

Ek M, Gierer J, Jansbo K. Study on the selectivity of bleaching with oxygen-containing species. Holzforschung, 1989; 43: 391–396.

Eklöf J, Planas A. Glycoside Hydrolase Family 16 in *CAZypedia*. 2011. Available at: http://www.cazypedia.org/ (accessed October 31, 2011).

El Gogary S, Leite A, Crivellaro O, El Dorry H, Eveleigh DE. *Trichoderma reesei* cellulases—from mutant to induction. In: Kubicek CP, Eveleigh DE, Esterbauer H, Steiner W, Kubicek-Pranz EM, (eds.), *Trichoderma reesei Cellulases: Biodiversity, Genetics, Physiology and Applications*, 1990, pp. 200–211. Royal Society of Chemistry, Cambridge.

El-Gogary S, Leite A, Crivellaro O, Eveleigh DE, El-Dorry H. Mechanism by which cellulose triggers cellobiohydrolase I gene expression in *Trichoderma reesei*. Proc Natl Acad Sci USA, 1989; 86: 6138–6141.

Elkins JG, Raman B, Keller M. Engineered microbial systems for enhanced conversion of lignocellulosic biomass. Curr Opin Biotechnol, 2010; 21: 657–662.

Elliott CG. *Reproduction in Fungi: Genetical and Physiological Aspects*, 1994, p. 324. Chapman & Hall, London.

Elshafei AM, Mohawed SM, Ammar MS, Abdel-Fatah OM. Properties of enzymes involved in D-galactonate catabolism in fungi. Antonie Van Leeuwenhoek, 1995; 67: 211–216.

Embley TM, van der Giezen M, Horner DS, Dyal PL, Bell S, Foster PG. Hydrogenosomes, mitochondria and early eukaryotic evolution. IUBMB Life, 2003; 55: 387–395.

Endler A, Persson S. Cellulose synthases and synthesis in Arabidopsis. Mol Plant, 2011; 4: 199–211.

Endo Y, Yokoyama M, Morimoto M, Shirai K, Chikamatsu G, Kato N, Tsukagoshi N, Kato M, Kobayashi T. Novel promoter sequence required for inductive expression of the *Aspergillus nidulans* endoglucanase gene *eglA*. Biosci Biotechnol Biochem, 2008; 72: 312–320.

Eneyskaya EV, Sundqvist G, Golubev AM, Ibatullin FM, Ivanen DR, Shabalin KA, Brumer H, Kulminskaya AA. Transglycosylating and hydrolytic activities of the beta-mannosidase from *Trichoderma reesei*. Biochimie, 2009,; 91: 632–638.

Enoki A, Tanaka H, Itakura S. Physical and chemical characteristics of glycopeptide from wood decay fungi. In: (Goodell B, Nicholas DD, Schultz TP, (eds): *Wood Deterioration and Preservation*, 1994, pp. 140–153. Oxford University Press, Washington, DC.

Entian KD, Fröhlich KU. *Saccharomyces cerevisiae* mutants provide evidence of hexokinase PII as a bifunctional enzyme with catalytic and regulatory domains for triggering carbon catabolite repression. J Bacteriol, 1984; 158: 29–35.

Epstein SS (ed). *Uncultivated Microorganisms*, 2009. Springer-Verlag, New York.

Eriksson KE, Blanchette RA, Ander P. *Microbial and Enzymatic Degradation of Wood and Wood Components*, 1990, p. 406. Springer, Berlin-Heidelberg-New York.

Eriksson T, Stals I, Collén A, Tjerneld F, Claeyssens M, Stålbrand H, Brumer H. Heterogeneity of homologously expressed *Hypocrea jecorina* (*Trichoderma reesei*) Cel7B catalytic module. Eur J Biochem, 2004; 271: 1266–1276.

Escutia MR, Bowater L, Edwards A, Bottrill AR, Burrell MR, Polanco R, Vicuña R, Bornemann S. Cloning and sequencing of two *Ceriporiopsis subvermispora* bicupin oxalate oxidase allelic isoforms: implications for the reaction specificity of oxalate oxidases and decarboxylases. Appl Environ Microbiol, 2005; 71: 3608–3616.

Esterbauer H, Steiner W, Labudova I, Hermann A, Hayn M. Production of *Trichoderma* cellulase in laboratory and pilot plant scale. Bioresource Technol, 1991; 36: 51–65.

Eveleigh DE, Montenecourt BS. Increasing yields of extracellular enzymes. Adv Appl Microbiol, 1979; 25: 57–74.

Ezeji TC, Qureshi N, Blaschek HP. Bioproduction of butanol from biomass: from genes to bioreactors. Curr Opin Biotechnol, 2007; 18: 220–227.

Fagerstam LG, Pettersson LG. The 1.4-β-glucan cellobiohydrolases of *Trichoderma reesei* QM 9414: a new type of cellulolytic synergism. FEBS Letters, 1980; 119: 97–100.

Faison BD, Kirk TK. Factors involved in the regulation of ligninase activity in *Phanerochaete chrysosporium*. Appl Environ Microbiol, 1985; 49: 299–304.

Fang X, Yano S, Inoue H, Sawayama S. Lactose enhances cellulase production by the filamentous fungus *Acremonium cellulolyticus*. J Biosci Bioeng, 2008; 106: 115–120.

Fass D. Hunting for alternative disulfide bond formation pathways: endoplasmic reticulum janitor turns professor and teaches a lesson. Mol Cell, 2010; 40: 685–686.

Fekete E, Seiboth B, Kubicek CP, Szentirmai A, Karaffa L. Lack of aldose 1-epimerase in *Hypocrea jecorina* (anamorph *Trichoderma reesei*): a key to cellulase gene expression on lactose. Proc Natl Acad Sci USA, 2008; 105: 7141–7146.

Felenbok B, Flipphi M, Nikolaev I. Ethanol catabolism in *Aspergillus nidulans*: a model system for studying gene regulation. Prog Nucleic Acid Res Mol Biol, 2001; 69: 149–204.

Fenn P, Kirk TK. Relationship of nitrogen to the onset and suppression of ligninolytic activity and secondary metabolism in *Phanerochaete chrysosporium*. Arch Microbiol, 1981; 130: 59–65.

Fiaux J, Cakar ZP, Sonderegger M, Wüthrich K, Szyperski T, Sauer U. Metabolic-flux profiling of the yeasts *Saccharomyces cerevisiae* and *Pichia stipitis*. Eukaryot Cell, 2003; 2: 170–180.

Fierro F, Kosalková K, Gutiérrez S, Martin JF. Autonomously replicating plasmids carrying the AMA1 region in *Penicillium chrysogenum*. Curr Genet, 1996; 29: 482–489.

FitzPatrick M, Champagne P, Cunningham MF, Whitney RA. A biorefinery processing perspective: treatment of lignocellulosic materials for the production of value-added products. Bioresour Technol, 2010; 101: 8915–8922.

Flipphi M, Sun J, Robellet X, Karaffa L, Fekete E, Zeng AP, Kubicek CP. Biodiversity and evolution of primary carbon metabolism in *Aspergillus nidulans* and other *Aspergillus* spp. Fungal Genet Biol, 2009; 46: S19–S44.

Flipphi M, van de Vondervoort PJ, Ruijter GJ, Visser V, Arst HN Jr, Felenbok B. Onset of carbon catabolite repression in *Aspergillus nidulans*. Parallel involvement of hexokinase and glucokinase in sugar signaling. J Biol Chem, 2003; 278: 11849–11857.

Foreman P, Van Solingen P, Goedegebuur F, Ward M. CIP1 polypeptides and their uses. 2010; US patent: 7923235.

Foreman PK, Brown D, Dankmeyer L, Dean R, Diener S, Dunn-Coleman NS, Goedegebuur F, Houfek TD, England GJ, Kelley AS, Meerman HJ, Mitchell T, Mitchinson C, Olivares HA, Teunissen PJ, Yao J, Ward M. Transcriptional regulation of biomass-degrading enzymes in the filamentous fungus *Trichoderma reesei*. J Biol Chem, 2003; 278: 31988–31997.

Forment JV, Ramón D, MacCabe AP. Consecutive gene deletions in *Aspergillus nidulans*: application of the Cre/loxP system. Curr Genet, 2006; 50: 217–224.

Fort DA, Remsing RC, Swatloski RP, Moyna P, Moyna G, Rogers RD. Can ionic liquids dissolve wood? Processing and analysis of lignocellulosic materials with 1-*n*-butyl-3-methylimidazolium chloride. Green Chem, 2007; 9: 63–69.

Fowler T, Brown RD Jr. The *bgl1* gene encoding extracellular β-glucosidase from *Trichoderma reesei* is required for rapid induction of the cellulase complex. Mol Microbiol, 1992; 6: 3225–3235.

Fox JM, Levine SE, Clark DS, Blanch HW. Initial- and processive-cut products from cellobiohydrolase-catalyzed hydrolysis of cellulose reveal rate-limiting steps and role of companion enzymes. Biochemistry, 2011, Nov 21 [Epub ahead of print].

Freudenberg K, Nash AC (eds). *Constitution and Biosynthesis of Lignin*, 1968. Springer-Verlag: Berlin.

Fritscher C, Messner R, Kubicek CP. Cellobiose metabolism and cellobiohydrolase I biosynthesis by *Trichoderma reesei*. Exp Mycol, 1990; 14: 451–461.

Fromme JC, Orci L, Schekman R. Coordination of COPII vesicle trafficking by Sec23. Trends Cell Biol, 2008; 18: 330–336.

Fry SC. Cross-linking of matrix polymers in the growing cell walls of angiosperms. Annu Rev Plant Physiol, 1986; 37: 165–186.

Fujii T, Fang X, Inoue H, Murakami K, Sawayama S. Enzymatic hydrolyzing performance of *Acremonium cellulolyticus* and *Trichoderma reesei* against three lignocellulosic materials. Biotechnol Biofuels, 2009; 2: 24.

Fujimoto Z, Kaneko S, Kuno A, Kobayashi H, Kusakabe I, Mizuno H. Crystal structures of decorated xylooligosaccharides bound to a family 10 xylanase from *Streptomyces olivaceoviridis* E-86. J Biol Chem, 2004; 279: 9606–9614.

Galazka JM, Tian C, Beeson WT, Martinez B, Glass NL, Cate JH. Cellodextrin transport in yeast for improved biofuel production. Science, 2010; 330: 84–86.

Galbe M, Sassner P, Wingren A, Zacchi G. Process engineering economics of bioethanol production. Adv Biochem Eng Biotechnol, 2007; 108: 303–327.

Galbe M, Zacchi G. Pretreatment of lignocellulosic materials for efficient bioethanol production. Adv Biochem Eng Biotechnol, 2007; 108: 41–65.

Gamauf C, Marchetti M, Kallio J, Puranen T, Vehmaanperä J, Allmaier G, Kubicek CP, Seiboth B. Characterization of the bga1-encoded glycoside hydrolase family 35 beta-galactosidase of *Hypocrea jecorina* with galacto-beta-D-galactanase activity. FEBS J, 2007; 274: 1691–1700.

Gao D, Chundawat SP, Krishnan C, Balan V, Dale BE. Mixture optimization of six core glycosyl hydrolases for maximizing saccharification of ammonia fiber expansion (AFEX) pretreated corn stover. Bioresour Technol, 2010; 101: 2770–2781.

García R, Cremata JA, Quintero O, Montesino R, Benkestock K, Ståhlberg J. Characterization of protein glycoforms with N-linked neutral and phosphorylated oligosaccharides: studies on the glycosylation of endoglucanase 1 (Cel7B) from *Trichoderma reesei*. Biotechnol Appl Biochem, 2001; 33: 141–152.

García-Cubero MA, González-Benito G, Indacoechea I, Coca M, Bolado S. Effect of ozonolysis pretreatment on enzymatic digestibility of wheat and rye straw. Bioresour Technol, 2009; 100: 1608–1613.

Garcia-Sandoval R, Wang Z, Binder M, Hibbett DS. Molecular phylogenetics of the *Gloeophyllales* and relative ages of clades of *Agaricomycotina* producing a brown rot. Mycologia, 2011; 103: 510–524.

Gardiner ES, Sarko A. Packing analysis of carbohydrates and polysaccharides. 16. The crystal strucrures of cellulose IVI and IVII. Can J Chem, 1985; 63: 173–180.

Gasser B, Saloheimo M, Rinas U, Dragosits M, Rodríguez-Carmona E, Baumann K, Giuliani M, Parrilli E, Branduardi P, Lang C, Porro D, Ferrer P, Tutino ML, Mattanovich D, Villaverde A. Protein folding and conformational stress in microbial cells producing recombinant proteins: a host comparative overview. Microb Cell Fact, 2008; 7: 11.

Gerini MF, Roccatano D, Baciocchi E, Nola AD. Molecular dynamics simulations of lignin peroxidase in solution. Biophys J, 2003; 84: 3883–3893.

Gerst JE. Message on the web: mRNA and ER co-trafficking. Trends Cell Biol, 2008; 18: 68–76.

Geysens S, Pakula T, Uusitalo J, Dewerte I, Penttilä M, Contreras R. Cloning and characterization of the glucosidase II alpha subunit gene of *Trichoderma reesei*: a frameshift mutation results in the aberrant glycosylation profile of the hypercellulolytic strain Rut-C30. Appl Environ Microbiol, 2005; 71: 2910–2924.

Giardina P, Faraco V, Pezzella C, Piscitelli A, Vanhulle S, Sannia G. Laccases: a never-ending story. Cell Mol Life Sci, 2010; 67: 369–385.

Gibson DM, King BC, Hayes ML, Bergstrom GC. Plant pathogens as a source of diverse enzymes for lignocellulose digestion. Curr Opin Microbiol, 2011; 14: 264–270.

Gielkens M, González-Candelas L, Sánchez-Torres P, van de Vondervoort P, de Graaff L, Visser J, Ramón D. The *abfB* gene encoding the major α-L-arabinofuranosidase of *Aspergillus nidulans*: nucleotide sequence, regulation and construction of a disrupted strain. Microbiol UK, 1999a; 145: 735–741.

Gielkens MM, Dekkers E, Visser J, de Graaff LH. Two cellobiohydrolase-encoding genes from *Aspergillus niger* require D-xylose and the xylanolytic transcriptional activator XlnR for their expression. Appl Environ Microbiol, 1999b; 65: 4340–4345.

Gilbert HJ. Glycoside Hydrolase Family 26 in *CAZypedia*. 2011a. Available at: http://www.cazypedia.org/(accessed October 31, 2011).

Gilbert HJ. Glycoside Hydrolase Family 43 in *CAZypedia*. 2011b. Available at: http://www.cazypedia.org/(accessed October 31, 2011).

Gilbert HJ. Glycoside Hydrolase Family 67 in *CAZypedia*. 2011c. Available at: http://www.cazypedia.org/(accessed October 31, 2011).

Gilbert HJ, Hazlewood GP, Laurie JI, Orpin CG, Xue GP. Homologous catalytic domains in a rumen fungal xylanase: evidence for gene duplication and prokaryotic origin. Mol Microbiol, 1992; 6: 2065–2072.

Gilbert HJ, Stålbrand H, Brumer H. How the walls come crumbling down: recent structural biochemistry of plant polysaccharide degradation. Curr Opin Plant Biol, 2008; 11: 338–348.

Godfrey T, West ST (eds.). *Industrial Enzymology*, 2nd edn, 1996. Macmillan Press, London.

Goedegebuur F, Fowler T, Phillips J, van der Kley P, van Solingen P, Dankmeyer L, Power SD. Cloning and relational analysis of 15 novel fungal endoglucanases from family 12 glycosyl hydrolase. Curr Genet, 2002; 41: 89–98.

Golan G, Shallom D, Teplitsky A, Zaide G, Shulami S, Baasov T, Stojanoff V, Thompson A, Shoham Y, Shoham G. Crystal structures of *Geobacillus stearothermophilus* alpha-glucuronidase complexed with its substrate and products: mechanistic implications. J Biol Chem, 2004; 279: 3014–3024.

Gold MH, Alic M. Molecular biology of the lignin-degrading basidiomycete *Phanerochaete chrysosporium*. Microbiol Rev, 1993; 57: 605–622.

Goll CM, Pastore A, Nilges M. The three-dimensional structure of a type I module from titin: a prototype of intracellular fibronectin type III domains. Structure, 1998; 6: 1291–1302.

Golubev AM, Nagem RA, Brandão Neto JR, Neustroev KN, Eneyskaya EV, Kulminskaya AA, Shabalin KA, Savel'ev AN, Polikarpov I. Crystal structure of alpha-galactosidase from *Trichoderma reesei* and its complex with galactose: implications for catalytic mechanism. J Mol Biol, 2004; 339: 413–422.

Goode D, Lewis ME, Crabbe MJ. Glucuronate metabolism in the mammalian eye lens—origin of xylitol. Biochem Soc Trans, 1996; 24: 453S.

Goodwin SB, M'barek SB, Dhillon B, Wittenberg AH, Crane CF, Hane JK, Foster AJ, Van der Lee TA, Grimwood J, Aerts A, Antoniw J, Bailey A, Bluhm B, Bowler J, Bristow J, van der Burgt A, Canto-Canché B, Churchill AC, Conde-Ferràez L, Cools HJ, Coutinho PM, Csukai M, Dehal P, De Wit P, Donzelli B, van de Geest HC, van Ham RC, Hammond-Kosack KE, Henrissat B, Kilian A, Kobayashi AK, Koopmann E, Kourmpetis Y, Kuzniar A, Lindquist E, Lombard V, Maliepaard C, Martins N, Mehrabi R, Nap JP, Ponomarenko A, Rudd JJ, Salamov A, Schmutz J, Schouten HJ, Shapiro H, Stergiopoulos I, Torriani SF, Tu H, de Vries RP, Waalwijk C, Ware SB, Wiebenga A, Zwiers LH, Oliver RP, Grigoriev IV, Kema GH. Finished genome of the fungal wheat pathogen *Mycosphaerella graminicola* reveals dispensome structure, chromosome plasticity, and stealth pathogenesis. PLoS Genet, 2011; 7: e1002070.

Gordon TR, Leveau JH. Plant pathology: a story about biology. Annu Rev Phytopathol, 2010; 48: 293–309.

Gouka RJ, Van Hartingsveldt W, Bovenberg RAL, Van Zeijl CMJ, Van Den Hondel CAMJJ, Van Gorcom RFM. Development of a new transformant selection system for *Penicillium chrysogenum*: isolation and characterization of the *P. chrysogenum* acetyl-coenzyme A synthetase gene (*facA*) and its use as a homologous selection marker. Appl Microbiol Biotechnol, 1993; 38: 514–519.

Gould JM. Alkaline peroxide delignification of agricultural residues to enhance enzymatic saccharification. Biotechnol Bioeng, 1984; 26: 46–52.

Grabar TB, Zhou S, Shanmugam KT, Yomano LP, Ingram LO. Methylglyoxal bypass identified as source of chiral contamination in L(+) and D(−)-lactate fermentations by recombinant *Escherichia coli*. Biotechnol Letts, 2006; 28: 1527–1535.

Grabber JH, Hatfield RD, Lu F, Ralph J. Coniferyl ferulate incorporation into lignin enhances the alkaline delignification and enzymatic degradation of cell walls. Biomacromolecules, 2008; 9: 2510–2516.

Grassick A, Murray PG, Thompson R, Collins CM, Byrnes L, Birrane G, Higgins TM, Tuohy MG. Three-dimensional structure of a thermostable native cellobiohydrolase, CBH IB, and molecular characterization of the cel7 gene from the filamentous fungus, *Talaromyces emersonii*. Eur J Biochem, 2004; 271: 4495–4506.

Green F, Larsen MJ, Winandy JE, Highley TL. Role of oxalic acid in incipient brown rot decay. Mater Organ, 1991; 26: 191–213.

Gregg DJ, Saddler JN. Factors affecting cellulose hydrolysis and the potential of enzyme recycle to enhance the efficiency of an integrated wood to ethanol process. Biotechnol Bioeng, 1996; 51: 375–383.

Grethlein HE. Pretreatment for enhanced hydrolysis of cellulosic biomass. Biotechnol Adv, 1984; 2: 43–62.

Griffith GW, Baker SE, Fliegerova K, Liegenstorfer A, van der Giezen M, Voigt K, Beakes, G. Anaerobic fungi *Neocallimastigomycota*. IMA Fungus, 2010; 1: 181–185.

Grishutin SG, Gusakov AV, Markov AV, Ustinov BB, Semenova MV, Sinitsyn AP. Specific xyloglucanases as a new class of polysaccharide-degrading enzymes. Biochim Biophys Acta, 2004; 1674: 268–281.

Guerriero G, Fugelstad J, Bulone V. What do we really know about cellulose biosynthesis in higher plants? J Integr Plant Biol, 2010; 52: 161–175.

Guillaumie S, Mzid R, Méchin V, Léon C, Hichri I, Destrac-Irvine A, Trossat-Magnin C, Delrot S, Lauvergeat V. The grapevine transcription factor WRKY2 influences the lignin pathway and xylem development in tobacco. Plant Mol Biol, 2010; 72: 215–234.

Guillén D, Sánchez S, Rodríguez-Sanoja R. Carbohydrate-binding domains: multiplicity of biological roles. Appl Microbiol Biotechnol, 2010; 85: 1241–1249.

Guo W, Jia W, Li Y, Chen S. Performances of *Lactobacillus brevis* for producing lactic acid from hydrolysate of lignocellulosics. Appl Biochem Biotechnol, 2010; 161: 124–136.

Gusakov AV. Alternatives to *Trichoderma reesei* in biofuel production. Trends Biotechnol, 2011; 29: 419–425.

Haakana H, Miettinen-Oinonen A, Joutsjoki V, Mantyla A, Suominen P, Vehmaanpera J. Cloning of cellulase genes from *Melanocarpus albomyces* and their efficient expression in *Trichoderma reesei*. Enzyme Microb Technol, 2004; 34: 159–167.

Hägglund P, Eriksson T, Collén A, Nerinckx W, Claeyssens M, Stålbrand H. A cellulose-binding module of the *Trichoderma reesei* beta-mannanase Man5A increases the mannan-hydrolysis of complex substrates. J Biotechnol, 2003; 101: 37–48.

Hahn-Hägerdal B, Galbe M, Gorwa-Grauslund MF, Lidén G, Zacchi G. Bio-ethanol—the fuel of tomorrow from the residues of today. Trends Biotechnol, 2006; 24: 549–556.

Hahn-Hägerdal B, Karhumaa K, Fonseca C, Spencer-Martins I, Gorwa-Grauslund MF. Towards industrial pentose-fermenting yeast strains. Appl Microbiol Biotechnol, 2007; 74: 937–953.

Halliwell B, Gutteridge JMC (eds). *Free Radicals in Biology and Medicine*, 1999. Oxford University Press, Oxford.

Halliwell G. Hydrolysis of fibrous cotton and reprecipitated cellulose by cellulolytic enzymes from microorganisms. Biochem J, 1965; 95: 270–281.

Hamada N, Okumura R, Fuse N, Kodaira R, Shimosaka M, Kanda T, Okazaki M. Isolation and transcriptional analysis of a cellulase gene (*cell*) from the basidiomycete *Irpex lacteus*. J Biosci Bioeng, 1999; 87: 97–102.

Hammel KE, Cullen D. Role of fungal peroxidases in biological ligninolysis. Curr Opin Plant Biol, 2008; 11: 349–355.

Hampton RY. ER-associated degradation in protein quality control and cellular regulation. Curr Opin Cell Biol, 2002; 14: 476–482.

Hanai T, Atsumi S, Liao JC. Engineered synthetic pathway for isopropanol production in *Escherichia coli*. Appl Environ Microbiol, 2007; 73: 7814–7818.

Hantus S, Pauly M, Darvill AG, Albersheim P, York WS. Structural characterization of novel L-galactose-containing oligosaccharide subunits of jojoba seed xyloglucans. Carbohydr Res, 1997; 304:;11–20.

Hara Y, Hinoki Y, Shimoi H, Ito K. Cloning and sequence analysis of endoglucanase genes from an industrial fungus, *Aspergillus kawachii*. Biosci Biotechnol Biochem, 2003; 67: 2010–2013.

Harding HP, Zhang Y, Bertolotti A, Zeng H, Ron D. Perk is essential for translational regulation and cell survival during the unfolded protein response. Mol Cell, 2000; 5: 897–904.

Harding HP, Zhang Y, Ron D. Protein translation and folding are coupled by an endoplasmic reticulum-resident kinase. Nature, 1999; 397: 271–274.

Harholt J, Suttangkakul A, Vibe Scheller H. Biosynthesis of pectin. Plant Physiol, 2010; 153: 384–395.

Harper AD, Bar-Peled M. Biosynthesis of UDP-xylose. Cloning and characterization of a novel *Arabidopsis* gene family, UXS, encoding soluble and putative membrane-bound UDP-glucuronic acid decarboxylase isoforms. Plant Physiol, 2002; 130: 2188–2198.

Harrier LA, Millam S. Biolistic transformation of arbuscular mycorrhizal fungi. Progress and perspectives. Mol Biotechnol, 2001; 18: 25–33.

Harris D, DeBolt S. Synthesis, regulation and utilization of lignocellulosic biomass. Plant Biotechnol J, 2010; 8: 244–262.

Harris JF. Acid hydrolysis and dehydration reactions for utilizing plant carbohydrates. Appl Poly Symp, 1975; 28: 131–144.

Harris PV, Welner D, McFarland KC, Re E, Navarro Poulsen JC, Brown K, Salbo R, Ding H, Vlasenko E, Merino S, Xu F, Cherry J, Larsen S, Lo Leggio L. Stimulation of lignocellulosic biomass hydrolysis by proteins of glycoside hydrolase family 61: structure and function of a large, enigmatic family. Biochemistry, 2010; 49: 3305–3316.

Harrison MJ, Nouwens AS, Jardine DR, Zachara NE, Gooley AA, Nevalainen H, Packer NH. Modified glycosylation of cellobiohydrolase I from a high cellulase-producing mutant strain of *Trichoderma reesei*. Eur J Biochem, 1998; 256: 119–127.

Hart DO, He S, Chany CJ 2nd, Withers SG, Sims PF, Sinnott ML, Brumer H 3rd. Identification of Asp-130 as the catalytic nucleophile in the main α-galactosidase from *Phanerochaete chrysosporium*, a family 27 glycosyl hydrolase. Biochemistry, 2000; 39: 9826–9836.

Hartl L, Gastebois A, Aimanianda V, Latgé JP. Characterization of the GPI-anchored endo β-1,3-glucanase Eng2 of *Aspergillus fumigatus*. Fungal Genet Biol, 2011; 48: 185–191.

Hartl L, Kubicek CP, Seiboth B. Induction of the *gal* pathway and cellulase genes involves no transcriptional inducer function of the galactokinase in *Hypocrea jecorina*. J Biol Chem, 2007; 282: 18654–18659.

Hartl L, Seiboth B. Sequential gene deletions in *Hypocrea jecorina* using a single blaster cassette. Curr Genet, 2005; 48: 204–211.

Harvey PJ, Schomaker HE, Palmer JM. Veratryl alcohol as a mediator and the role of radical cations in lignin biodegradation by *Phanerochaete chrysosporium*. FEBS Letters, 1986; 195: 242–246.

Hashimoto H. Recent structural studies of carbohydrate-binding modules. Cell Mol Life Sci, 2006,; 63: 2954–2967.

Hasper AA, Dekkers E, van Mil M, van de Vondervoort PJI, de Graaff LH. EglC, a new endoglucanase from *Aspergillus niger* with major activity towards xyloglucan. Appl Environ Microbiol, 2002; 68: 1556–1560.

Hasper AA, Trindade LM, van der Veen D, van Ooyen AJ, de Graaff LH. Functional analysis of the transcriptional activator XlnR from *Aspergillus niger*. Microbiol UK, 2004; 150: 1367–1375.

Hayakawa Y, Ishikawa E, Shoji JY, Nakano H, Kitamoto K. Septum-directed secretion in the filamentous fungus *Aspergillus oryzae*. Mol Microbiol, 2011; 81: 40–55.

Hayashi T, Kaida R. Functions of xyloglucan in plant cells. Mol Plant, 2011; 4: 17–24.

Hazelwood LA, Daran JM, van Maris AJ, Pronk JT, Dickinson JR. The Ehrlich pathway for fusel alcohol production: a century of research on *Saccharomyces cerevisiae* metabolism. Appl Environ Microbiol, 2008; 74: 2259–2266.

Heer D, Heine D, Sauer U. Resistance of *Saccharomyces cerevisiae* to high concentrations of furfural is based on NADPH-dependent reduction by at least two oxireductases. Appl Environ Microbiol, 2009; 75: 7631–7638.

Heilmann-Clausen J, Boddy L. Inhibition and stimulation effects in communities of wood decay fungi: exudates from colonized wood influence growth by other species. Microb Ecol, 2005; 49: 399–406.

Heinzelman P, Snow CD, Smith MA, Yu X, Kannan A, Boulware K, Villalobos A, Govindarajan S, Minshull J, Arnold FH. SCHEMA recombination of a fungal cellulase uncovers a single mutation that contributes markedly to stability. *J Biol Chem*, 2009; 284:26229–26233.

Helenius A, Aebi M. Roles of *N*-linked glycans in the endoplasmic reticulum. Annu Rev Biochem, 2004; 73: 1019–1049.

Hendriks AT, Zeeman G. Pretreatments to enhance the digestibility of lignocellulosic biomass. Bioresour Technol, 2009; 100: 10–18.

Henriksson G, Nutt A, Henriksson H, Petterson B, Stahlberg J, Johansson G, Pettersson G. Endoglucanase 28 (cel12A) a new *Phanerochaete chrysosporium* cellulase. Eur J Biochem, 1999; 259: 88–95.

Henrissat B. A classification of glycosyl hydrolases based on amino acid sequence similarities. Biochem J, 1991; 280: 309–316.

Henrissat B, Bairoch A. New families in the classification of glycosyl hydrolases based on amino acid sequence similarities. Biochem J, 1993; 293: 781–788.

Henrissat B, Claeyssens M, Tomme P, Lemesle L, Mornon JP. Cellulase families revealed by hydrophobic cluster analysis. Gene, 1989; 81: 83–95.

Hermann T. Industrial production of amino acids by coryneform bacteria. J Biotechnol, 2003; 104: 155–172.

Hermoso JA, Sanz-Aparicio J, Molina R, Juge N, González R, Faulds CB. The crystal structure of feruloyl esterase A from *Aspergillus niger* suggests evolutive functional convergence in feruloyl esterase family. J Mol Biol, 2004; 338: 495–506.

Herpoël-Gimbert I, Margeot A, Dolla A, Jan G, Mollé D, Lignon S, Mathis H, Sigoillot JC, Monot F, Asther M. Comparative secretome analyses of two *Trichoderma reesei* RUT-C30 and CL847 hypersecretory strains. Biotechnol Biofuels, 2008; 23: 18.

Herrmann MC, Vrsanska M, Jurickova M, Hirsch J, Biely P, Kubicek CP. The β-D-xylosidase of *Trichoderma reesei* is a multifunctional β-D-xylan xylohydrolase. Biochem J, 1997; 321: 375–381.

Hess K, Kissig J. Zur kenntnis der hochtemperatur-modifikation der cellulose (cellulose IV). Z Physikal Chem B, 1941; 49: 235–239.

Hildén K, Martinez AT, Hatakka A, Lundell T. The two manganese peroxidases Pr-MnP2 and Pr-MnP3 of Phlebia radiata, a lignin-degrading basidiomycete, are phylogenetically and structurally divergent. Fungal Genet Biol, 2005; 42: 403–419.

Hilditch S, Berghäll S, Kalkkinen N, Penttilä M, Richard P. The missing link in the fungal D-galacturonate pathway: identification of the L-threo-3-deoxy-hexulosonate aldolase. J Biol Chem, 2007; 282: 26195–26201.

Hinz SWA, Pouvreau L, Joosten R, Bartels J, Jonathan MC, Wery J, Schols HA. Hemicellulase production in *Chrysosporium lucknowense* C1. J Cereal Sci, 2009; 50: 318–323.

Hirano T, Enoki A, Tanaka H. Immunogold labeling of an extracellular substance producing hydroxyl radicals in wood degraded by brown-rot fungus *Tyromyces palustris*. J Wood Sci, 2000; 46: 45–51.

Hirano T, Tanaka H, Enoki A. Extracellular substance from the brown rot basidiomycete *Tyromyces palustris* that reduces molecular oxygen to hydroxyl radicals and ferric iron to ferrous iron. Mokuzai Gakkaishi, 1995; 41: 334–341.

Ho NW, Chen Z, Brainard AP. Genetically engineered *Saccharomyces* yeast capable of effective cofermentation of glucose and xylose. Appl Environ Microbiol, 1998; 64: 1852–1859.

Hoegger PJ, Kilaru S, James TY, Thacker JR, Kües U. Phylogenetic comparison and classification of laccase and related multicopper oxidase protein sequences. FEBS J, 2006; 273: 2308–2326.

Hofrichter M, Ullrich R, Pecyna MJ, Liers C, Lundell T. New and classic families of secreted fungal heme peroxidases. Appl Microbiol Biotechnol, 2010; 87: 871–897.

Hoffman M, Jia Z, Peña MJ, Cash M, Harper A, Blackburn AR 2nd, Darvill A, York WS. Structural analysis of xyloglucans in the primary cell walls of plants in the subclass *Asteridae*. Carbohydr Res, 2005; 340: 1826–1840.

Hogg D, Pell G, Dupree P, Goubet F, Martín-Orúe SM, Armand S, Gilbert HJ. The modular architecture of *Cellvibrio japonicus* mannanases in glycoside hydrolase families 5 and 26 points to differences in their role in mannan degradation. Biochem J, 2003; 371: 1027–1043.

Honda S, Selker EU. Tools for fungal proteomics: multifunctional *Neurospora* vectors for gene replacement, protein expression and protein purification. Genetics, 2009; 182: 11–23.

Hong J, Tamaki H, Yamamoto K, Kumagai H. Cloning of a gene encoding thermostable cellobiohydrolase from *Thermoascus aurantiacus* and its expression in yeast. Appl Microbiol Biotechnol, 2003; 63: 42–50.

Horbach R, Navarro-Quesada AR, Knogge W, Deising HB. When and how to kill a plant cell: infection strategies of plant pathogenic fungi. J Plant Physiol, 2011; 168: 51–62.

Hou J, Vemuri GN, Bao X, Olsson L. Impact of overexpressing NADH kinase on glucose and xylose metabolism in recombinant xylose-utilizing *Saccharomyces cerevisiae*. Appl Microbiol Biotechnol, 2009; 82: 909–919.

Hou Y, Wang T, Long H, Zhu H. Cloning, sequencing and expression analysis of the first cellulase gene encoding cellobiohydrolase 1 from a cold-adaptive *Penicillium chrysogenum* FS010. Acta Biochim Biophys Sin (Shanghai), 2007; 39: 101–107.

Hrmova M, De Gori R, Smith BJ, Fairweather JK, Driguez H, Varghese JN, Fincher GB. Structural basis for broad substrate specificity in higher plant β-D-glucan glucohydrolases. Plant Cell, 2002; 14: 1033–1052.

Hrmova M, MacGregor EA, Biely P, Stewart RJ, Fincher GB. Substrate binding and catalytic mechanism of a barley β-D-glucosidase(1,4)-β-D-glucan exohydrolase. J Biol Chem, 1998; 273: 11134–11143.

Hua D, Ma C, Song L, Lin S, Zhang Z, Deng Z, Xu P. Enhanced vanillin production from ferulic acid using adsorbent resin. Appl Microbiol Biotechnol, 2007; 74: 783–790.

Huang TT, Mitchinson C, Sala RF. Swollenin compositions and methods of increasing the efficiency of a cellulase. 2011; US Patent: 20110136182.

Huber GW, Dale BE. Grassoline at the pump. Sci Am Ind, 2009; 4: 40–45.

Hubert EE. *An Outline of Forest Pathology*, 1931, p. 543. John Wiley & Sons, New York, UK.

Hui JP, Lanthier P, White TC, McHugh SG, Yaguchi M, Roy R, Thibault P. Characterization of cellobiohydrolase I (Cel7A) glycoforms from extracts of *Trichoderma reesei* using capillary isoelectric focusing and electrospray mass spectrometry. J Chromatogr B Biomed Sci Appl, 2001; 752: 349–368.

Hui JP, White TC, Thibault P. Identification of glycan structure and glycosylation sites in cellobiohydrolase II and endoglucanases I and II from *Trichoderma reesei*. Glycobiol, 2002; 12: 837–849.

Ichinose H, Yoshida M, Kotake T, Kuno A, Igarashi K, Tsumuraya Y, Samejima M, Hirabayashi J, Kobayashi H, Kaneko S. An exo-β-1,3-galactanase having a novel β-1,3-galactan-binding module from *Phanerochaete chrysosporium*. J Biol Chem, 2005; 280: 25820–25829.

Igarashi K, Koivula A, Wada M, Kimura S, Penttilä M, Samejima M. High speed atomic force microscopy visualizes processive movement of *Trichoderma reesei* cellobiohydrolase I on crystalline cellulose. J Biol Chem, 2009; 284: 36186–36190.

Ilmén M, Saloheimo A, Onnela ML, Penttilä ME. Regulation of cellulase gene expression in the filamentous fungus *Trichoderma reesei*. Appl Environ Microbiol, 1997; 63: 1298–1306.

Ilmén M, Thrane C, Penttilä M. The glucose repressor gene cre1 of *Trichoderma*: isolation and expression of a full-length and a truncated mutant form. Mol Gen Genet, 1996; 251: 451–460.

Ingram LO, Gomez PF, Lai X, Moniruzzaman M, Wood BE, Yomano LP, York SW. Metabolic engineering of bacteria for ethanol production. Biotechnol Bioeng, 1998; 58: 204–214.

Inui M, Kawaguchi H, Murakami S, Vertès AA, Yukawa H. Metabolic engineering of *Corynebacterium glutamicum* for fuel ethanol production under oxygen-deprivation conditions. J Mol Microbiol Biotechnol, 2004, 8: 243–254.

Inui M, Murakami S, Okino S, Kawaguchi H, Vertes AA, Yukawa H. Metabolic analysis of *Corynebacterium glutamicum* during lactate and succinate productions under oxygen deprivation conditions. J Mol Microbiol Biotechnol, 2004; 7: 182–196.

Ishida T, Fujimoto Z, Ichinose H, Igarashi K, Kaneko S, Samejima M. Crystallization of selenomethionyl exo-β-1,3-galactanase from the basidiomycete *Phanerochaete chrysosporium*. Acta Crystallogr Sect F Struct Biol Cryst Commun, 2009; 65: 1274–1276.

Ishii T. *O*-acetylated oligosaccharides from pectins of potato tuber cell walls. Plant Physiol, 1997; 113: 1265–1272.

Ito T, Yokoyama E, Sato H, Ujita M, Funaguma T, Furukawa K, Hara A. Xylosidases associated with the cell surface of *Penicillium herquei* IFO 4674. J Biosci Bioeng, 2003; 96: 354–359.

Itoh H, Wada M, Honda Y, Kuwahara M, Watanabe T. Bioorganosolve pretreatments for simultaneous saccharification and fermentation of beech wood by ethanolysis and white rot fungi. J Biotechnol, 2003; 103: 273–280.

Ivanova C, Baath JA, Seibioth B, Kubicek CP. Systemic analysis of lactose metabolism in *Trichoderma reesei* reveals that uptake of lactose is dispensable for cellulase induction. 2012. (Submitted)

Iwaki T, Takegawa K. A set of loxP marker cassettes for Cre-mediated multiple gene disruption in *Schizosaccharomyces pombe*. Biosci Biotechnol Biochem, 2004; 68: 545–550.

Iyayi CB, Bruchmann E-E, Kubicek CP. Induction of cellulase formation in *Trichoderma reesei* by cellobiono-1,5-lactone. Arch Microbiol, 1989; 151: 326–330.

Jäger G, Wulfhorst H, Zeithammel EU, Elinidou E, Spiess AC, Büchs J. Screening of cellulases for biofuel production: online monitoring of the enzymatic hydrolysis of insoluble cellulose using high-throughput scattered light detection. Biotechnol J, 2011; 6: 74–85.

Jaklitsch WM, Samuels GJ, Dodd SL, Lu BS, Druzhinina IS. *Hypocrea rufa/Trichoderma viride*: a reassessment, and description of five closely related species with and without warted conidia. Stud Mycol, 2006; 56: 135–177.

Jeng WY, Wang NC, Lin MH, Lin CT, Liaw YC, Chang WJ, Liu CI, Liang PH, Wang AH. Structural and functional analysis of three β-glucosidases from bacterium *Clostridium cellulovorans*, fungus *Trichoderma reesei* and termite *Neotermes koshunensis*. J Struct Biol, 2011; 173: 46–56.

Jenkins J, Mayans O, Pickersgill R. Structure and evolution of parallel β-helix proteins. J Struct Biol, 1998; 122: 236–246.

Jensen D, Schekman R. COPII-mediated vesicle formation at a glance. J Cell Sci, 2011; 124: 1–4.

Jensen MH, Otten H, Christensen U, Borchert TV, Christensen LL, Larsen S, Leggio LL. Structural and biochemical studies elucidate the mechanism of rhamnogalacturonan lyase from *Aspergillus aculeatus*. J Mol Biol, 2010; 404: 100–111.

Jeppsson H, Alexander NJ, Hahn-Hagerdal B. Existence of cyanide-insensitive respiration in the yeast *Pichia stipitis* and its possible influence on product formation during xylose utilization. Appl Environ Microbiol, 1995; 61: 2596–2600.

Jeppsson M, Bengtsson O, Franke K, Lee H, Hahn-Hägerdal B, Gorwa-Grauslund MF. The expression of a *Pichia stipitis* xylose reductase mutant with higher K_M for NADPH increases ethanol production from xylose recombinant *Saccharomyces cerevisiae*. Biotechnol Bioeng, 2006; 93: 665–673.

Jeppsson M, Johansson B, Hahn-Hägerdal B, Gorwa-Grauslund MF. Reduced oxidative pentose phosphate pathway flux in recombinant xylose-utilizing *Saccharomyces cerevisiae* strains improves the ethanol yield from xylose. Appl Environ Microbiol, 2002; 68: 1604–1609.

Jin YS, Alper H, Yang YT, Stephanopoulos G. Improvement of xylose uptake and ethanol production in recombinant *Saccharomyces cerevisiae* through an inverse metabolic engineering approach. Appl Environ Microbiol, 2005; 71: 8249–8256.

Jin YS, Ni H, Laplaza JM, Jeffries TW. Optimal growth and ethanol production from xylose by recombinant *Saccharomyces cerevisiae* require moderate D-xylulokinase activity. Appl Environ Microbiol, 2003; 69: 495–503.

John RP, Sukumaran RK, Nampoothiri KM, and Pandey A, Statistical optimization of simultaneous saccharification and L(+)-lactic acid fermentation from cassava bagasse using mixed culture of lactobacilli by response surface methodology. Biochem Eng J, 2007; 36: 262–267.

Jojima T, Omumasaba CA, Inui M, Yukawa H. Sugar transporters in efficient utilization of mixed sugar substrates: current knowledge and outlook. Appl Microbiol Biotechnol, 2010; 85: 471–480.

Jorgensen H, Vibe-Pedersen J, Larsen J, Felby C. Liquefaction of lignocellulose at high-solids concentrations. Biotechnol Bioeng, 2007; 96: 862–870.

Jouhten P, Pitkänen E, Pakula T, Saloheimo M, Penttilä M, Maaheimo H. 13C-metabolic flux ratio and novel carbon path analyses confirmed that *Trichoderma reesei* uses primarily the respirative pathway also on the preferred carbon source glucose. BMC Syst Biol, 2009; 3: 104.

Kacher Y, Brumshtein B, Boldin-Adamsky S, Toker L, Shainskaya A, Silman I, Sussman JL, Futerman AH. Acid β-glucosidase: insights from structural analysis and relevance to Gaucher disease therapy. Biol Chem, 2008; 389: 1361–1369.

Kainz E, Gallmetzer A, Hatzl C, Nett JH, Li H, Schinko T, Pachlinger R, Berger H, Reyes-Dominguez Y, Bernreiter A, Gerngross T, Wildt S, Strauss J. *N*-glycan modification in *Aspergillus* species. Appl Environ Microbiol, 2008; 74: 1076–1086.

Kalnenieks U. Physiology of *Zymomonas mobilis*: some unanswered questions. Adv Microb Physiol, 2006; 51: 73–117.

Kalsner I, Hintz W, Reid LS, Schachter H. Insertion into *Aspergillus nidulans* of functional UDP-GlcNAc: alpha 3-D-mannoside β-1,2-*N*-acetylglucosaminyl-transferase I, the enzyme catalysing the first committed step from oligomannose to hybrid and complex *N*-glycans. Glycoconj J, 1995; 12: 360–370.

Kapich AN, Steffen KT, Hofrichter M, Hatakka A. Involvement of lipid peroxidation in the degradation of a non-phenolic lignin model compound by manganese peroxidase of the litter-decomposing fungus *Stropharia coronilla*. Biochem Biophys Res Commun, 2005; 330: 371–377.

Kapoor M. Gene transfer by electroporation of filamentous fungi. Methods Mol Biol, 1995; 47: 279–289.

Karhumaa K, Fromanger R, Hahn-Hägerdal B, Gorwa-Grauslund MF. High activity of xylose reductase and xylitol dehydrogenase improves xylose fermentation by recombinant *Saccharomyces cerevisiae*. Appl Microbiol Biotechnol, 2007; 73: 1039–1046.

Karkehabadi S, Hansson H, Kim S, Piens K, Mitchinson C, Sandgren M. The first structure of a glycoside hydrolase family 61 member, Cel61B from *Hypocrea jecorina*, at 1.6 Å resolution. J Mol Biol, 2008; 383: 144–154.

Kärkönen A, Koutaniemi S. Lignin biosynthesis studies in plant tissue cultures. J Integr Plant Biol, 2010; 52: 176–185.

Karlsson J, Saloheimo M, Siika-Aho M, Tenkanen M, Penttilä M, Tjerneld F. Homologous expression and characterization of Cel61A (EG IV) of *Trichoderma reesei*. Eur J Biochem, 2001; 268: 6498–6507.

Katayama T. Glycoside Hydrolase Family 95 in *CAZypedia*. 2011. Available at: http://www.cazypedia.org/ (accessed October 31, 2011).

Kato M. An overview of the CCAAT-box binding factor in filamentous fungi: assembly, nuclear translocation, and transcriptional enhancement. Biosci Biotechnol Biochem, 2005; 69: 663–672.

Katz ME, Kelly JM. Glucose. In: Borkovich KA, Ebbole DJ (eds) *Cellular and Molecular Biology of Filamentous Fungi*, 2010, pp. 291–311. ASM Press, Washington, DC.

Kauppinen S, Christgau S, Kofod LV, Halkier T, Dörreich K, Dalbøge H. Molecular cloning and characterization of a rhamnogalacturonan acetylesterase from *Aspergillus aculeatus*. Synergism between rhamnogalacturonan degrading enzymes. J Biol Chem, 1995; 270: 27172–27178.

Kawaguchi H, Vertes AA, Okino S, Inui M, Yukawa H. Engineering of a xylose metabolic pathway in *Corynebacterium glutamicum*. Appl Environ Microbiol, 2007; 72: 3418–3428.

Keegstra K, Walton J. Plant science. Beta-glucans–brewer's bane, dietician's delight. Science, 2006; 311: 1872–1873.

Keller NP, Hohn TM. Metabolic pathway gene clusters in filamentous fungi. Fungal Genet Biol, 1997; 21: 17–29.

Keränen S, Penttilä M. Production of recombinant proteins in the filamentous fungus *Trichoderma reesei*. Curr Opin Biotechnol, 1995; 6: 534–537.

Kersten PJ, Kalyanaraman B, Hammel KE, Reinhammar B. Comparison of lignin peroxidase, horseradish peroxidase and laccase in the oxidation of methoxybenzene. Biochem J, 1990; 268: 475–480.

Kersten P, Cullen D. Extracellular oxidative systems of the lignin-degrading Basidiomycete *Phanerochaete chrysosporium*. Fungal Genet Biol, 2007; 44: 77–87.

Kerstetter FD, Lyons JK. Wheat straw for ethanol production in Washington: a resource, technical, and economic assessment. Washington State University, Cooperative Extension Energy Program, 2001, 89 pp.

Keshwani DR, Cheng JJ. Modeling changes in biomass composition during microwave-based alkali pretreatment of switchgrass. Biotechnol Bioeng, 2010; 105: 88–97.

Kester HCM, Esteban Warren M, Orlando R, Benen JAE, Bergmann B, Visser J. Tandem mass spectrometric analysis of *Aspergillus niger* pectin methyl esterase; mode of action on fully methylesterified oligogalacturonates. Biochem J, 2000; 346: 469–474.

Khandke KM, Vithayathil PJ, Murthy SK. Purification and characterization of an α-D-glucuronidase from a thermophilic fungus, *Thermoascus aurantiacus*. Arch Biochem Biophys, 1989; 274: 511–517.

Khindaria A, Grover TA, Aust SD. Oxalate-dependent reductive activity of manganese peroxidase from *Phanerochaete chrysosporium*. Arch Biochem Biophys, 1994; 314: 301–306.

Kilby NJ, Snaith MR, Murray JA. Site-specific recombinases: tools for genome engineering. Trends Genet, 1993; 9: 413–421.

Kilpelainen I, Xie H, King A, Granstrom M, Heikkinen S, Argyropoulos DS. Dissolution of wood in ionic liquids. J Agric Food Chem, 2007; 55: 9142–9148.

Kim KK, Fravel DR, Papavizas GC. Glucose oxidase as the antifungal principle of talaron from *Talaromyces flavus*. Can J Microbiol, 1990; 36: 760–764.

Kim S, Dale BE. Global potential bioethanol production from wasted crops and crop residues. Biomass Bioenergy, 2004; 4: 361–375.

Kim S, Holtzapple MT. Lime pretreatment and enzymatic hydrolysis of corn stover. Bioresour Technol, 2005; 96: 1994–2006.

Kim TH, Lee YY. Pretreatment and fractionation of corn stover by ammonia recycle percolation process. Biores Technol, 2005; 96: 2007–2013.

Kimura M, Kamakura T, Tao QZ, Kaneko I, Yamaguchi I. Cloning of the blasticidin S deaminase gene (BSD) from *Aspergillus terreus* and its use as a selectable marker for *Schizosaccharomyces pombe* and *Pyricularia oryzae*. Mol Gen Genet, 1994; 242: 121–129.

King BC, Waxman KD, Nenni NV, Walker LP, Bergstrom GC, Gibson DM. Arsenal of plant cell wall degrading enzymes reflects host preference among plant pathogenic fungi. Biotechnol Biofuels, 2011; 4: 4.

Kirchhausen T. Making COPII coats. Cell, 2007; 129: 1251–1252.

Kirk TK, Ibach R, Mozuch MD, Conner AH, Highley TL. Characteristics of cotton cellulose depolymerized by a brown-rot fungus, by acid, or by chemical oxidants. Holzforschung, 1991; 45: 239–244.

Kishi K, Wariishi H, Marquez L, Dunford HB, Gold MH. Mechanism of manganese peroxidase compound II reduction. Effect of organic acid chelators and pH. Biochemistry, 1994; 33: 8694–8701.

Kleizen B, Braakman I. Protein folding and quality control in the endoplasmic reticulum. Curr Opin Cell Biol, 2004; 16: 343–349.

Klemm D, Heublein B, Fink H, Bohn A. Cellulose: fascinating biopolymer/sustainable raw material. Ang Chemie (Intl Edn), 2004; 44: 3358–3364.

Klinke HB, Ahring BK, Schmidt AS, Thomsen AB. Characterization of degradation products from alkaline wet oxidation of wheat straw. Bioresour Technol, 2002; 82: 15–26.

Klionsky DJ. Protein transport from the cytoplasm into the vacuole. J Membr Biol, 1997; 157: 105–115.

Koch A, Weigel CT, Schulz G. Cloning, sequencing, and heterologous expression of a cellulase-encoding cDNA (*cbh1*) from *Penicillium janthinellum*. Gene, 1993; 124: 57–65.

Koenigs JW. Hydrogen peroxide and iron: a proposed system for decomposition of wood by brown-rot basidiomycetes. Wood Fiber, 1974; 6: 66–80

Koivula A, Reinikainen T, Ruohonen L, Valkeajärvi A, Claeyssens M, Teleman O, Kleywegt GJ, Szardenings M, Rouvinen J, Jones TA, Teeri TT. The active site of *Trichoderma reesei* cellobiohydrolase II: the role of tyrosine 169. Protein Eng, 1996; 9: 691–699.

Koivula A, Ruohonen L, Wohlfahrt G, Reinikainen T, Teeri TT, Piens K, Claeyssens M, Weber M, Vasella A, Becker D, Sinnott ML, Zou JY, Kleywegt GJ, Szardenings M, Ståhlberg J, Jones TA. The active site of cellobiohydrolase Cel6A from *Trichoderma reesei*: the roles of aspartic acids D221 and D175. J Am Chem Soc, 2002; 124: 10015–10024.

Kolenová K, Ryabova O, Vrsanská M, Biely P. Inverting character of family GH115 α-glucuronidases. FEBS Lett, 2010; 584: 4063–4068.

Konno N, Igarashi K, Habu N, Samejima M, Isogai A. Cloning of the *Trichoderma reesei* cDNA encoding a glucuronan lyase belonging to a novel polysaccharide lyase family. Appl Environ Microbiol, 2009; 75: 101–107.

Kopke K, Hoff B, Kück U. Application of the *Saccharomyces cerevisiae* FLP/FRT recombination system in filamentous fungi for marker recycling and construction of knockout strains devoid of heterologous genes. Appl Environ Microbiol, 2010; 76: 4664–4674.

Koseki T, Fushinobu S, Ardiansyah, Shirakawa H, Komai M. Occurrence, properties, and applications of feruloyl esterases. Appl Microbiol Biotechnol, 2009; 84: 803–810.

Koseki T, Mese Y, Nishibori N, Masaki K, Fujii T, Handa T, Yamane Y, Shiono Y, Murayama T, Iefuji H. Characterization of an α-L-rhamnosidase from *Aspergillus kawachii* and its gene. Appl Microbiol Biotechnol, 2008; 80: 1007–1013.

Koshland DE. Stereochemistry and mechanism of enzymatic reactions. Biol Rev, 1953; 28: 416–436.

Kotake T, Hirata N, Degi Y, Ishiguro M, Kitazawa K, Takata R, Ichinose H, Kaneko S, Igarashi K, Samejima M, Tsumuraya Y. Endo-{beta}-1,3-galactanase from Winter mushroom *Flammulina velutipes*. J Biol Chem, 2011; 286: 27848–27854.

Kremer SM, Wood PM. Production of fenton reagent by cellobiose oxidase from cellulolytic cultures of *Phanerochaete chrysosporium*. Eur J Biochem, 1992; 208: 807–814.

Kroll J, Klinter S, Schneider C, Voss I, Steinbüchel A. Plasmid addiction systems: perspectives and applications in biotechnology. Microb Biotechnol, 2010; 3: 634–657.

Kruszewska JS, Saloheimo M, Migdalski A, Orlean P, Penttilä M, Palamarczyk G. Dolichol phosphate mannose synthase from the filamentous fungus *Trichoderma reesei* belongs to the human and *Schizosaccharomyces pombe* class of the enzyme. Glycobiology, 2000; 10: 983–991.

Kuan IC, Tien M. Phosphorylation of lignin peroxidases from *Phanerochaete chrysosporium*. Identification of mannose 6-phosphate. J Biol Chem, 1989; 264: 20350–20355.

Kubicek CP. Release of carboxymethyl-cellulase and β-glucosidase from the cell walls of *Trichoderma reesei*. Eur J Appl Microbiol Biotechnol, 1981; 13: 226–231.

Kubicek CP, Bölzlbauer UM, Kovacs W, Mach RL, Kuhls K, Lieckfeldt E, Börner T, Samuels GJ. Cellulase formation by species of *Trichoderma* sect. *Longibrachiatum* and of *Hypocrea* spp. with anamorphs referable to *Trichoderma* sect. *Longibrachiatum*. Fungal Genet Biol, 1996; 20: 105–114.

Kubicek CP, Herrera-Estrella A, Seidl-Seiboth V, Martinez DA, Druzhinina IS, Thon M, Zeilinger S, Casas-Flores S, Horwitz BA, Mukherjee PK, Mukherjee M, Kredics L, Alcaraz LD, Aerts A, Antal Z, Atanasova L, Cervantes-Badillo MG, Challacombe J, Chertkov O, McCluskey K, Coulpier F, Deshpande N, von Döhren H, Ebbole DJ, Esquivel-Naranjo EU, Fekete E, Flipphi M, Glaser F, Gómez-Rodríguez EY, Gruber S, Han C, Henrissat B, Hermosa R, Hernández-Oñate M, Karaffa L, Kosti I, Le Crom S, Lindquist E, Lucas S, Lübeck M, Lübeck PS, Margeot A, Metz B, Misra M, Nevalainen H, Omann M, Packer N, Perrone G, Uresti-Rivera EE, Salamov A, Schmoll M, Seiboth B, Shapiro H, Sukno S, Tamayo-Ramos JA, Tisch D, Wiest A, Wilkinson HH, Zhang M, Coutinho PM, Kenerley CM, Monte E, Baker SE, Grigoriev IV. Comparative genome sequence analysis underscores mycoparasitism as the ancestral life style of *Trichoderma*. Genome Biol, 2011; 12: R40.

Kubicek CP, Messner R, Gruber F, Mandels M, Kubicek-Pranz EM. Triggering of cellulase biosynthesis by cellulose in *Trichoderma reesei*. Involvement of a constitutive, sophorose-inducible, glucose-inhibited beta-diglucoside permease. J Biol Chem, 1993; 268: 19364–19368.

Kubicek CP, Penttilä M. Regulation of production of plant polysaccharide degrading enzymes by *Trichoderma*. In: Harman GE, Kubicek CP (eds) *Trichoderma & Gliocladium*, 1998, pp. 49–72. Taylor & Francis, London.

Kubicek CP, Mikus M, Schuster A, Schmoll M, Seiboth B. Metabolic engineering strategies for the improvement of cellulase production by *Hypocrea jecorina*. Biotechnol Biofuels, 2009; 2: 19.

Kubicek CP, Mühlbauer G, Grotz M, John E, Kubicek-Pranz EM. Properties of a conidial bound cellulase enzyme system from *Trichoderma reesei*. J Gen Microbiol, 1988; 134: 1215–1222.

Kück U, Hoff B. New tools for the genetic manipulation of filamentous fungi. Appl Microbiol Biotechnol, 2010; 86: 51–62.

Kuhls K, Lieckfeldt E, Samuels GJ, Kovacs W, Meyer W, Petrini O, Gams W, Börner T, Kubicek CP. Molecular evidence that the asexual industrial fungus *Trichoderma reesei* is a clonal derivative of the ascomycete *Hypocrea jecorina*. Proc Natl Acad Sci USA, 1996; 93: 7755–7760.

Kühnel S, Westphal Y, Hinz SW, Schols HA, Gruppen H. Mode of action of *Chrysosporium lucknowense* C1 α-L-arabinohydrolases. Bioresour Technol, 2011; 102: 1636–1643.

Kulmburg P, Mathieu M, Dowzer C, Kelly J, Felenbok B. Specific binding sites in the alcR and *alcA* promoters of the ethanol regulon for the CREA repressor mediating carbon catabolite repression in *Aspergillus nidulans*. Mol Microbiol, 1993; 7: 847–857.

Kulminskaya A, Maksimainen M, Rouvinen J. Glycoside Hydrolase Family 35 in *CAZypedia*. 2011. Available at: http://www.cazypedia.org/ (accessed October 31, 2011).

Kuorelahti S, Jouhten P, Maaheimo H, Penttilä M, Richard P. L-galactonate dehydratase is part of the fungal path for D-galacturonic acid catabolism. *Mol Microbiol*. 2006; 61: 1060–1068.

Kurasin M, Väljamäe P. Processivity of cellobiohydrolases is limited by the substrate. J Biol Chem, 2011; 286: 169–177.

Kuratsu M, Taura A, Shoji JY, Kikuchi S, Arioka M, Kitamoto K. Systematic analysis of SNARE localization in the filamentous fungus *Aspergillus oryzae*. Fungal Genet Biol, 2007; 44: 1310–1323.

Kurek B, Gaudard F. Oxidation of spruce wood sawdust by MnO_2 plus oxalate: a biochemical investigation. J Agri Food Chem, 2000; 48: 3058–3062.

Kurzatkowski W, Törrönen A, Filipek J, Mach RL, Herzog P, Sowka S, Kubicek CP. Glucose-induced secretion of *Trichoderma reesei* xylanases. Appl Environ Microbiol, 1996,; 62: 2859–2865.

Lafon A, Han KH, Seo JA, Yu JH, d'Enfert C. G-protein and cAMP-mediated signaling in *Aspergilli*: a genomic perspective. Fungal Genet Biol, 2006; 43: 490–502.

Langston JA, Shaghasi T, Abbate E, Xu F, Vlasenko E, Sweeney MD. Oxidoreductive cellulose depolymerization by the enzymes cellobiose dehydrogenase and glycoside hydrolase 61. Appl Environ Microbiol, 2011; 77: 7007–7015.

Lara-Márquez A, Zavala-Páramo MG, López-Romero E, Camacho HC. Biotechnological potential of pectinolytic complexes of fungi. Biotechnol Lett, 2011; 33: 859–868.

Larsson AM, Anderson L, Xu B, Munõz IG, Usón I, Janson J-C, Stålbrand H, Ståhlberg J. Three-dimensional crystal structure and enzymic characterization of β-mannanase Man5A from blue mussel *Mytilus edulis*. J Mol Biol, 2006; 357: 1500–1510.

Lavigne JA, Hill CMD, Tremblay A, St. Pierre P, Tomashek JJ. Cellulase variants with reduced inhibition by glucose. 2009; US Patent: 0186381 A1.

Lawrence CW. Classical mutagenesis techniques. Methods Enzymol, 2002; 350: 189–199.

Lazar T, Götte M, Gallwitz D. Vesicular transport: how many Ypt/Rab-GTPases make a eukaryotic cell? Trends Biochem Sci, 1997; 22: 468–472.

Le Crom S, Schackwitz W, Pennacchio L, Magnuson JK, Culley DE, Collett JR, Martin J, Druzhinina IS, Mathis H, Monot F, Seiboth B, Cherry B, Rey M, Berka R, Kubicek CP, Baker SE, Margeot A. Tracking the roots of cellulase hyperproduction by the fungus *Trichoderma reesei* using massively parallel DNA sequencing. Proc Natl Acad Sci USA, 2009; 106: 16151–16156.

Le Nours J, Ryttersgaard C, Lo Leggio L, Østergaard PR, Borchert TV, Christensen LL, Larsen S. Structure of two fungal beta-1,4-galactanases: searching for the basis for temperature and pH optimum. Protein Sci, 2003; 12: 1195–1204.

Le Nours J, De Maria L, Welner D, Jørgensen CT, Christensen LL, Borchert TV, Larsen S, Lo Leggio L. Investigating the binding of β-1,4-galactan to *Bacillus licheniformis* β-1,4-galactanase by crystallography and computational modeling. Proteins, 2009; 75: 977–989.

Leandro MJ, Fonseca C, Gonçalves P. Hexose and pentose transport in ascomycetous yeasts: an overview. FEMS Yeast Res, 2009; 9: 511–525.

Lee K, Tirasophon W, Shen X, Michalak M, Prywes R, Okada T, Yoshida H, Mori K, Kaufman RJ. IRE1-mediated unconventional mRNA splicing and S2P-mediated ATF6 cleavage merge to regulate XBP1 in signaling the unfolded protein response. Genes Dev, 2002; 16: 452–466.

Lee PC, Lee SY, Hong SH, Chang HN, Park SC. Biological conversion of wood hydrolysate to succinic acid by *Anaerobiospirillum succiniciproducens*. Biotechnol Lett, 2003a; 25: 111–114.

Lee PC, Lee WG, Kwon S, Lee SY, Chang HN. Batch and continuous cultivation of *Anaerobiospirillum succiniciproducens* for the production of succinic acid from whey. Appl Microbiol Biotechnol, 2000a; 54: 23–27.

Lee RC, Hrmova M, Burton RA, Lahnstein J, Fincher GB. Bifunctional family 3 glycoside hydrolases from barley with alpha-L-arabinofuranosidase and beta-D-xylosidase activity. Characterization, primary structures, and COOH-terminal processing. J Biol Chem, 2003b; 278: 5377–5387.

Lee SS, Ha JK, Cheng K. Relative contributions of bacteria, protozoa, and fungi to *in vitro* degradation of orchard grass cell walls and their interactions. Appl Environ Microbiol, 2000b; 66: 3807–3813.

Lee SY. Deciphering bioplastic production. Nat Biotechnol, 2006; 24: 1227–1229.

Lee SY, Choi J, Wong HH. Recent advances in polyhydroxyalkanoate production by bacterial fermentation: mini-review. Int J Biol Macromol, 1999; 25: 31–36.

Lehtio J, Sugiyama J, Gustavsson M, Fransson L, Linder M, Teeri TT. The binding specificity and affinity determinants of family 1 and family 3 cellulose binding modules. Proc Natl Acad Sci USA, 2003; 100: 484–489.

Lejeune R, Baron GV. Simulation of growth of a filamentous fungus in 3 dimensions. Biotechnol Bioeng, 1997; 53: 139–150.

Lejeune R, Nielsen J, Baron GV. Morphology of *Trichoderma reesei* QM 9414 in submerged cultures. Biotechnol Bioeng, 1995; 47: 609–615.

Lerouxel O, Cavalier DM, Liepman AH, Keegstra K. Biosynthesis of plant cell wall polysaccharides—a complex process. Curr Opin Plant Biol, 2006; 9: 621–630.

Lev S, Horwitz BA. A mitogen-activated protein kinase pathway modulates the expression of two cellulase genes in *Cochliobolus heterostrophus* during plant infection. Plant Cell, 2003; 15: 835–844.

Levasseur A, Saloheimo M, Navarro D, Andberg M, Monot F, Nakari-Setälä T, Asther M, Record E. Production of a chimeric enzyme tool associating the *Trichoderma reesei* swollenin with the *Aspergillus niger* feruloyl esterase A for release of ferulic acid. Appl Microbiol Biotechnol, 2006; 73: 872–880.

Levine SE, Fox JM, Blanch HW, Clark DS. A mechanistic model of the enzymatic hydrolysis of cellulose. Biotechnol Bioeng, 2010; 107: 37–51.

Levine SE, Fox JM, Clark DS, Blanch HW. A mechanistic model for rational design of optimal cellulase mixtures. Biotechnol Bioeng, 2011, doi: 10.1002/bit.23249.

Levy S, Staehelin LA. Synthesis, assembly and function of plant cell wall macromolecules. Curr Opin Cell Biol, 1992; 4: 856–862.

Li H, Zhang X, Shen Y, Dong Y, Bao X. Inhibitors and their effects on *Saccharomyces cerevisiae* and relevant countermeasures in bioprocess of ethanol production from lignocelluloses—a review. Sheng Wu Gong Cheng Xue Bao, 2009a; 25: 1321–1328.

Li K, Horanyi PS, Collins R, Phillips RS, Eriksson KL. Investigation of the role of 3-hydroxyanthranilic acid in the degradation of lignin by white-rot fungus *Pycnoporus cinnabarinus*. Enzyme Microb Technol, 2001; 28: 301–307.

Li LL, McCorkle S, Monchy S, Taghavi S, van der Lelie D. Bioprospecting metagenomes: glycosyl hydrolases for converting biomass. Biotechnol Biofuel, 2009b; 2: 10.

Li LL, Taghavi S, McCorkle SM, Zhang YB, Blewitt MG, Brunecky R, Adney WS, Himmel ME, Brumm P, Drinkwater C, Mead DA, Tringe SG, Lelie D. Bioprospecting metagenomics of decaying wood: mining for new glycoside hydrolases. Biotechnol Biofuels, 2011; 4: 23.

Li XL, Skory CD, Cotta MA, Puchart V, Biely P. Novel family of carbohydrate esterases, based on identification of the *Hypocrea jecorina* acetyl esterase gene. Appl Environ Microbiol, 2008; 74: 7482–7492.

Li XL, Spániková S, de Vries RP, Biely P. Identification of genes encoding microbial glucuronoyl esterases. FEBS Letts, 2007; 581: 4029–4035.

Liggenstoffer AS, Youssef NH, Couger MB, Elshahed MS. Phylogenetic diversity and community structure of anaerobic gut fungi (phylum *Neocallimastigomycota*) in ruminant and non-ruminant herbivores. ISME J, 2010; 4: 1225–1235.

Limón MC, Pakula T, Saloheimo M, Penttilä M. The effects of disruption of phosphoglucose isomerase gene on carbon utilisation and cellulase production in *Trichoderma reesei* Rut-C30. Microb Cell Fact, 2011; 10: 40.

Lin H, Bennett GN, San KY. Fed-batch culture of a metabolically engineered *Escherichia coli* strain designed for high-level succinate production and yield under aerobic conditions. Biotechnol Bioeng, 2005; 90: 775–779.

Lin L, Wang F, Wei D. Chlorimuron ethyl as a new selectable marker for disrupting genes in the insect-pathogenic fungus *Metarhizium robertsii*. J Microbiol Methods, 2011 Aug 6 [Epub ahead of print]

Linder M, Lindeberg G, Reinikainen T, Teeri TT, Pettersson G. The difference in affinity between two fungal cellulose-binding domains is dominated by a single amino acid substitution. FEBS Lett, 1995; 372: 96–98.

Lipschütz JH, Mostov KE. Exocytosis: the many masters of the exocyst. Curr Biol, 2002; 12: R212–R214.

Liu H, Xu Y, Zheng Z, Liu D. 1,3-Propanediol and its copolymers: research, development and industrialization. Biotechnol J, 2010a; 5: 1137–1148.

Liu Y, Yoshida M, Kurakata Y, Miyazaki T, Igarashi K, Samejima M, Fukuda K, Nishikawa A, Tonozuka T. Crystal structure of a glycoside hydrolase family 6 enzyme, CcCel6C, a cellulase constitutively produced by *Coprinopsis cinerea*. FEBS J, 2010b; 277: 1532–1542.

Liu ZL. Molecular mechanisms of yeast tolerance and in situ detoxification of lignocellulose hydrolysates. Appl Microbiol Biotechnol, 2011; 90: 809–825.

Ljungdahl LG. The cellulase/hemicellulase system of the anaerobic fungus *Orpinomyces* PC-2 and aspects of its applied use. Ann NY Acad Sci, 2008; 1125: 308–321.

Lo Leggio L. Glycoside Hydrolase Family 53 in *CAZypedia*. 2011. Available at: http://www.cazypedia.org/ (accessed October 31, 2011).

Lo Leggio L, Kalogiannis S, Eckert K, Teixeira SC, Bhat MK, Andrei C, Pickersgill RW, Larsen S. Substrate specificity and subsite mobility in *Thermomyces aurantiacus* xylanase 10A. FEBS Lett, 2001; 509: 303–308.

Lockhart RJ, Van Dyke MI, Beadle IR, Humphreys P, McCarthy AJ. Molecular biological detection of anaerobic gut fungi (*Neocallimastigales*) from landfill sites. Appl Environ Microbiol, 2006; 72: 5659–5661.

Lockington RA, Rodbourn L, Barnett S, Carter CJ, Kelly JM. Regulation by carbon and nitrogen sources of a family of cellulases in *Aspergillus nidulans*. Fungal Genet Biol, 2002; 37: 190–196.

Lombard V, Bernard T, Rancurel C, Brumer H, Coutinho PM, Henrissat B. A hierarchical classification of polysaccharide lyases for glycogenomics. Biochem J, 2010; 432: 437–444.

López-Cortés N, Reyes-Duarte D, Beloqui A, Polaina J, Ghazi I, Golyshina OV, Ballesteros A, Golyshin PN, Ferrer M. Catalytic role of conserved HQGE motif in the CE6 carbohydrate esterase family. FEBS Lett, 2007; 581: 4657–4662.

Lovering AL, Lee SS, Kim YW, Withers SG, Strynadka NC. Mechanistic and structural analysis of a family 31 alpha-glycosidase and its glycosyl-enzyme intermediate. J Biol Chem, 2005; 280: 2105–2115.

Lubertozzi D, Keasling JD. Developing *Aspergillus* as a host for heterologous expression. Biotechnol Adv, 2009; 27: 53–75.

Lucia LA, Argyropoulos DS, Adamopoulos L, Gaspar AR. Chemicals and energy from biomass. Can J Chem, 2006; 84: 960–970.

Ludvigsen S, Poulsen FM. Three-dimensional structure in solution of barwin, a protein from barley seed. Biochemistry, 1992; 31: 8783–8789.

Lundell TK, Mäkelä MR, Hildén K. Lignin-modifying enzymes in filamentous basidiomycetes: ecological, functional and phylogenetic review. J Basic Microbiol, 2010; 50: 1–16.

Luonteri E, Alatalo E, Siika-Aho M, Penttilä M, Tenkanen M. α-Galactosidases of *Penicillium simplicissimum*: production, purification and characterization of the gene encoding AGLI. Biotechnol Appl Biochem, 1998; 28: 179–188.

Lynd LR, Weimer PJ, van Zyl WH, Pretorius IS. Microbial cellulose utilization: fundamentals and biotechnology. Microbiol Mol Biol Rev, 2002; 66: 506–577.

Ma B, Mayfield MB, Godfrey BJ, Gold MH. Novel promoter sequence required for manganese regulation of manganese peroxidase isozyme 1 gene expression in *Phanerochaete chrysosporium*. Eukaryot Cell, 2004; 3: 579–588.

Mabee WE, Saddler JN. Bioethanol from lignocellulosics: status and perspectives in Canada. Bioresour Technol, 2010; 101: 4806–4813.

Mabee WE, Fraser ED, McFarlane PN, Saddler JN. Canadian biomass reserves for biorefining. Appl Biochem Biotechnol, 2006; 132: 22–40.

Mach-Aigner AR, Grosstessner-Hain K, Poças-Fonseca MJ, Mechtler K, Mach RL. From an electrophoretic mobility shift assay to isolated transcription factors: a fast genomic-proteomic approach. BMC Genomics, 2010a; 11: 644.

Mach-Aigner AR, Pucher ME, Mach RL. D-Xylose as a repressor or inducer of xylanase expression in *Hypocrea jecorina* (*Trichoderma reesei*). Appl Environ Microbiol, 2010b; 76: 1770–1776.

Mach-Aigner AR, Pucher ME, Steiger MG, Bauer GE, Preis SJ, Mach RL. Transcriptional regulation of *xyr1*, encoding the main regulator of the xylanolytic and cellulolytic enzyme system in *Hypocrea jecorina*. Appl Environ Microbiol, 2008; 74: 6554–6562.

MacLeod AM, Lindhorst T, Withers SG, Warren RA. The acid/base catalyst in the exoglucanase/xylanase from *Cellulomonas fimi* is glutamic acid 127: evidence from detailed kinetic studies of mutants. Biochemistry, 1994; 33: 6371–6376.

Maeda Y, Kinoshita T. Dolichol-phosphate mannose synthase: structure, function and regulation. Biochim Biophys Acta, 2008; 1780: 861–868.

Magliano P, Flipphi M, Sanglard D, Poirier Y. Characterization of the *Aspergillus nidulans* biotin biosynthetic gene cluster and use of the *bioDA* gene as a new transformation marker. Fungal Genet Biol, 2011; 48: 208–215.

Makarova K, Slesarev A, Wolf Y, Sorokin A, Mirkin B, Koonin E, Pavlov A, Pavlova N, Karamychev V, Polouchine N, Shakhova V, Grigoriev I, Lou Y, Rohksar D, Lucas S, Huang K, Goodstein DM, Hawkins T, Plengvidhya V, Welker D, Hughes J, Goh Y, Benson A, Baldwin K, Lee JH, Díaz-Muñiz I, Dosti B, Smeianov V, Wechter W, Barabote R, Lorca G, Altermann E, Barrangou R, Ganesan B, Xie Y, Rawsthorne H, Tamir D, Parker C, Breidt F, Broadbent J, Hutkins R, O'Sullivan D, Steele J, Unlu G, Saier M, Klaenhammer T, Richardson P, Kozyavkin S, Weimer B, Mills D. Comparative genomics of the lactic acid bacteria. Proc Natl Acad Sci USA, 2006; 103: 15611–15616.

Mäki-Arvela P, Anugwom I, Virtanen P, Sjöholm R, Mikkola JP. Dissolution of lignocellulosic materials and its constituents using ionic liquids—a review. Ind Crops Prod, 2010, 32: 175–201.

Maksimainen M, Hakulinen N, Kallio JM, Timoharju T, Turunen O, Rouvinen J. Crystal structures of *Trichoderma reesei* beta-galactosidase reveal conformational changes in the active site. J Struct Biol, 2011; 174: 156–163.

Malkin R, Malmstrom BG. The state and function of copper in biological systems. Adv Enzymol, 1970; 33: 177–244.

Malsam J, Söllner TH. Organization of SNAREs within the Golgi Stack. Cold Spring Harb Perspect Biol, 2011; pii: a005249v1.

Manavalan A, Adav SS, Sze SK. iTRAQ-based quantitative secretome analysis of *Phanerochaete chrysosporium*. J Proteomics, 2011, Sep 13 [Epub ahead of print].

Mandels M, Parrish FW, Reese ET. Sophorose as an inducer of cellulase in *Trichoderma viride*. J Bacteriol, 1962; 83: 400–408.

Mansfield SD, DeJong E, Saddler JN. Cellobiose dehydrogenase, an active agent in cellulose depolymerization. Appl Environ Microbiol, 1997; 63: 3804–3809.

Manubens A, Avila M, Canessa P, Vicuña R. Differential regulation of genes encoding manganese peroxidase (MnP) in the basidiomycete *Ceriporiopsis subvermispora*. Curr Genet, 2003; 43: 433–438.

Manzanares P, de Graaff LH, Visser J. Purification and characterization of an alpha-L-rhamnosidase from *Aspergillus niger*. FEMS Microbiol Lett, 1997; 157: 279–283.

Manzanares P, Orejas M, Ibañez E, Vallés S, Ramón D. Purification and characterization of an alpha-L-rhamnosidase from *Aspergillus nidulans*. Lett Appl Microbiol, 2000; 31: 198–202.

Manzanares P, van den Broeck HC, de Graaff LH, Visser J. Purification and characterization of two different alpha-L-rhamnosidases, RhaA and RhaB, from *Aspergillus aculeatus*. Appl Environ Microbiol, 2001; 67: 2230–2234.

Maras M, De Bruyn A, Vervecken W, Uusitalo J, Penttila M, Busson P, Herdewijn P, Contreras R. *In vivo* synthesis of complex *N*-glycans by expression of human *N*-acetylglucosaminyltransferase I in the filamentous fungus *Trichoderma reesei*. FEBS Lett, 1999; 452: 365–370.

Marcinowski M, Höller M, Feige MJ, Baerend D, Lamb DC, Buchner J. Substrate discrimination of the chaperone BiP by autonomous and cochaperone-regulated conformational transitions. Nat Struct Mol Biol, 2011; 18: 150–158.

Margolles-Clark E, Tenkanen M, Luonteri E, Penttilä M. Three alpha-galactosidase genes of *Trichoderma reesei* cloned by expression in yeast. Eur J Biochem, 1996; 240: 104–111.

Marita JM, Vermerris W, Ralph J, Hatfield RD. Variations in the cell wall composition of maize brown midrib mutants. J Agric Food Chem, 2003; 51: 1313–1321.

Maruyama K, Goto C, Numata M, Suzuki T, Nakagawa Y, Hoshino T, Uchiyama T. O-acetylated xyloglucan in extracellular polysaccharides from cell-suspension cultures of Mentha. Phytochemistry, 1996; 41: 1309–1314.

Markovic O, Janecek S. Pectin methylesterases: sequence-structural features and phylogenetic relationships. Carbohydr Res, 2004; 339: 2281–2295.

Martínez AT, Speranza M, Ruiz-Dueñas FJ, Ferreira P, Camarero S, Guillén F, Martínez MJ, Gutiérrez A, del Río JC. Biodegradation of lignocellulosics: microbial, chemical, and enzymatic aspects of the fungal attack of lignin. *Int Microbiol.* 2005; 8: 195–204.

Martinez D, Berka RM, Henrissat B, Saloheimo M, Arvas M, Baker SE, Chapman J, Chertkov O, Coutinho PM, Cullen D, Danchin EG, Grigoriev IV, Harris P, Jackson M, Kubicek CP, Han CS, Ho I, Larrondo LF, de Leon AL, Magnuson JK, Merino S, Misra M, Nelson

B, Putnam N, Robbertse B, Salamov AA, Schmoll M, Terry A, Thayer N, Westerholm-Parvinen A, Schoch CL, Yao J, Barabote R, Nelson MA, Detter C, Bruce D, Kuske CR, Xie G, Richardson P, Rokhsar DS, Lucas SM, Rubin EM, Dunn-Coleman N, Ward M, Brettin TS. Genome sequencing and analysis of the biomass-degrading fungus *Trichoderma reesei* (syn. *Hypocrea jecorina*). Nat Biotechnol, 2008; 26: 553–560.

Martinez D, Challacombe J, Morgenstern I, Hibbett D, Schmoll M, Kubicek CP, Ferreira P, Ruiz-Duenas FJ, Martinez AT, Kersten P, Hammel KE, Vanden Wymelenberg A, Gaskell J, Lindquist E, Sabat G, Bondurant SS, Larrondo LF, Canessa P, Vicuna R, Yadav J, Doddapaneni H, Subramanian V, Pisabarro AG, Lavín JL, Oguiza JA, Master E, Henrissat B, Coutinho PM, Harris P, Magnuson JK, Baker SE, Bruno K, Kenealy W, Hoegger PJ, Kües U, Ramaiya P, Lucas S, Salamov A, Shapiro H, Tu H, Chee CL, Misra M, Xie G, Teter S, Yaver D, James T, Mokrejs M, Pospisek M, Grigoriev IV, Brettin T, Rokhsar D, Berka R, Cullen D. Genome, transcriptome, and secretome analysis of wood decay fungus Postia placenta supports unique mechanisms of lignocellulose conversion. Proc Natl Acad Sci USA, 2009; 106: 1954–1959.

Maruyama J, Kikuchi S, Kitamoto K. Differential distribution of the endoplasmic reticulum network as visualized by the BipA-EGFP fusion protein in hyphal compartments across the septum of the filamentous fungus, *Aspergillus oryzae*. Fungal Genet Biol, 2006; 43: 642–654.

Mathew S, Abraham E. Bioconversions of ferulic acid and hydroxycinnamic acid. Crit Rev Microbiol, 2006; 32: 115–125.

Mathew S, Abraham E, Sudheesh S. Rapid conversion of ferulic acid to 4-vinylguaiacol and vanillin metabolites by *Debaryomyces hansenii*. J Mol Catal B, 2007; 44: 48–52.

Matsunaga T, Ishii T, Matsumoto S, Higuchi M, Darvill A, Albersheim P, O'Neill MA. Occurrence of the primary cell wall polysaccharide rhamnogalacturonan II in pteridophytes, lycophytes, and bryophytes. Implication for the evolution of vascular plants, Plant Physiol, 2004; 134: 339–351.

Matsushika A, Inoue H, Kodaki T, Sawayama S. Ethanol production from xylose in engineered *Saccharomyces cerevisiae* strains: current state and perspectives. Appl Microbiol Biotechnol, 2009; 84: 37–53.

Matsushita J, Kato Y, Matsuda K. Purification and properties of an alpha-D-xylosidase from *Aspergillus niger*. J Biochem, 1985; 98: 825–832.

Mauseth JD. *Plant Anatomy*, 1988. The Benjamin/Cummings Publishing Company.

Mayans O, Scott M, Connerton I, Gravesen T, Benen J, Visser J, Pickersgill R, Jenkins J. Two crystal structures of pectin lyase A from *Aspergillus* reveal a pH driven conformational change and striking divergence in the substrate-binding clefts of pectin and pectate lyases. Structure, 1997; 5: 677–689.

McCarter JD, Withers SG. Mechanisms of enzymatic glycoside hydrolysis. Curr Opin Struct Biol, 1994; 4: 885–892.

McHardy J, Sawan SP. *Supercritical Fluid Cleaning: Fundamentals, Technology and Applications*, 1998. Noyes Publications, Park Ridge, NJ.

McKinlay JB, Vieille C, Zeikus JG. Prospects for a bio-based succinate industry. Appl Microbiol Biotechnol, 2007; 76: 727–740.

McLaughlin SB, Kszos LA. Development of switchgrass (*Panicum virgatum*) as a bioenergy feedstock in the United States. Biomass Bioenergy, 2005; 28: 515–535.

McQueen-Mason S, Cosgrove DJ. Disruption of hydrogen bonding between plant cell wall polymers by proteins that induce wall extension. Proc Natl Acad Sci USA, 1994; 91: 6574–6578.

Meinke A, Damude HG, Tomme P, Kwan E, Kilburn DG, Miller RC Jr, Warren RA, Gilkes NR. Enhancement of the endo-beta-1,4-glucanase activity of an exocellobiohydrolase by deletion of a surface loop. J Biol Chem, 1995; 270: 4383–4386.
Mellerowicz EJ, Immerzeel P, Hayashi T. Xyloglucan: the molecular muscle of trees. Ann Bot, 2008; 102: 659–665.
Merino ST, Cherry J. Progress and challenges in enzyme development for biomass utilization. Adv Biochem Eng Biotechnol., 2007; 108: 95–120.
Mertens JA, Bowman MJ. Expression and characterization of fifteen *Rhizopus oryzae* 99–880 polygalacturonase enzymes in *Pichia pastoris*. Curr Microbiol, 2011; 62: 1173–1178.
Metz B, de Vries RP, Polak S, Seidl V, Seiboth B. The *Hypocrea jecorina* (syn. *Trichoderma reesei*) *lxr1* gene encodes a D-mannitol dehydrogenase and is not involved in L-arabinose catabolism. FEBS Lett, 2009; 583: 1309–1313.
Metz B, Seidl-Seiboth V, Haarmann T, Kopchinskiy A, Lorenz P, Seiboth B, Kubicek CP. Expression of biomass degrading enzymes is a major event during conidia development in *Trichoderma reesei*. Eukaryot Cell, 2011; Sep 2, [Epub ahead of print].
Meyer KH, Misch L. 31. Position des atomes dans le noveau modele spatiale de la cellulose. Helv Chim Acta, 1937; 20: 232–245.
Meyer MM, Silberg JJ, Voigt CA, Endelman JB, Mayo SL, Wang ZG, Arnold FH. Library analysis of SCHEMA-guided protein recombination. Protein Sci, 2003; 12: 1686–1693.
Mi S, Meng K, Wang Y, Bai Y, Yuan T, Luo H, Yao B. Molecular cloning and characterization of a novel alpha-galactosidase gene from *Penicillium* sp. F63 CGMCC 1669 and expression in *Pichia pastoris*. Enzyme Microb Technol, 2007; 40: 1373–1380.
Micales JA, Highley TL. Factors associated with decay capacity of the brown-rot fungus *Postia placenta*. In: Llewllyn GG, O'Rear CE (eds) *Biodeterioration Research III*, 1991, pp. 285–302. Plenum, New York.
Mishra P, Singh A. Microbial pentose utilization. Adv Appl Microbiol, 1993; 39: 91–152.
Mitchell RA, Dupree P, Shewry PR. A novel bioinformatics approach identifies candidate genes for the synthesis and feruloylation of arabinoxylan. Plant Physiol, 2007; 144: 43–53.
Miyanaga A, Koseki T, Matsuzawa H, Wakagi T, Shoun H, Fushinobu S. Crystal structure of a family 54 alpha-L-arabinofuranosidase reveals a novel carbohydrate-binding module that can bind arabinose. J Biol Chem, 2004; 279: 44907–44914.
Mohnen D. Pectin structure and biosynthesis. Curr Opin Plant Biol, 2008; 11: 266–277.
Mojzita D, Penttilä M, Richard P. Identification of an l-arabinose reductase gene in *Aspergillus niger* and its role in L-arabinose catabolism. J Biol Chem, 2010a; 285: 23622–23628.
Mojzita D, Vuoristo K, Koivistoinen OM, Penttilä M, Richard P. The "true" L-xylulose reductase of filamentous fungi identified in *Aspergillus niger*. FEBS Lett, 2010b; 584: 3540–3544.
Mojzita D, Wiebe M, Hilditch S, Boer H, Penttilä M, Richard P. Metabolic engineering of fungal strains for conversion of D-galacturonate to *meso*-galactarate. Appl Environ Microbiol, 2010c; 76: 169–175.
Mølgaard A, Kauppinen S, Larsen S. Rhamnogalacturonan acetylesterase elucidates the structure and function of a new family of hydrolases. Structure, 2000; 8: 373–383.
Mølhøj M, Pagant S, Höfte H. Towards understanding the role of membrane-bound endo-beta-1,4-glucanases in cellulose biosynthesis. Plant Cell Physiol, 2002; 43: 1399–1406.
Molinari M, Galli C, Vanoni O, Arnold SM, Kaufman RJ. Persistent glycoprotein misfolding activates the glucosidase II/UGT1-driven calnexin cycle to delay aggregation and loss of folding competence. Mol Cell, 2005; 20: 503–512.
Montanier C, Money VA, Pires VM, Flint JE, Pinheiro BA, Goyal A, Prates JA, Izumi A, Stålbrand H, Morland C, Cartmell A, Kolenova K, Topakas E, Dodson EJ, Bolam DN,

Davies GJ, Fontes CM, Gilbert HJ. The active site of a carbohydrate esterase displays divergent catalytic and noncatalytic binding functions. PLoS Biol, 2009; 7: e71.

Moon HS, Abercrombie JM, Kausch AP, Stewart CN Jr. Sustainable use of biotechnology for bioenergy feedstocks. Environ Manage, 2010; 46: 531–538.

Moralejo FJ, Watson AJ, Jeenes DJ, Archer DB, Martín JF. A defined level of protein disulfide isomerase expression is required for optimal secretion of thaumatin by *Aspergillus awamori*. Mol Genet Genomics, 2001; 266: 246–253.

Mora-Pale M, Meli L, Doherty TV, Linhardt RJ, Dordick JS. Room temperature ionic liquids as emerging solvents for the pretreatment of lignocellulosic biomass. Biotechnol Bioeng, 2011; 108: 1229–1245.

Moreira LR, Filho EX. An overview of mannan structure and mannan-degrading enzyme systems. Appl Microbiol Biotechnol, 2008; 79: 165–178.

Morelle W, Bernard M, Debeaupuis JP, Buitrago M, Tabouret M, Latgé JP. Galactomannoproteins of *Aspergillus fumigatus*. Eukaryot Cell, 2005; 4: 1308–1316.

Moreno F, Herrero P. The hexokinase 2-dependent glucose signal transduction pathway of *Saccharomyces cerevisiae*. FEMS Microbiol Rev, 2002; 26: 83–90.

Morschhäuser J, Michel S, Staib P. Sequential gene disruption in *Candida albicans* by FLP-mediated site-specific recombination. Mol Microbiol, 1999; 32: 547–556.

Moser F, Irwin D, Chen SL, Wilson DB. Regulation and characterization of *Thermobifida fusca* carbohydrate-binding module proteins E7 and E8. Biotechnol Bioeng, 2008; 100: 1066–1077.

Mosier N, Wyman C, Dale B, Elander R, Lee YY, Holtzapple M, Ladisch M. Features of promising technologies for pretreatment of lignocellulosic biomass. Bioresour Technol, 2005; 96: 673–686.

Mueller SC, Brown RM Jr. Evidence for an intramembraneous component associated with a cellulose microfibril synthesizing complex in higher plants. J Cell Biol, 1980; 84: 315–326.

Mulakala C, Reilly PJ. *Hypocrea jecorina* (*Trichoderma reesei*) Cel7A as a molecular machine: a docking study. Proteins, 2005; 60: 598–605.

Munasinghe PC, Khanal SK. Biomass-derived syngas fermentation into biofuels: opportunities and challenges. Bioresour Technol, 2010; 10: 5013–5022.

Muñoz IG, Ubhayasekera W, Henriksson H, Szabó I, Pettersson G, Johansson G, Mowbray SL, Ståhlberg J. Family 7 cellobiohydrolases from *Phanerochaete chrysosporium*: Crystal structure of the catalytic module of Cel7D (CBH58) at 1.32 Å resolution and homology models of the isozymes. J Mol Biol, 2001; 314: 1097–1111.

Munson M, Novick P. The exocyst defrocked, a framework of rods revealed. Nat Struct Mol Biol, 2006; 13: 577–581.

Murphy C, Powlowski J, Wu M, Butler G, Tsang A. Curation of characterized glycoside hydrolases of fungal origin. Database, 2011a; bar020.

Murphy L, Cruys-Bagger N, Damgaard HD, Baumann MJ, Nymand Olsen S, Borch K, Flensted Lassen S, Sweeney M, Tatsumi H, Westh P. Origin of the initial burst in activity for *Trichoderma reesei* endoglucanases hydrolyzing insoluble cellulose. J Biol Chem, 2011b, Nov 22 [Epub ahead of print].

Mutter M, Colquhoun IJ, Schols HA, Beldman G, Voragen AG. Rhamnogalacturonase B from *Aspergillus aculeatus* is a rhamnogalacturonan alpha-L-rhamnopyranosyl-(1–>4)-alpha-D-galactopyranosyluronide lyase. Plant Physiol, 1996; 110: 73–77.

Mutter M, Beldman G, Pitson SM, Schols HA, Voragen AG. Rhamnogalacturonan alpha-D-galactopyranosyluronohydrolase. An enzyme that specifically removes the terminal nonreducing galacturonosyl residue in rhamnogalacturonan regions of pectin. Plant Physiol, 1998; 117: 153–163.

Myneni RB, Dong J, Tucker CJ, Kaufmann RK, Kauppi PE, Liski J, Zhou L, Alexeyev V, Hughes MK. A large carbon sink in the woody biomass of Northern forests. Proc Natl Acad Sci USA, 2001; 98: 14784–14789.

Nagae M, Tsuchiya A, Katayama T, Yamamoto K, Wakatsuki S, Kato R. Structural basis of the catalytic reaction mechanism of novel 1,2-alpha-L-fucosidase from *Bifidobacterium bifidum*. J Biol Chem, 2007; 282: 18497–18509.

Nagy T, Tunnicliffe RB, Higgins LD, Walters C, Gilbert HJ, Williamson MP. Characterization of a double dockerin from the cellulosome of the anaerobic fungus *Piromyces equi*. J Mol Biol, 2007; 373: 612–622.

Nakamura CE, Whited GM. Metabolic engineering for the microbial production of 1,3-propanediol. Curr Opin Biotechnol, 2003; 14: 454–459.

Nakazawa H, Kawai T, Ida N, Shida Y, Kobayashi Y, Okada H, Tani S, Sumitani JI, Kawaguchi T, Morikawa Y, Ogasawara W. Construction of a recombinant *Trichoderma reesei* strain expressing *Aspergillus aculeatus* β-glucosidase 1 for efficient biomass conversion. Biotechnol Bioeng, 2011, Aug 9, doi: 10.1002/bit.23296.

Nakazawa H, Okada K, Onodera T, Ogasawara W, Okada H, Morikawa Y. Directed evolution of endoglucanase III (Cel12A) from *Trichoderma reesei*. Appl Microbiol Biotechnol, 2009; 83: 649–657.

Naumoff DG. Hierarchical classification of glycoside hydrolases. Biochemistry (Mosc), 2011; 76: 622–635.

Neustroev KN, Golubev AM, Firsov LM, Ibatullin FM, Protasevich II, Makarov AA. Effect of modification of carbohydrate component on properties of glucoamylase. FEBS Lett, 1993; 316: 157–160.

Nevoigt E. Progress in metabolic engineering of *Saccharomyces cerevisiae*. Microbiol Mol Biol Rev, 2008; 72: 379–412.

Ni H, Laplaza M, Jeffries TW. Transposon mutagenesis to improve the growth of recombinant *Saccharomyces cerevisiae* on D-xylose. Appl Environ Microbiol, 2007; 73: 2061–2066.

Nickel W, Rabouille C. Mechanisms of regulated unconventional protein secretion. Nat Rev Mol Cell Biol, 2009; 10: 148–155.

Nijikken Y, Tsukada T, Igarashi K, Samejima M, Wakagi T, Shoun H, Fushinobu S. Crystal structure of intracellular family 1 beta-glucosidase BGL1A from the basidiomycete *Phanerochaete chrysosporium*. FEBS Lett, 2007; 581: 1514–1520.

Nilvebrant N-O, Persson P, Reimann A, De Sousa F, Gorton L, Jonsson LJ. Limits for alkaline detoxification of dilute-acid lignocellulose hydrolysates. Appl Biochem Biotechnol, 2003; 107: 615–628.

Ninomiya Y, Suzuki K, Ishii C, Inoue H. Highly efficient gene replacements in *Neurospora* strains deficient for nonhomologous end-joining. Proc Natl Acad Sci USA, 2004; 101: 12248–12253.

Noguchi Y, Sano M, Kanamaru K, Ko T, Takeuchi M, Kato M, Kobayashi T. Genes regulated by AoXlnR, the xylanolytic and cellulolytic transcriptional regulator, in *Aspergillus oryzae*. Appl Microbiol Biotechnol, 2009; 85: 141–154.

Noguchi Y, Tanaka H, Kanamaru K, Kato M, Kobayashi T. Xylose triggers reversible phosphorylation of XlnR, the fungal transcriptional activator of xylanolytic and cellulolytic genes in *Aspergillus oryzae*. Biosci Biotechnol Biochem, 2011; 75: 953–959.

Nomura CT, Taguchi S. PHA synthase engineering toward superbiocatalysts for custom-made biopolymers. Appl Microbiol Biotechnol, 2007; 73: 969–979.

Nothwehr SF, Stevens TH. Sorting of membrane proteins in the yeast secretory pathway. J Biol Chem, 1994; 269: 10185–10188.

Nurizzo D, Turkenburg JP, Charnock SJ, Roberts SM, Dodson EJ, McKie VA, Taylor EJ, Gilbert HJ, Davies GJ. *Cellvibrio japonicus* alpha-L-arabinanase 43A has a novel five-blade beta-propeller fold. Nat Struct Biol, 2002; 9: 665–668.

Nyon MP, Rice DW, Berrisford JM, Hounslow AM, Moir AJ, Huang H, Nathan S, Mahadi NM, Bakar FD, Craven CJ. Catalysis by *Glomerella cingulata* cutinase requires conformational cycling between the active and inactive states of its catalytic triad. J Mol Biol, 2009; 385: 226–235.

Nystrom JM, Allen AL. Pilot scale investigations and economics of cellulase production. Biotechnol Bioengin Symp, 1976; 6: 55–74.

Ogawa N, Mori K. Autoregulation of the HAC1 gene is required for sustained activation of the yeast unfolded protein response. Genes Cells, 2004; 9: 95–104.

Ohgren K, Bengtsson O, Gorwa-Grauslund MF, Galbe M, Hahn-Hägerdal B, Zacchi G. Simultaneous saccharification and co-fermentation of glucose and xylose in steam-pretreated corn stover at high fiber content with *Saccharomyces cerevisiae* TMB3400. J Biotechnol, 2006; 126: 488–498.

Oka T, Nemoto T, Jigami Y. Functional analysis of *Arabidopsis thaliana* RHM2/MUM4, a multidomain protein involved in UDP-D-glucose to UDP-L-rhamnose conversion. J Biol Chem, 2007; 282: 5389–5403.

Okabe M, Lies D, Kanamasa S, Park EY. Biotechnological production of itaconic acid and its biosynthesis in *Aspergillus terreus*. Appl Microbiol Biotechnol, 2009; 84: 597–606.

Okemoto K, Uekita T, Tsumuraya Y, Hashimoto Y, Kasama T. Purification and characterization of an endo-beta-(1,6)-galactanase from *Trichoderma viride*. Carbohydr Res, 2003; 338: 219–230.

Olivares-Hernández R, Sunner H, Frisvad JC, Olsson L, Nielsen J, Panagiotou G. Combining substrate specificity analysis with support vector classifiers reveals feruloyl esterase as a phylogenetically informative protein group. PLoS One, 2010; 5: e12781.

Oosterveld A, Beldman G, Voragen AG. Oxidative cross-linking of pectic polysaccharides from sugar beet pulp. Carbohydr Res, 2000; 328: 199–207.

Oppenheimer M, Poulin MB, Lowary TL, Helm RF, Sobrado P. Characterization of recombinant UDP-galactopyranose mutase from *Aspergillus fumigatus*. Arch Biochem Biophys, 2010; 502: 31–38.

Orbach MJ, Porro EB, Yanofsky C. Cloning and characterization of the gene for β-tubulin from a benomyl-resistant mutant of *Neurospora crassa* and its use as a dominant selectable marker. Mol Cell Biol, 1986; 6: 2452–2461.

Orejas M, MacCabe AP, Peréz González JA, Kumar S, Ramón D. Carbon catabolite repression of the *Aspergillus nidulans xlnA* gene. Mol Microbiol, 1999; 31: 177–184.

Orejas M, MacCabe AP, Peréz González JA, Kumar S, Ramón D. The wide-domain carbon catabolite repressor CreA indirectly controls expression of the *Aspergillus nidulans xlnB* gene, encoding the acidic endo-β-(1,4)-xylanase X(24). J Bacteriol, 2001; 183: 1517–1523.

Orfila C, Sørensen SO, Harholt J, Geshi N, Crombie H, Truong HN, Reid JS, Knox JP, Scheller HV. QUASIMODO1 is expressed in vascular tissue of *Arabidopsis thaliana* inflorescence stems, and affects homogalacturonan and xylan biosynthesis. Planta, 2005; 222: 613–622.

Orpin CG. Invasion of plant tissue in the rumen by the flagellate *Neocallimastix frontalis*. J Gen Microbiol, 1977; 98: 423–430.

Orpin CG. Studies on the rumen flagellate *Neocallimastix frontalis*. J Gen Microbiol, 1975; 92: 249–262.

O'Sullivan AC. Cellulose: the structure slowly unravels. Cellulose, 1997; 4: 173–207.

Ou MS, Ingram LO, Shanmugam KT. L(+)-Lactic acid production from non-food carbohydrates by thermotolerant *Bacillus coagulans*. J Ind Microbiol Biotechnol, 2011; 38: 599–605.

Paës G, Skov LK, O'Donohue MJ, Rémond C, Kastrup JS, Gajhede M, Mirza O. The structure of the complex between a branched pentasaccharide and *Thermobacillus xylanolyticus* GH-51 arabinofuranosidase reveals xylan-binding determinants and induced fit. Biochemistry, 2008; 47: 7441–7451.

Paetzel M, Karla A, Strynadka NC, Dalbey RE. Signal peptidases. Chem Rev, 2002; 102: 4549–4580.

Pakula TM, Laxell M, Huuskonen A, Uusitalo J, Saloheimo M, Penttilä M. The effects of drugs inhibiting protein secretion in the filamentous fungus *Trichoderma reesei*. Evidence for down-regulation of genes that encode secreted proteins in the stressed cells. J Biol Chem, 2003; 278: 45011–45020.

Palmer JM, Keller NP. Secondary metabolism in fungi: does chromosomal location matter? Curr Opin Microbiol, 2010; 13: 431–436.

Palmqvist B, Wiman M, Lidén G. Effect of mixing on enzymatic hydrolysis of steam-pretreated spruce: a quantitative analysis of conversion and power consumption. Biotechnol Biofuels, 2011; 4: 10.

Palonen H, Thomsen AB, Tenkanen M, Schmidt AS, Viikari L. Evaluation of wet oxidation pretreatment for enzymatic hydrolysis of softwood. Appl Biochem Biotechnol, 2004; 117: 1–17.

Palonen H, Viikari L. Role of oxidative enzymatic treatments on enzymatic hydrolysis of softwood. Biotechnol Bioeng, 2004; 86: 550–557.

Pan X, Arato C, Gilkes N, Gregg D, Mabee W, Pye K, Xiao Z, Zhang X, Saddler J. Biorefining of softwoods using ethanol organosolv pulping: preliminary evaluation of process streams for manufacture of fuel-grade ethanol and co-products. Biotechnol Bioeng, 2005; 90: 473–481.

Panagiotou G, Christakopoulos P, Olsson L. Simultaneous saccharification and fermentation of cellulose by *Fusarium oxysporum* F3-growth characteristics and metabolite profiling. Enzyme Microb Technol, 2005; 36: 693–699.

Pang X, Zhuang X, Tang Z, Chen X. Polylactic acid (PLA): research, development and industrialization. Biotechnol J, 2010; 5: 1125–1136.

Pao SS, Paulsen IT, Saier MH Jr. Major facilitator superfamily. Microbiol Mol Biol Rev, 1998; 62: 1–34.

Park BH, Karpinets TV, Syed MH, Leuze MR, Uberbacher EC. CAZymes Analysis Toolkit (CAT): web service for searching and analyzing carbohydrate-active enzymes in a newly sequenced organism using CAZy database. Glycobiology, 2010; 20: 1574–1584.

Parkkinen T, Koivula A, Vehmaanperä J, Rouvinen J. Crystal structures of *Melanocarpus albomyces* cellobiohydrolase Cel7B in complex with cello-oligomers show high flexibility in the substrate binding. Protein Sci, 2008; 17: 1383–1394.

Passardi F, Theiler G, Zamocky M, Cosio C, Rouhier N, Teixera F, Margis-Pinheiro M, Ioannidis V, Penel C, Falquet L, Dunand C. PeroxiBase: the peroxidase database. Phytochemistry, 2007a; 68: 1605–1611.

Passardi F, Bakalovic N, Teixeira FK, Margis-Pinheiro M, Penel C, Dunand C. Prokaryotic origins of the non-animal peroxidase superfamily and organelle-mediated transmission to eukaryotes. Genomics, 2007b; 89: 567–579.

Paszczynski A, Huynh V-B, Crawford R. Comparison of ligninase-I and peroxidase-M2 from the white-rot fungus *Phanerochaete chrysosporium*. Arch Biochem Biophys, 1986; 244: 750–765.

Patil C, Walter P. Intracellular signaling from the endoplasmic reticulum to the nucleus: the unfolded protein response in yeast and mammals. Curr Opin Cell Biol, 2001; 13: 349–355.

Patil CK, Li H, Walter P. Gcn4p and novel upstream activating sequences regulate targets of the unfolded protein response. PLoS Biol, 2004; 2: E246.

Patrick WM, Firth AE. Strategies and computational tools for improving randomized protein libraries. Biomol Eng, 2005; 22: 105–112.

Pauly M, Keegstra K. Plant cell wall polymers as precursors for biofuels. Curr Opin Plant Biol, 2010; 13: 304–311.

Payan F, Leone P, Porciero S, Furniss C, Tahir T, Williamson G, Durand A, Manzanares P, Gilbert HJ, Juge N, Roussel A. The dual nature of the wheat xylanase protein inhibitor XIP-I: structural basis for the inhibition of family 10 and family 11 xylanases. J Biol Chem, 2004; 279: 36029–36037.

Pazur JH, Tominaga Y, Forsberg LS, Simpson DL. Glycoenzymes: an unusual type of glycoprotein structure for a glucoamylase. Carbohydr Res, 1980; 84: 103–114.

Pedersen M, Johansen KS, Meyer AS. Low temperature lignocelluloses pretreatment: effects and interactions of pretreatment pH are critical for maximizing enzymatic monosaccharide yields from wheat straw. Biotechnol Biofuels, 2011; 4: 11.

Pedersen M, Meyer AS. Influence of substrate particle size and wet oxidation on physical surface structures and enzymatic hydrolysis of wheat straw. Biotechnol Prog, 2009; 25: 399–408.

Pedersen M, Meyer AS. Lignocellulose pretreatment severity—relating pH to biomatrix opening. N Biotechnol, 2010; 27: 739–750.

Pell G, Taylor EJ, Gloster TM, Turkenburg JP, Fontes CM, Ferreira LM, Nagy T, Clark SJ, Davies GJ, Gilbert HJ. The mechanisms by which family 10 glycoside hydrolases bind decorated substrates. J Biol Chem, 2004; 279: 9597–9605.

Pelloux J, Rustérucci C, Mellerowicz EJ. New insights into pectin methylesterase structure and function. Trends Plant Sci, 2007; 12: 267–277.

Peñalva MA, Arst HN Jr. Recent advances in the characterization of ambient pH regulation of gene expression in filamentous fungi and yeasts. Ann Rev Microbiol, 2004; 58: 425–451.

Peñalva MA, Tilburn J, Bignell E, Arst HN Jr. Ambient pH gene regulation in fungi: making connections. Trends Microbiol, 2008; 16: 291–300.

Peng L, Kawagoe Y, Hogan P, Delmer D. Sitosterol-beta-glucoside as primer for cellulose synthesis in plants. Science, 2002; 295: 147–150.

Peralta-Yahya PP, Keasling JD. Advanced biofuel production in microbes. Biotechnol J, 2010; 5: 147–162.

Percival Zhang YH, Himmel ME, Mielenz JR. Outlook for cellulase improvement: screening and selection strategies. Biotechnol Adv, 2006; 24: 452–481.

Perez-Boada M, Ruiz-Duenas FJ, Pogni R, Basosi R, Choinowski T, Martinez M, Piontek K, Martínez AT. Versatile peoxidase oxidation of high redox potential aromatic compounds: site-directed mutagenesis, spectroscopic and crystallographic investigation of three long-range electron transfer pathways. J Mol Biol, 2005; 354: 385–402.

Peterbauer CK, Volc J. Pyranose dehydrogenases: biochemical features and perspectives of technological applications. Appl Microbiol Biotechnol, 2010; 85: 837–848.

Petersen TN, Kauppinen S, Larsen S. The crystal structure of rhamnogalacturonase A from *Aspergillus aculeatus*: a right-handed parallel beta helix. Structure, 1997; 5: 533–544.

Petschacher B, Nidetzky B. Altering the coenzyme preference of xylose reductase to favor utilization of NADH enhances ethanol yield from xylose in a metabolically engineered strain of *Saccharomyces cerevisiae*. Microb Cell Fact, 2008; 7: 9.

Phalip V, Delalande F, Carapito C, Goubet F, Hatsch D, Leize-Wagner E, Dupree P, Dorsselaer AV, Jeltsch JM. Diversity of the exoproteome of *Fusarium graminearum* grown on plant cell wall. Curr Genet, 2005; 48: 366–379.

Phillips CM, Beeson WT, Cate JH, Marletta MA. Cellobiose dehydrogenase and a copper-dependent polysaccharide monooxygenase potentiate cellulose degradation by *Neurospora crassa*. ACS Chem Biol, 2011; Oct 25 [Epub ahead of print].

Pickersgill R, Smith D, Worboys K, Jenkins J. Crystal structure of polygalacturonase from *Erwinia carotovora* ssp. *carotovora*. J Biol Chem, 1998; 273: 24660–24664.

Piens K, Davies G. Glycoside Hydrolase Family 6 in *CAZypedia*. 2011. Available at: http://www.cazypedia.org/ (accessed October 31, 2011).

Plemper RK, Wolf DH. Retrograde protein translocation: ERADication of secretory proteins in health and disease. Trends Biochem Sci, 1999; 24: 266–270.

Pokkuluri PR, Duke NE, Wood SJ, Cotta MA, Li XL, Biely P, Schiffer M. Structure of the catalytic domain of glucuronoyl esterase Cip2 from *Hypocrea jecorina*. Proteins, 2011; 79: 2588–2592.

Polanco R, Canessa P, Rivas A, Larrondo LF, Lobos S, Vicuña R. Cloning and functional characterization of the gene encoding the transcription factor Ace1 in the basidiomycete *Phanerochaete chrysosporium*. Biol Res, 2006; 39: 641–648.

Polizeli ML, Rizzatti AC, Monti R, Terenzi HF, Jorge JA, Amorim DS. Xylanases from fungi: properties and industrial applications. Appl Microbiol Biotechnol, 2005; 67: 577–591.

Pollet A, Lagaert S, Eneyskaya E, Kulminskaya A, Delcour JA, Courtin CM. Mutagenesis and subsite mapping underpin the importance for substrate specificity of the aglycon subsites of glycoside hydrolase family 11 xylanases. Biochim Biophys Acta, 2010; 1804: 977–985.

Popp JL, Kirk TK. Oxidation of methoxybenzenes by manganese peroxidase and by Mn^{3+}. Arch Biochem Biophys, 1991; 268: 145–148.

Popper ZA. Evolution and diversity of green plant cell walls. Curr Opin Plant Biol, 2008; 11: 286–292.

Poulos TL, Edwards SL, Wariishi H, Gold MH. Crystallographic refinement of lignin peroxidase at 2Δ. J Biol Chem, 1993; 268: 4429–4440.

Powlowski J, Mahajan S, Schapira M, Master ER. Substrate recognition and hydrolysis by a fungal xyloglucan-specific family 12 hydrolase. Carbohydr Res, 2009; 344: 1175–1179.

Prakasham RS, Rao RS, Hobbs PJ. Current trends in biotechnological production of xylitol and future prospects. Curr Trends Biotechnol Pharm, 2009; 3: 8–36.

Priefert H, Rabenhorst J, Steinbüchel A. Biotechnological production of vanillin. Appl Microbiol Biotechnol, 2001; 56: 296–314.

Pu YQ, Jiang N, Ragauskas AJ. Ionic liquid as a green solvent for lignin. J Wood Chem Technol, 2007; 27: 23–33.

Puchart V, Gariépy MC, Shareck F, Dupont C. Identification of catalytically important amino acid residues of *Streptomyces lividans* acetylxylan esterase A from carbohydrate esterase family 4. Biochim Biophys Acta, 2006; 1764: 263–274.

Puchart V, Vršanská M, Mastihubová M, Topakas E, Vafiadi C, Faulds CB, Tenkanen M, Christakopoulos P, Biely P. Substrate and positional specificity of feruloyl esterases for monoferuloylated and monoacetylated 4-nitrophenyl glycoside. J Biotechnol, 2007; 127: 235–243.

Punt PJ, Drint-Kuijvenhoven A, Lokman BC, Spencer JA, Jeenes D, Archer DA, van den Hondel CA. The role of the *Aspergillus niger* furin-type protease gene in processing of fungal proproteins and fusion proteins. Evidence for alternative processing of recombinant (fusion-)proteins. J Biotechnol, 2003; 106: 23–32.

Punt PJ, Seiboth B, Weenink XO, van Zeijl C, Lenders M, Konetschny C, Ram AF, Montijn R, Kubicek CP, van den Hondel CA. Identification and characterization of a family of secretion-related small GTPase-encoding genes from the filamentous fungus *Aspergillus niger*: a putative SEC4 homologue is not essential for growth. Mol Microbiol, 2001; 41: 513–525.

Quan LJ, Zhang B, Shi WW, Li HY. Hydrogen peroxide in plants: a versatile molecule of the reactive oxygen species network. J Integr Plant Biol, 2008; 50: 2–18.

Quijano G, Couvert A, Amrane A. Ionic liquids: applications and future trends in bioreactor technology. Bioresour Technol, 2010; 101: 8923–8930.

Ragauskas AJ, Williams CK, Davison BH, Britovsek G, Cairney J, Eckert CA, Submit WJ, Hallett JP, Leak DJ, Liotta CL, Mielenz JR, Murphy R, Templer R, Tschaplinski T. The path forward for biofuels and biomaterials. Science, 2006; 311: 484–489.

Ralph J, Brunow G, Harris PJ, Dixon RA, Schatz PF, Boerjan W. Lignification: Are lignins biosynthesized via simple combinatorial chemistry or via proteinaceous control and template replication?. In: F. Daayf and V. Lattanzio (eds) *Recent Advances in Polyphenol Research*, 2008, vol. 1, pp. 36-66. Blackwell Publishing Ltd.

Ramakrishnan S, Collier J, Oyetunji R, Stutts B, Burnett R. Enzymatic hydrolysis of cellulose dissolved in *N*-methyl morpholine oxide/water solutions. Bioresour Technol, 2010; 101: 4965–4970.

Rath J, Messner R, Kosma P, Altmann F, März L, Kubicek CP. The alpha-D-mannan core of a complex cell-wall heteroglycan of *Trichoderma reesei* is responsible for beta-glucosidase activation. Arch Microbiol, 1995; 164: 414–419.

Raven PH et al., *Biology of Plants*, 6th edn, 1999. W.H. Freeman and Company.

Read ND. Exocytosis and growth do not occur only at hyphal tips. Mol Microbiol, 2011; 81: 4–7.

Reese ET. History of the cellulase program at the U.S. army Natick Development Center. Biotechnol Bioeng Symp, 1976; 6: 9–20.

Reese ET, Maguire AH, Parrish FW. Glucosidases and exo-glucanases. Can J Biochem, 1968; 46: 25–34.

Reese ET, Siu RGH, Levinson HS. The biological degradation of soluble cellulose derivatives and its relationship to the mechanism of cellulose hydrolysis. J Bacteriol, 1950; 59: 485–497.

Reiter WD. Biosynthesis and properties of the plant cell wall. Curr Opin Plant Biol, 2002; 5: 536–342.

Rempel B, St John F. Glycoside Hydrolase Family 30 in *CAZypedia*. 2011. Available at: http://www.cazypedia.org/ (accessed October 31, 2011).

Ren C, Chen T, Zhang J, Liang L, Lin Z. An evolved xylose transporter from *Zymomonas mobilis* enhances sugar transport in *Escherichia coli*. Microb Cell Fact, 2009; 8: 66.

Richard P, Hilditch S. D-Galacturonic acid catabolism in microorganisms and its biotechnological relevance. Appl Microbiol Biotechnol, 2009; 82: 597–604.

Richard P, Putkonen M, Väänänen R, Londesborough J, Penttilä M. The missing link in the fungal l-arabinose catabolic pathway, identification of the L-xylulose reductase gene. Biochemistry, 2002; 41: 6432–6437.

Richmond TA, Somerville CR. The cellulose synthase superfamily. Plant Physiol, 2000; 124: 495–498.

Rietveld A, Simons K. The differential miscibility of lipids as the basis for the formation of functional membrane rafts. Biochim Biophys Acta, 1998; 1376: 467–479.

Rifai MA. A revision of the genus *Trichoderma*. Mycol Pap, 1969; 116: 1–56.

Riley GL, Tucker KG, Paul GC, Thomas CR. Effect of biomass concentration and mycelial morphology on fermentation broth rheology. Biotechnol Bioeng, 2000; 68: 160–172.

Rizk SE, Abdel-Massih RM, Baydoun EA, Brett CT. Protein- and pH-dependent binding of nascent pectin and glucuronoarabinoxylan to xyloglucan in pea. Planta, 2000; 211: 423–429.

Rizzi M, Erlemann P, Bui-Thanh NA, Dellweg H. Xylose fermentation by yeasts. 4. Purification and kinetic studies of xylose reductase from *Pichia stipitis*. Appl Microbiol Biotechnol, 1988; 29: 148–154.

Rizzi M, Harwart K, Erlemann P, Bui-Thanh NA, Dellweg H. Purification and properties of the NAD^+-xylitol-dehydrogenase from the yeast *Pichia stipitis*. J Ferment Bioeng, 1989; 67: 20–24.

Roberts CJ, Nothwehr SF, Stevens TH. Membrane protein sorting in the yeast secretory pathway: evidence that the vacuole may be the default compartment. J Cell Biol, 1992; 119: 69–83.

Roche CM, Dibble CJ, Stickel JJ. Laboratory-scale method for enzymatic saccharification of lignocellulosic biomass at high-solids loadings. Biotechnol Biofuels, 2009; 2: 28.

Rockwell NC, Krysan DJ, Komiyama T, Fuller RS. Precursor processing by kex2/furin proteases. Chem Rev, 2002; 102: 4525–4548.

Rockwood DL, Rudie AW, Ralph SA, Zhu JY, Winandy JE. Energy product options for eucalyptus species grown as short rotation woody crops. Int J Mol Sci, 2008; 9: 1361–1378.

Rodrigues-Pousada CA, Nevitt T, Menezes R, Azevedo D, Pereira J, Amaral C. Yeast activator proteins and stress response: an overview. FEBS Lett, 2004; 567: 80–85.

Rogers PLK, Lee J, Skotnicki ML, Tribe DE. Ethanol production by *Zymononas mobilis*. Adv Biochem Eng, 1982; 23: 37–84.

Rojas AL, Fischer H, Eneiskaya EV, Kulminskaya AA, Shabalin KA, Neustroev KN, Craievich AF, Golubev AM, Polikarpov I. Structural insights into the beta-xylosidase from *Trichoderma reesei* obtained by synchrotron small-angle X-ray scattering and circular dichroism spectroscopy. Biochemistry, 2005; 44: 15578–15584.

Rose JK, Saladié M, Catalá C. The plot thickens: new perspectives of primary cell wall modification. Curr Opin Plant Biol, 2004; 7: 296–301.

Rosenau T, Potthast A, Sixta H, Kosma P. The chemistry of side reactions and byproduct formation in the system NMMO/cellulose (Lyocell process). Prog Polymer Sci, 2001; 26: 1763–1837.

Rosgaard L, Pedersen S, Cherry JR, Harris P, Meyer AS. Efficiency of new fungal cellulase systems in boosting enzymatic degradation of barley straw lignocellulose. Biotechnol Prog, 2006; 22: 493–498.

Roth J, Zuber C, Park S, Jang I, Lee Y, Kysela KG, Le Fourn V, Santimaria R, Guhl B, Cho JW. Protein *N*-glycosylation, protein folding, and protein quality control. Mol Cells, 2010; 30: 497–506.

Rouvinen J, Bergfors T, Teeri T, Knowles JKC, Jones TA. Three dimensional structure of cellobiohydrolase II from *Trichoderma reesei*. Science, 1990; 249: 380–386.

Rubio-Texeira M. Endless versatility in the biotechnological applications of *Kluyveromyces* LAC genes. Biotechnol Adv, 2006; 24: 212–225.

Ruijte GJ, Vanhanen SA, Gielkens MM, van de Vondervoort PJ, Visser J. Isolation of *Aspergillus niger creA* mutants and effects of the mutations on expression of arabinases and L-arabinose catabolic enzymes. Microbiol UK, 1997; 143: 2991–2998.

Ruiz-Dueñas FJ, Martínez MJ, Martínez AT. Molecular characterization of a novel peroxidase isolated from the ligninolytic fungus *Pleurotus eryngii*. Mol Microbiol, 1999; 31: 223–235.

Rüsch genannt Klaas M, Schöne H. Direct, high-yield conversions of cellulose into biofuel and platform chemicals-on the way to a sustainable biobased economy. Chem Sus Chem, 2009; 2: 127–128.

Russell JB, Muck RE, Weimer PJ. Quantitative analysis of cellulose degradation and growth of cellulolytic bacteria in the rumen. FEMS Microbiol Ecol, 2009; 67: 183–197.

Russell JB, Rychlik JL. Factors that alter rumen microbial ecology. Science, 2001; 292: 1119–1122.

Ryabova O, Vrsanská M, Kaneko S, van Zyl WH, Biely P. A novel family of hemicellulolytic alpha-glucuronidase. FEBS Letts, 2009; 583: 1457–1462.

Ryttersgaard C, Le Nours J, Lo Leggio L, Jørgensen CT, Christensen LL, Bjørnvad M, and Larsen S. The structure of endo-beta-1,4-galactanase from *Bacillus licheniformis* in complex with two oligosaccharide products. J Mol Biol, 2004,; 341: 107–117.

Sabini E, Wilson KS, Siika-aho M, Boisset C, Chanzy H. Digestion of single crystals of mannan I by an endo-mannanase from *Trichoderma reesei*. Eur J Biochem, 2000; 267: 2340–2344.

Saier MH Jr, Crasnier M. Inducer exclusion and the regulation of sugar transport. Res Microbiol, 1996; 147: 482–489.

Saier MH Jr, Tran CV, Barabote RD. TCDB: the transporter classification database for membrane transport protein analyses and information. Nucl Acids Res, 2006; 34: D181–D186.

Sakai S, Nishide T, Munir E, Baba K, Inui H, Nakano Y, Hattori T, Shimada M. Subcellular localization of glyoxylate cycle key enzymes involved in oxalate biosynthesis of wood-destroying basidiomycete *Fomitopsis palustris* grown on glucose. Microbiology, 2006; 152: 1857–1866.

Sakamoto T, Fujita T, Kawasaki H. Transglycosylation catalyzed by a *Penicillium chrysogenum* exo-1,5-alpha-L-arabinanase. Biochim Biophys Acta, 2004; 1674: 85–90.

Sakamoto T, Kitaura H, Minami M, Honda Y, Watanabe T, Ueda A, Suzuki K, Irie T. Transcriptional effect of a calmodulin inhibitor, W-7, on the ligninolytic enzyme genes in *Phanerochaete chrysosporium*. Curr Genet, 2010; 56: 401–410.

Salghetti SE, Caudy AA, Chenoweth JG, Tansey JW. Regulation of transcriptional activation domain function by ubiquitin. Science, 2001; 293: 1651–1653.

Saloheimo A, Aro N, Ilmén M, Penttilä M. Isolation of the ace1 gene encoding a Cys(2)-His(2) transcription factor involved in regulation of activity of the cellulase promoter *cbh1* of *Trichoderma reesei*. J Biol Chem, 2000; 275: 5817–5825.

Saloheimo A, Henrissat B, Hoffrén AM, Teleman O, Penttilä M. A novel, small endoglucanase gene, *egl5*, from *Trichoderma reesei* isolated by expression in yeast. Mol Microbiol, 1994; 13: 219–228.

Saloheimo A, Rauta J, Stasyk OV, Sibirny AA, Penttilä M, Ruohonen L. Xylose transport studies with xylose-utilizing *Saccharomyces cerevisiae* strains expressing heterologous and homologous permeases. Appl Microbiol Biotechnol, 2007; 74: 1041–1052.

Saloheimo M, Lehtovaara P, Penttilä M, Teeri TT, Ståhlberg J, Johansson G, Pettersson G, Claeyssens M, Tomme P, Knowles JK. EGIII, a new endoglucanase from *Trichoderma reesei*: the characterization of both gene and enzyme. Gene, 1988; 63: 11–22.

Saloheimo M, Paloheimo M, Hakola S, Pere J, Swanson B, Nyyssönen E, Bhatia A, Ward M, Penttilä M. Swollenin, a *Trichoderma reesei* protein with sequence similarity to the plant expansins, exhibits disruption activity on cellulosic materials. Eur J Biochem, 2002; 269: 4202–4211.

Saloheimo M, Siika-Aho M, Tenkanen M, Penttila ME. Xylanase from *Trichoderma reesei*, method for production thereof, and methods employing this enzyme. 2003; US Patent: 6555335.

Salvachúa D, Prieto A, López-Abelairas M, Lu-Chau T, Martínez AT, Martínez MJ. Fungal pretreatment: an alternative in second-generation ethanol from wheat straw. Bioresour Technol, 2011; 102: 7500–7506.

Samayam IP, Hanson BL, Langan P, Schall CA. Ionic-liquid induced changes in cellulose structure associated with enhanced biomass hydrolysis. Biomacromolecules, 2011; 12: 3091–3098.

Sampedro J, Cosgrove DJ. The expansin superfamily. Genome Biol, 2005; 6: 242.

Samuelov NS, Datta R, Jain MK, Zeikus JG. Whey fermentation by *Anaerobiospirillum succiniciproducens* for production of a succinate-based animal feed additive. Appl Environ Microbiol, 1999; 65: 2260–2263.

Sanderson MA, Reed RL, McLaughlin SB, Wullschleger SD, Conger BV, Parrish DJ, Wolf DD, Taliaferro C, Hopkins AA, Ocumpaugh WR, Hussey MA, Read JC, Tischler CR. Switchgrass as a sustainable bioenergy crop. Bioresource Technology, 1996; 56: 83–93.

Sandgren M, Shaw A, Ropp TH, Wu S, Bott R, Cameron AD, Ståhlberg J, Mitchinson C, Jones TA. The X-ray crystal structure of the *Trichoderma reesei* family 12 endoglucanase 3, Cel12A, at 1.9 A resolution. J Mol Biol, 2001; 308: 295–310.

Sandgren M, Ståhlberg J, Mitchinson C. Structural and biochemical studies of GH family 12 cellulases: improved thermal stability, and ligand complexes. Prog Biophys Mol Biol, 2005; 89: 246–291.

Sandhu AP, Randhawa GS, Dhugga KS. Plant cell wall matrix polysaccharide biosynthesis. Mol Plant, 2009; 2: 840–850.

Savageau MA. Design of gene circuitry by natural selection: analysis of the lactose catabolic system in *Escherichia coli*. Biochem Soc Trans, 1999; 27: 264–270.

Scandellari F, Hobbie EA, Ouimette AP, Stucker VK. Tracing metabolic pathways of lipid biosynthesis in ectomycorrhizal fungi from position-specific 13C-labelling in glucose. Environ Microbiol, 2009; 11: 3087–3095.

Scazzocchio C. The fungal GATA factors. Curr Opin Microbiol, 2000; 3: 126–131.

Schacht C, Zetzl C, Brunner G. From plant materials to ethanol by means of supercritical fluid technology. J Supercrit Fluids, 2008; 46: 299–321.

Scheel T, Hofer M, Ludwig S, Holker U. Differential expression of manganese peroxidase and laccase in white-rot fungi in the presence of manganese or aromatic compounds. Appl Microbiol Biotechnol, 2000; 54: 686–691.

Schekman R. Charting the secretory pathway in a simple eukaryote. Mol Biol Cell, 2010; 21: 3781–3784.

Scheller HV, Ulvskov P. Hemicelluloses. Annu Rev Plant Biol, 2010; 61: 263–89.

Schenk B, Fernandez F, Waechter CJ. The in(side) and out(side) of dolichyl phosphate biosynthesis and recycling in the endoplasmic reticulum. Glycobiology, 2001; 11: 61R–70R.

Schmid J, Müller-Hagen D, Bekel T, Funk L, Stahl U, Sieber V, Meyer V. Transcriptome sequencing and comparative transcriptome analysis of the scleroglucan producer *Sclerotium rolfsii*. BMC Genomics, 2010; 11: 329.

Schmidt A, Schlacher A, Steiner W, Schwab H, Kratky C. Structure of the xylanase from *Penicillium simplicissimum*. Protein Sci, 1998,; 7: 2081–2088.

Schmidt E. *Wood and Tree Fungi: Biology, Damage, Protection, and Use*, 2006, p. 334. Springer, Berlin-Heidelberg-New York.

Schmitz RF, Gibson KE. USDA Forest Services Forest Insect & Disease Leaflet 5. Available at: http://www.barkbeetles.org/douglasfir/dfir.html.

Schmoll M, Esquivel-Naranjo EU, Herrera-Estrella A. *Trichoderma* in the light of day—physiology and development. Fungal Genet Biol, 2010; 47: 909–916.

Schmoll M, Schuster A, Silva Rdo N, Kubicek CP. The G-alpha protein GNA3 of *Hypocrea jecorina* (anamorph *Trichoderma reesei*) regulates cellulase gene expression in the presence of light. Eukaryot Cell, 2009; 8: 410–420.

Schneider H. Conversion of pentoses to ethanol by yeasts and fungi. Crit Rev Biotechnol, 1989; 9: 1–40.

Schnorr KM, Landvik S, Spendler T, Christensen LLH. Family GH-61 polypeptides. 2007; US Patent: 7273738.

Schoemaker HE, Leisola MSA. Degradation of lignin by *Phanerochaete chrysosporium*. J Biotechnol, 1990; 13: 101–109.

Schols HA, Posthumus MA, Voragen AGJ. Structural features of hairy regions of pectins isolated from apple juice produced by the liquefaction process. Carbohydr Res, 1990; 206: 117–129.

Schröder M, Chang JS, Kaufman RJ. The unfolded protein response represses nitrogen-starvation induced developmental differentiation in yeast. Genes Dev, 2000; 14: 2962–2975.

Schülein M, Kauppinen M, Lange L, Lassen S, Andersen L, Klysner S, Nielsen J. Characterization of fungal cellulases for fiber modification. ACS Symp Series, 1998; 687: 66–74.

Schulein M, Wu W, Kauppinen MS, Stringer MA. Xyloglucanase from *Malbranchea*. 2002; US Patent: 6500658.

Schülein M. Enzymatic properties of cellulases from *Humicola insolens*. J Biotechnol, 1997; 57: 71–81.

Schuster A, Bruno KS, Collett JR, Baker SE, Seiboth B, Kubicek CP, Schmoll M. A versatile toolkit for high throughput functional genomics with *Trichoderma reesei*. Biotechnol Biofuels, 2012; 5: 1.

Scott BR, St. Pierre P, Lavigne L, Masri JN, White TC, Tomashek JJ. Novel lignin-resistant cellulase enzymes. 2010; US Patent: 7375197.

Sedlak M, Ho NW. Characterization of the effectiveness of hexose transporters for transporting xylose during glucose and xylose co-fermentation by a recombinant *Saccharomyces* yeast. Yeast, 2004; 21: 671–684.

Seibel C, Gremel G, do Nascimento Silva R, Schuster A, Kubicek CP, Schmoll M. Light-dependent roles of the G-protein alpha subunit GNA1 of *Hypocrea jecorina* (anamorph *Trichoderma reesei*). BMC Biol, 2009; 7: 58.

Seiboth B, Gamauf C, Pail M, Hartl L, Kubicek CP. The d-xylose reductase of *Hypocrea jecorina* is the major aldose reductase in pentose and d-galactose catabolism and necessary for beta-galactosidase and cellulase induction by lactose. Mol Microbiol, 2007a; 66: 890–900.

Seiboth B, Hartl L, Pail M, Fekete E, Karaffa L, Kubicek CP. The galactokinase of *Hypocrea jecorina* is essential for cellulase induction by lactose but dispensable for growth on D-galactose. Mol Microbiol, 2004; 51: 1015–1025.

Seiboth B, Hartl L, Salovuori N, Lanthaler K, Robson GD, Vehmaanperä J, Penttilä ME, Kubicek CP. Role of the bga1-encoded extracellular β-galactosidase of *Hypocrea jecorina* in cellulase induction by lactose. Appl Environ Microbiol, 2005; 71: 851–857.

Seiboth B, Karaffa L, Sándor E, Kubicek C. The *Hypocrea jecorina gal10* (uridine 5′-diphosphate-glucose 4-epimerase-encoding) gene differs from yeast homologues in structure, genomic organization and expression. Gene, 2002; 295: 143–149.

Seiboth B, Karimi R, Phatale PA, Linke R, Sauer DG, Smith KM, Baker SE, Freitag M, Kubicek CP. The putative protein methyltransferase LAE1 controls cellulase gene expression in *Trichoderma reesei*. Mol Microbiol, 2012 May 3. doi: 10.1111/j.1365-2958.2012.08083.x. [Epub ahead of print].

Seiboth B, Metz B. Fungal arabinan and L-arabinose metabolism. Appl Microbiol Biotechnol, 2011; 89: 1665–1673.

Seiboth B, Pakdaman BS, Hartl L, Kubicek CP. Lactose metabolism in filamentous fungi: how to deal with an unknown substrate. Fungal Biol Rev, 2007b; 21: 42–48.

Seidl V, Seibel C, Kubicek CP, Schmoll M. Sexual development in the industrial workhorse *Trichoderma reesei*. Proc Natl Acad Sci USA, 2009,; 106: 13909–13914.

Seidl V, Seiboth B. *Trichoderma reesei*: genetic approaches to improving strain efficiency. Biofuels, 2010; 1: 343–354.

Seifert GJ, Barber C, Wells B, Dolan L, Roberts K. Galactose biosynthesis in *Arabidopsis*: genetic evidence for substrate channeling from UDP-D-galactose into cell wall polymers. Curr Biol, 2002; 12: 1840–1845.

Seifert GJ, Barber C, Wells B, Roberts K. Growth regulators and the control of nucleotide sugar flux. Plant Cell, 2004; 16: 723–730.

Semenova MV, Drachevskaya MI, Sinitsyna OA, Gusakov AV, Sinitsyn AP. Isolation and properties of extracellular beta-xylosidases from fungi *Aspergillus japonicas* and *Trichoderma reesei*. Biochemistry, 2009; 74: 1002–1008.

Sestak S, Farkas V. Metabolic regulation of endoglucanase synthesis in *Trichoderma reesei*: participation of cyclic AMP and glucose-6-phosphate. Can J Microbiol, 1993; 39: 342–347.

Shibuya H, Nagasaki H, Kaneko S, Yoshida S, Park GG, Kusakabe I, Kobayashi H. Cloning and high-level expression of alpha-galactosidase cDNA from *Penicillium purpurogenum*. Appl Environ Microbiol, 1998; 64: 4489–4494.

Shima Y, Ito Y, Kaneko S, Hatabayashi H, Watanabe Y, Adachi Y, Yabe K. Identification of three mutant loci conferring carboxin-resistance and development of a novel transformation system in *Aspergillus oryzae*. Fungal Genet Biol, 2009; 46: 67–76.

Shimada M, Akamtsu Y, Tokimatsu T, Mii K, Hattori T. Possible biochemical roles of oxalic acid as a low molecular weight compound involved in brown-rot and white-rot wood decays. J Biotechnol, 1997; 53: 103–113.

Shishido K. Filamentous ascomycetes and basidiomycetes: development of host-vector systems. Bioprocess Technol, 1994; 19: 683–700.

Shoham Y. Glycoside Hydrolase Family 51 in *CAZypedia*. 2011. Available at: http://www.cazypedia.org/ (accessed October 31, 2011).

Shoji JY, Arioka M, Kitamoto K. Dissecting cellular components of the secretory pathway in filamentous fungi: insights into their application for protein production. Biotechnol Letts, 2008; 30: 7–14.

Shoseyov O, Shani Z, Levy I. Carbohydrate binding modules: biochemical properties and novel applications. Microbiol Mol Biol Rev, 2006; 70: 283–295.

Sigl C, Handler M, Sprenger G, Kürnsteiner H, Zadra I. A novel homologous dominant selection marker for genetic transformation of *Penicillium chrysogenum*: overexpression of squalene epoxidase-encoding *ergA*. J Biotechnol, 2010; 150: 307–311.

Simmons BA, Loque D, Blanch HW. Next-generation biomass feedstocks for biofuel production. Genome Biol, 2008; 9: 242.

Simmons BA, Loqué D, Ralph J. Advances in modifying lignin for enhanced biofuel production. Curr Opin Plant Biol, 2010; 13: 313–320.

Simmons EG. Classification of some cellulase-producing *Trichoderma* species. In: *Book of Abstracts: Second International Mycological Congress*, Tampa, FL, 1977, p. 618

Sims I, Craik DJ, Bacic A. Structural characterisation of galactoglucomannan secreted by suspension-cultured cells of *Nicotiana plumbaginifolia*. Carbohydr Res, 1997; 303: 79–92.

Singh D, Chen S. The white-rot fungus *Phanerochaete chrysosporium*: conditions for the production of lignin-degrading enzymes. Appl Microbiol Biotechnol, 2008; 81: 399–417.

Singhania RR, Sukumaran RK, Patel AK, Larroche C, Pandey A. Advancement and comparative profiles in the production technologies using solid-state and submerged fermentation for microbial cellulases. Enzyme Microb Technol, 2010; 46: 541–549.

Skomarovsky AA, Gusakov AV, Okunev ON, Solov'eva IV, Bubnova TV, Kondrat'eva EG, Sinitsyn AP. Studies of hydrolytic activity of enzyme preparations of *Penicillium* and *Trichoderma* fungi. Appl Biochem Microbiol, 2005; 41: 182–184.

Slininger PJ, Dien BS, Gorsich SW, Liu ZL. Nitrogen source and mineral optimization enhance D-xylose conversion to ethanol by the yeast *Pichia stipitis* NRRL Y-7124. Appl Microbiol Biotechnol, 2006; 72: 1285–1296.

Slot JC, Rokas A. Multiple GAL pathway gene clusters evolved independently and by different mechanisms in fungi. Proc Natl Acad Sci USA, 2010; 107: 10136–10141.

Sogabe Y, Kitatani T, Yamaguchi A, Kinoshita T, Adachi H, Takano K, Inoue T, Mori Y, Matsumura H, Sakamoto T, Tada T. High-resolution structure of exo-arabinanase from *Penicillium chrysogenum*. Acta Crystallogr D Biol Crystallogr, 2011; 67: 415–422.

Somerville C. Cellulose synthesis in higher plants. Annu Rev Cell Dev Biol, 2006; 22: 53–78.

Sonderegger M, Jeppsson M, Hahn-Hägerdal B, Sauer U. Molecular basis for anaerobic growth of *Saccharomyces cerevisiae* on xylose, investigated by global gene expression and metabolic flux analysis. Appl Environ Microbiol, 2004; 70: 2307–2317.

Sonderegger M, Sauer U. Evolutionary engineering of *Saccharomyces cerevisiae* for anaerobic growth on xylose. Appl Environ Microbiol, 2003; 69: 1990–1998.

Sørensen A, Teller PJ, Lübeck PS, Ahring BK. Onsite enzyme production during bioethanol production from biomass: screening for suitable fungal strains. Appl Biochem Biotechnol, 2011; 164: 1058–1070.

Sørensen HR, Jørgensen CT, Hansen CH, Jørgensen CI, Pedersen S, Meyer AS. A novel GH43 alpha-L-arabinofuranosidase from *Humicola insolens*: mode of action and synergy with GH51 alpha-L-arabinofuranosidases on wheat arabinoxylan. Appl Microbiol Biotechnol, 2006; 73: 850–861.

Špániková S, Biely P. Glucuronoyl esterase—novel carbohydrate esterase produced by *Schizophyllum commune*. FEBS Lett, 2006; 580: 4597–4601.

Špániková S, Poláková M, Joniak D, Hirsch J, Biely P. Synthetic esters recognized by glucuronoyl esterase from *Schizophyllum commune*. Arch Microbiol, 2007; 188: 185–189.

Sposato P, Ahn JH, Walton JD. Characterization and disruption of a gene in the maize pathogen *Cochliobolus carbonum* encoding a cellulase lacking a cellulose binding domain and hinge region. Mol Plant Microbe Interact, 1995; 8: 602–609.

St John FJ, González JM, Pozharski E. Consolidation of glycosyl hydrolase family 30: a dual domain 4/7 hydrolase family consisting of two structurally distinct groups. FEBS Letts, 2010; 584: 4435–4441.

Stahlberg J. Glycoside Hydrolase Family 7 in *CAZypedia*. 2011. Available at: http://www.cazypedia.org/ (accessed October 31, 2011).

Ståhlberg J, Divne C, Koivula A, Piens K, Claeyssens M, Teeri TT, Jones TA. Activity studies and crystal structures of catalytically deficient mutants of cellobiohydrolase I from *Trichoderma reesei*. J Mol Biol, 1996; 264: 337–349.

Steen EJ, Kang Y, Bokinsky G, Hu Z, Schirmer A, McClure A, Del Cardayre SB, Keasling JD. Microbial production of fatty-acid-derived fuels and chemicals from plant biomass. Nature, 2010; 463: 559–562.

Steenbakkers PJ, Irving JA, Harhangi HR, Swinkels WJ, Akhmanova A, Dijkerman R, Jetten MS, van der Drift C, Whisstock JC, Op den Camp HJ. A serpin in the cellulosome of the anaerobic fungus *Piromyces* sp. strain E2. Mycol Res, 2008; 112: 999–1006.

Steiger MG, Vitikainen M, Uskonen P, Brunner K, Adam G, Pakula T, Penttilä M, Saloheimo M, Mach RL, Mach-Aigner AR. Transformation system for *Hypocrea jecorina* (*Trichoderma reesei*) that favors homologous integration and employs reusable bidirectionally selectable markers. Appl Environ Microbiol, 2011; 77: 114–121.

Sterling JD, Atmodjo MA, Inwood SE, Kumar Kolli VS, Quigley HF, Hahn MG, Mohnen D. Functional identification of an Arabidopsis pectin biosynthetic homogalacturonan galacturonosyltransferase. Proc Natl Acad Sci USA, 2006; 103: 5236–5241.

Sternberg D, Mandels GR. Induction of cellulolytic enzymes in *Trichoderma reesei* by sophorose. J Bacteriol, 1979; 139: 761–769.

Straathof AJ, Sie S, Franco TT, van der Wielen LA. Feasibility of acrylic acid production by fermentation. Appl Microbiol Biotechnol, 2005; 67: 727–734.

Strauss J, Horvath HK, Abdallah BM, Kindermann J, Mach RL, Kubicek CP. The function of CreA, the carbon catabolite repressor of *Aspergillus nidulans*, is regulated at the transcriptional and post-transcriptional level. Mol Microbiol, 1999; 32: 169–178.

Strauss J, Reyes-Dominguez J. Regulation of secondary metabolism by chromatin structure and epigenetic codes. Fungal Genet Biol, 2011; 48: 62–69.

Stricker AR, Grosstessner-Hain K, Würleitner E, Mach RL. Xyr1 (xylanase regulator 1) regulates both the hydrolytic enzyme system and D-xylose metabolism in *Hypocrea jecorina*. Eukaryot Cell, 2006; 5: 2128–2137.

Stricker AR, Mach RL, de Graaff LH. Regulation of transcription of cellulases- and hemicellulases-encoding genes in *Aspergillus niger* and *Hypocrea jecorina* (*Trichoderma reesei*). Appl Microbiol Biotechnol, 2008; 78: 211–220.

Stricker AR, Steiger MG, Mach RL. Xyr1 receives the lactose induction signal and regulates lactose metabolism in *Hypocrea jecorina*. FEBS Letts, 2007; 581: 3915–3920.

Sugano Y. DyP-type peroxidases comprise a novel heme peroxidase family. Cell Mol Life Sci, 2009; 66: 1387–1403.

Sugano Y, Muramatsu R, Ichiyanagi A, Sato T, Shoda M. DyP, a unique dye-decolorizing peroxidase, represents a novel heme peroxidase family: ASP171 replaces the distal histidine of classical peroxidases. J Biol Chem, 2007; 282: 36652–36658.

Sulzenbacher G, Mackenzie LF, Wilson KS, Withers SG, Dupont C, Davies GJ. The crystal structure of a 2-fluorocellotriosyl complex of the *Streptomyces lividans* endoglucanase CelB2 at 1.2 A resolution. Biochemistry, 1999; 38: 4826–4833.

Sundaramoorthy M, Kishi K, Gold MH, Poulos TL. Preliminary crystallographic analysis of manganese peroxidase from *Phanerochaete chrysosporium*. J Mol Biol, 1994; 238: 845–848.

Sundaramoorthy M, Youngs HL, Gold MH, Poulos TL. High-resolution crystal structure of manganese peroxidase: substrate and inhibitor complexes. Biochemistry, 2005; 44: 6463–6470.

Suriyamongkol P, Weselake R, Narine S, Moloney M, Shah S. Biotechnological approaches for the production of polyhydroxyalkanoates in microorganisms and plants—a review. Biotechnol Adv, 2007; 25: 148–175.

Suzuki H, Igarashi K, Samejima M. Cellotriose and cellotetraose as inducers of the genes encoding cellobiohydrolases in the basidiomycete *Phanerochaete chrysosporium*. Appl Environ Microbiol, 2010; 76: 6164–6170.

Swatloski RP, Spear SK, Holbrey JD, Rogers RD. Dissolution of cellose with ionic liquids. J Am Chem Soc, 2002; 124: 4974–4975.

Szakmary K, Wottawa A, Kubicek CP. Origin of oxidized cellulose degradation products and mechanism of their promotion of cellobiohydrolase I biosynthesis in *Trichoderma reesei*. J Gen Microbiol, 1991; 137: 2873–2878.

Szijártó N, Horan E, Zhang J, Puranen T, Siika-Aho M, Viikari L. Thermostable endoglucanases in the liquefaction of hydrothermally pretreated wheat straw. Biotechnol Biofuels, 2011; 4: 2.

Taherzadeh MJ, Niklasson C, Liden G. On-line control of fed-batch fermentation of dilute-acid hydrolyzates. Biotechnol Bioengin, 2000; 69: 330–338.

Takada G, Kawaguchi T, Sumitani J-I, Arai M. Cloning, nucleotide sequence, and transcriptional analysis of *Aspergillus aculeatus* no. F-50 cellobiohydrolase I (*cbhI*) gene. J Ferm Bioengin, 1998; 85: 1–9.

Takashima S, Iikura H, Nakamura A, Hidaka M, Masaki H, Uozumi T. Isolation of the gene and characterization of the enzymatic properties of a major exoglucanase of *Humicola grisea* without a cellulose-binding domain. J Biochem, 1998; 124: 717–725.

Takata R, Tokita K, Mori S, Shimoda R, Harada N, Ichinose H, Kaneko S, Igarashi K, Samejima M, Tsumuraya Y, Kotake T. Degradation of carbohydrate moieties of arabinogalactan-proteins by glycoside hydrolases from *Neurospora crassa*. Carbohydr Res, 2010; 345: 2516–2522.

Talebnia F, Karakashev D, Angelidaki I. Production of bioethanol from wheat straw: an overview on pretreatment, hydrolysis and fermentation. Bioresour Technol, 2010; 101: 4744–4753.

Tanaka H, Itakura S, Hirano T, Enoki A. An extracellular substance from the white-rot basidiomycete *Phanerochaete chrysosporium* for reducing molecular oxygen and ferric iron. Holzforschung, 1996; 50: 541–548.

Tanaka H, Yoshida G, Baba Y, Matsumura K, Wasada H, Murata J, Agawa M, Itakura S, Enoki A. Characterization of a hydroxyl-radical-producing glycoprotein and its presumptive genes from the white-rot basidiomycete *Phanerochaete chrysosporium*. J Biotechnol, 2007; 128: 500–511.

Tanaka N, Akamtsu Y, Hattori T, Shimada M. Effect of oxalic acid on the oxidative breakdown of cellulose by Fenton reaction. Wood Res, 1994; 81: 8–10.

Taniguchi M, Suzuki H, Watanabe D, Sakai K, Hoshino K, Tanaka T. Evaluation of pretreatment with *Pleurotus ostreatus* for enzymatic hydrolysis of rice straw. J Biosci Bioeng, 2005; 100: 637–643.

Teeri TT. Crystalline cellulose degradation: new insight into the function of cellobiohydrolases. Trends Biotechnol, 1997; 15: 160–167.

Teeri TT, Lehtovaara P, Kauppinen S, Salovuori I, Knowles J. Homologous domains in *Trichoderma reesei* cellulolytic enzymes: gene sequence and expression of cellobiohydrolase II. Gene, 1987; 51: 43–52.

Tenkanen M. Action of *Trichoderma reesei* and *Aspergillus oryzae* esterases in the deacetylation of hemicelluloses. Biotechnol Appl Biochem, 1998; 27: 19–24.

Tenkanen M, Siika-Aho M. An alpha-glucuronidase of *Schizophyllum commune* acting on polymeric xylan. J Biotechnol, 2000; 78: 149–161.

Teymouri F, Laureano-Pérez L, Alizadeh H, Dale BE. Ammonia fiber explosion treatment of corn stover. Appl Biochem Biotechnol, 2004; 115: 951–963.

Tian C, Beeson WT, Iavarone AT, Sun J, Marletta MA, Cate JH, Glass NL. Systems analysis of plant cell wall degradation by the model filamentous fungus *Neurospora crassa*. Proc Natl Acad Sci USA, 2009; 106: 22157–22162.

Tilburn J, Scazzocchio C, Taylor GG, Zabicky-Zissman JH, Lockington RA, Davies RW. Transformation by integration in *Aspergillus nidulans*. Gene, 1983; 26: 205–221.

Timberlake WE, Marshall MA. Genetic engineering of filamentous fungi. Science, 1989; 244: 1313–1317.

Timmers J, Vernhettes S, Desprez T, Vincken JP, Visser RG, Trindade LM. Interactions between membrane-bound cellulose synthases involved in the synthesis of the secondary cell wall. FEBS Lett, 2009; 583: 978–982.

Tinoco R, Verdin J, Vazquez-Duhalt R. Role of oxidizing mediators and tryptophan 172 in the decoloration of industrial dyes by the versatile peroxidases from *Bjerkandera adusta*. J Mol Catal B Enzym, 2007; 46: 1–7.

Toivari MH, Salusjarvi L, Ruohonen L, Penttilä M. Endogenous xylose pathway in *Saccharomyces cerevisiae*. Appl Environ Microbiol, 2004; 70: 3681–3686.

Tolan JS, Foody B. Cellulase from submerged fermentation. Adv Biochem Engin Biotechnol, 1999; 65: 41–67.

Torney F, Moeller L, Scarpa A, Wang K. Genetic engineering approaches to improve bioethanol production from maize. Curr Opin Biotechnol, 2007; 18: 193–199.

Torpenholt S, Le Nours J, Christensen U, Jahn M, Withers S, Ostergaard PR, Borchert TV, Poulsen JC, Lo Leggio L. Activity of three β-1,4-galactanases on small chromogenic substrates. Carbohydr Res, 2011; May 27 [Epub ahead of print].

Törrönen A, Harkki A, Rouvinen J. Three-dimensional structure of endo-1,4-beta-xylanase II from *Trichoderma reesei*: two conformational states in the active site. EMBO J, 1994; 13: 2493–2501.

Törrönen A, Kubicek CP, Henrissat B. Amino acid sequence similarities between low molecular weight endo-1,4-beta-xylanases and family H cellulases revealed by clustering analysis. FEBS Letts, 1993; 321: 135–139.

Törrönen A, Mach RL, Messner R, Gonzalez R, Kalkkinen N, Harkki A, Kubicek CP. The two major xylanases from *Trichoderma reesei*: characterization of both enzymes and genes. Bio/Technology, 1992; 10: 1461–1465.

Törrönen A, Rouvinen J. Structural comparison of two major endo-1,4-xylanases from *Trichoderma reesei*. Biochemistry, 1995; 34: 847–856.

Toyama H, Nakano M, Satake Y, Toyama N. Rapid isolation of the *Trichoderma* strain with higher degrading ability of a filter paper and superior proliferation characteristics using avicel plates and the double-layer selection medium. Appl Biochem Biotechnol, 2008; 145: 23–28.

Travers KJ, Patil CK, Wodicka L, Lockhart DJ, Weissman JS, Walter P. Functional and genomic analyses reveal an essential coordination between the unfolded protein response and ER-associated degradation. Cell, 2000; 101: 249–258.

Tummala SB, Junne SG, Papoutsakis ET. Antisense RNA downregulation of coenzyme A transferase combined with alcohol-aldehyde dehydrogenase overexpression leads to predominantly alcohologenic *Clostridium acetobutylicum* fermentations. J Bacteriol, 2003; 185: 3644–3653.

Turgeon BG, Condon B, Liu J, Zhang N. Protoplast transformation of filamentous fungi. Methods Mol Biol, 2010; 638: 3–19.

Ubhayasekera W, Muñoz IG, Vasella A, Ståhlberg J, Mowbray SL. Structures of *Phanerochaete chrysosporium* Cel7D in complex with product and inhibitors. FEBS J, 2005; 272: 1952–1964.

Umemura M, Yuguchi Y. Conformational folding of xyloglucan side chains in aqueous solution from molecular dynamics simulation. Carbohydr Res, 2005; 340: 2520–2532.

Ungermann C, Wickner W. Vam7p, a vacuolar SNAP-25 homolog, is required for SNARE complex integrity and vacuole docking and fusion. EMBO J, 1998; 17: 3269–3276.

Upton C, Buckley JT. A new family of lipolytic enzymes? Trends Biochem Sci, 1995; 20: 178–179.

Urbániková L, Vršanská M, Mørkeberg Krogh KB, Hoff T, Biely P. Structural basis for substrate recognition by *Erwinia chrysanthemi* GH30 glucuronoxylanase. FEBS J, 2011; doi: 10.1111/j.1742–4658.2011.08127.x [Epub ahead of print].

Urzua U, Kersten PJ, Vicuna R. Manganese peroxidase-dependent oxidation of glyoxylic and oxalic acids synthesized by *Ceriporiopsis subvermispora* produces extracellular hydrogen peroxide. Appl Environ Microbiol, 1998; 64: 68–73.

Vaaje-Kolstad G, Houston DR, Riemen AH, Eijsink VG, van Aalten DM. Crystal structure and binding properties of the *Serratia marcescens* chitin-binding protein CBP21. J Biol Chem, 2005,; 280: 11313–11319.

Vaheri M, Leisola M, Kauppinen V. Transglycosylation products of cellulase system of *Trichoderma reesei*. Biotechnol Letts, 1979; 1: 696–699.

Valkonen M, Kalkman ER, Saloheimo M, Penttilä M, Read ND, Duncan RR. Spatially segregated SNARE protein interactions in living fungal cells. J Biol Chem, 2007; 282: 22775–22785.

Valkonen M, Penttilä M, Saloheimo M. The *ire1* and *ptc2* genes involved in the unfolded protein response pathway in the filamentous fungus *Trichoderma reesei*. Mol Genet Genomics, 2004; 272: 443–451.

Valli K, Wariishi H, Gold MH. Oxidation of monomethoxylated aromatic compounds by lignin peroxidase: role of veratryl alcohol in lignin biodegradation. Biochem, 1990; 29: 8535–8539.

van den Brink J, de Vries RP. Fungal enzyme sets for plant polysaccharide degradation. Appl Microbiol Biotechnol, 2011; 91: 1477–1492.

van den Broek LA, Lloyd RM, Beldman G, Verdoes JC, McCleary BV, Voragen AG. Cloning and characterization of arabinoxylan arabinofuranohydrolase-D3 (AXHd3) from *Bifidobacterium adolescentis* DSM20083. Appl Microbiol Biotechnol, 2005; 67: 641–647.

Vanholme R, Morreel K, Ralph J, Boerjan W. Lignin engineering. Curr Opin Plant Biol, 2008; 11: 278–285.

van Peij NN, Visser J, de Graaff LH. Isolation and analysis of xlnR, encoding a transcriptional activator co-ordinating xylanolytic expression in *Aspergillus niger*. Mol Microbiol, 1998; 27: 131–142.

van Santen Y, Benen JA, Schroter KH, Kalk KH, Armand S, Visser J, Dijkstra BW. 1.68-Å crystal structure of endopolygalacturonase II from *Aspergillus niger* and identification of active site residues by site-directed mutagenesis. J Biol Chem, 1999; 274: 30474–30480.

Van Vleet JH, Jeffries TW. Yeast metabolic engineering for hemicellulosic ethanol production. Curr Opin Biotechnol, 2009; 20: 300–306.

van Zyl WH, Lynd LR, den Haan R, McBride JE. Consolidated bioprocessing for bioethanol production using *Saccharomyces cerevisiae*. Adv Biochem Eng Biotechnol, 2007; 108: 205–235.

Vanden Wymelenberg A, Covert S, Cullen D. Identification of the gene encoding the major cellobiohydrolase of the white rot fungus *Phanerochaete chrysosporium*. Appl Environ Microbiol, 1993; 59: 3492–3494.

Vanden Wymelenberg A, Sabat G, Mozuch M, Kersten PJ, Cullen D, Blanchette RA. Structure, organization, and transcriptional regulation of a family of copper radical oxidase genes in the lignin-degrading basidiomycete *Phanerochaete chrysosporium*. Appl Environ Microbiol, 2006; 72: 4871–4877.

Vandermarliere E, Bourgois TM, Winn MD, van Campenhout S, Volckaert G, Delcour JA, Strelkov SV, Rabijns A, Courtin CM. Structural analysis of a glycoside hydrolase family 43 arabinoxylan arabinofuranohydrolase in complex with xylotetraose reveals a different binding mechanism compared with other members of the same family. Biochem J, 2009; 418: 39–47.

Vardakou M, Dumon C, Murray JW, Christakopoulos P, Weiner DP, Juge N, Lewis RJ, Gilbert HJ, Flint JE. Understanding the structural basis for substrate and inhibitor recognition in eukaryotic GH11 xylanases. J Mol Biol, 2008; 375: 1293–1203.

Varghese JN, Hrmova M, Fincher GB. Three-dimensional structure of a barley beta-D-glucan exohydrolase, a family 3 glycosyl hydrolase. Structure, 1999; 7: 179–190.

Varrot A, Frandsen TP, von Ossowski I, Boyer V, Cottaz S, Driguez H, Schülein M, Davies GJ. Structural basis for ligand binding and processivity in cellobiohydrolase Cel6A from *Humicola insolens*. Structure, 2003; 11: 855–864.

Vautard-Mey G, Cotton P, Fèvre M. Expression and compartmentation of the glucose repressor CRE1 from the phytopathogenic fungus *Sclerotinia sclerotiorum*. Eur J Biochem, 1999; 266: 252–259.

Vautard-Mey G, Fèvre M. Mutation of a putative AMPK phosphorylation site abolishes the repressor activity but not the nuclear targeting of the fungal glucose regulator CRE1. Curr Genet, 2000; 37: 328–332.

Vemuri GN, Eiteman MA, Altman E. Succinate production in dual-phase *Escherichia coli* fermentations depends on the time of transition from aerobic to anaerobic conditions. J Ind Microbiol Biotechnol, 2002; 28: 325–332.

Verdoes JC, Punt PJ, van der Berg P, Debets F, Stouthamer AH, van den Hondel CA. Characterization of an efficient gene cloning strategy for *Aspergillus niger* based on an autonomously replicating plasmid: cloning of the *nicB* gene of *A. niger*. Gene, 1994; 146: 159–165.

Viikari L, Alapuranen M, Puranen T, Vehmaanperä J, Siika-Aho M. Thermostable enzymes in lignocellulose hydrolysis. Adv Biochem Eng Biotechnol, 2007; 108: 121–145.

Vincken JP, York WS, Beldman G, Voragen AGJ. Two general branching patterns of xyloglucan, XXXG and XXGG. Plant Physiol, 1997; 114: 9–13.

Virag A, Harris SD. The Spitzenkörper: a molecular perspective. Mycol Res, 2006; 110: 4–13.

Visser H, Joosten V, Punt PJ, Gusakov AV, Olson PT, Joosten R, Bartels J, Visser J, Sinitsyn AP, Emalfarb MA, Verdoes JC, Wery J. Development of a mature fungal technology and production platform for industrial enzymes based on a *Myceliophthora thermophila* isolate, previously known as *Chrysosporium lucknowense* C1. Ind Biotechnol, 2011; 7: 214–223.

Vitali J, Schick B, Kester HC, Visser J, Jurnak F. The tree-dimensional structure of *Aspergillus niger* pectin lyase B at 1.7-Å resolution. Plant Physiol, 1998; 116: 69–80.

Wada M, Sugiyama J, Okano T. Native celluloses on the basis of two crystalline phase (Iα/Iß) system. J Appl Polym Sci, 1993; 49: 1491–1496.

Walker JR, Corpina RA, Goldberg J. Structure of the Ku heterodimer bound to DNA and its implications for double-strand break repair. Nature, 2001; 412: 607–614.

Walter P, Johnson AE. Signal sequence recognition and protein targeting to the endoplasmic reticulum membrane. Annu Rev Cell Biol, 1994; 10: 87–119.

Wang B, Xia L. High efficient expression of cellobiase gene from *Aspergillus niger* in the cells of *Trichoderma reesei*. Bioresour Technol, 2011; 102: 4568–4572.

Wang L, Zhang Y, Gao P. A novel function for the cellulose-binding module of cellobiohydrolase I. Sci China C Life Sci, 2008; 51: 620–629.

Wang P, Nuss DL. Induction of a *Cryphonectria parasitica* cellobiohydrolase I gene is suppressed by hypovirus infection and regulated by a GTP-binding-protein-linked signaling pathway involved in fungal pathogenesis. Proc Natl Acad Sci USA, 1995; 92: 11529–11533.

Ward M, Wilkinson B, Turner G. Transformation of *Aspergillus nidulans* with a cloned oligomycin-resistant ATP synthase subunit 9 gene. Mol Gen Genet, 1986; 202: 289–293.

Wariishi H, Valli K, Gold MH. Manganese(II) oxidation by manganese peroxidase from the Basidiomycete *Phanerochaete chrysosporium*. J Biol Chem, 1992; 267: 23688–23695.

Warnick TA, Methé BA, Leschine SB. *Clostridium phytofermentans* sp. nov., a cellulolytic mesophile from forest soil. Int J Syst Evol Microbiol, 2002; 52: 1155–1160.

Watanabe T, Shirai N, Okada H, Honda Y, Kuwahara M. Production and chemiluminescent free radical reactions of glyoxal in lipid peroxidation of linoleic acid by the ligninolytic enzyme, manganese peroxidase. Eur J Biochem, 2001; 268: 6114–6122.

Watanabe T, Shitan N, Suzuki S, Umezawa T, Shimada M, Yazaki K, Hattori T. Oxalate efflux transporter from the brown rot fungus *Fomitopsis palustris*. Appl Environ Microbiol, 2010; 76: 7683–7690.

Wee Y-J, Kim J-N, Ryu H-W. Biotechnological production of lactic acid and its recent applications. Food Technol Biotechnol, 2006; 44: 163–172.

Wegener S, Ransom RF, Walton JD. A unique eukaryotic beta-xylosidase gene from the phytopathogenic fungus *Cochliobolus carbonum*. Microbiol, 1999; 145: 1089–1095.

Wei D, Houtman CJ, Kapich AN, Hunt CG, Cullen D, Hammel KE. Laccase and its role in production of extracellular reactive oxygen species during wood decay by the brown rot basidiomycete *Postia* placenta. Appl Environ Microbiol, 2010; 76: 2091–2097.

Weimer PJ, Chou YC. Anaerobic fermentation of woody biomass pretreated with supercritical ammonia. Appl Environ Microbiol, 1986; 52: 733–736.

Weld RJ, Plummer KM, Carpenter MA, Ridgway HJ. Approaches to functional genomics in filamentous fungi. Cell Res, 2006; 16: 31–44.

Wey TT, Hseu TH, Huang L. Molecular cloning and sequence analysis of the cellobiohydrolase I gene from *Trichoderma koningii* G-39. Curr Microbiol, 1994; 28: 31–39.

Whitman WB, Coleman DC, Wiebe WJ. Prokaryotes: the unseen majority. Proc Natl Acad Sci USA, 1998; 95: 6578–6583.

Whitney SE, Gidley MJ, McQueen-Mason SJ. Probing expansin action using cellulose/hemicellulose composites. Plant J, 2000; 22: 327–334.

Whittaker MM, Kersten PJ, Nakamura N, Sanders-Loehr J, Schweizer ES, Whittaker JW. Glyoxal oxidase from *Phanerochaete chrysosporium* is a new radical-copper oxidase. J Biol Chem, 1996; 271: 681–687.

Wierenga RK. The TIM-barrel fold: a versatile framework for efficient enzymes. FEBS Letts, 2001; 492: 193–198.

Wightman R, Turner S. Trafficking of the plant cellulose synthase complex. Plant Physiol, 2010; 153: 427–432.

Wild K, Halic M, Sinning I, Beckmann R. SRP meets the ribosome. Nat Struct Mol Biol, 2004; 11: 1049–1053.

Wildt S, Gerngross TU. The humanization of *N*-glycosylation pathways in yeast. Nature Rev Microbiol, 2005; 3: 119–128.

Wilkie JS. Carl Nägeli and the fine structure of living matter. Nature, 1961; 190: 1145–1150.

Willke T, Vorlop KD. Industrial bioconversion of renewable resources as an alternative to conventional chemistry. Appl Microbiol Biotechnol, 2004; 66: 131–142.

Williams S. "Glycoside Hydrolases" in *CAZypedia*. Available at: http://www.cazypedia.org/ (accessed October 31, 2011).

Williamson G, Kroon PA, Faulds CB. Hairy plant polysaccharides: a close shave with microbial esterases. Microbiol, 1998; 144: 2011–2023.

Wilson C, Venditti R, Rega LR, Colanzi A, D'Angelo G, De Matteis MA. The Golgi apparatus: an organelle with multiple complex functions. Biochem J, 2011; 433: 1–9.

Wilson DB. Cellulases and biofuels. Curr Opin Biotechnol, 2009; 20: 295–299.

Wilson DB. Microbial diversity of cellulose hydrolysis. Curr Opin Microbiol, 2011; 14: 259–263.

Withers S. Glycoside Hydrolase Family 1 in *CAZypedia*. 2011a. Available at: http://www.cazypedia.org/ (accessed October 31, 2011).

Withers S. Glycoside Hydrolase Family 11 in *CAZypedia*. 2011b. Available at: http://www.cazypedia.org/ (accessed October 31, 2011).

Wong CM, Wong KH, Chen XD. Glucose oxidase: natural occurrence, function, properties and industrial applications. Appl Microbiol Biotechnol, 2008a; 78: 927–938.

Wong DW. Structure and action mechanism of ligninolytic enzymes. Appl Biochem Biotechnol, 2009; 157: 174–209.

Wong KH, Hynes MJ, Davis MA. Recent advances in nitrogen regulation: a comparison between *Saccharomyces cerevisiae* and filamentous fungi. Eukaryot Cell, 2008b; 7: 917–925.

Wood SM, Layzell DB. *A Canadian Biomass Inventory: Feedstocks for a Bio-based Economy*, 2003. BIOCAP Canada Foundation, Kingston, ON, Canada.

Wood TM, McCrae SI. The purification and properties of the C1 component of *Trichoderma koningii* cellulase. Biochem J, 1972; 128: 1183–1192.

Wu SC, Halley JE, Luttig C, Fernekes LM, Gutiérrez-Sanchez G, Darvill AG, Albersheim P. Identification of an endo-beta-1,4-D-xylanase from *Magnaporthe grisea* by gene knockout analysis, purification, and heterologous expression. Appl Environ Microbiol, 2006; 72: 986–993.

Wu W, Lange L, Skovlund DA, Liu Y. Polypeptides having cellobiohydrolase II activity and polynucleotides encoding same. 2008; US Patent: 7348168.

Wyman CE, Dale BE, Elander RT, Holtzapple M, Ladisch MR, Lee YY. Comparative sugar recovery data from laboratory scale application of leading pretreatment technologies to corn stover. Bioresour Technol, 2005; 96: 2026–2032.

Ximenes EA, Chen H, Kataeva IA, Cotta MA, Felix CR, Ljungdahl LG, Li XL. A mannanase, ManA, of the polycentric anaerobic fungus *Orpinomyces* sp. strain PC-2 has carbohydrate binding and docking modules. Can J Microbiol, 2005; 51: 559–568.

Xu F. Oxidation of phenols, anilines, and benzenethiols by fungal laccases: correlation between activity and redox potentials as well as halide inhibition. Biochem, 1996; 35: 7608–7614.

Xu F, Shin W, Brown SH, Wahleithner JA, Sundaram UM, Solomon EI. A study of a series of recombinant fungal laccases and bilirubin oxidase that exhibit significant differences in redox potential, substrate specificity, and stability. Biochim Biophys Acta, 1996; 1292: 303–311.

Xu X, Lin J, Cen P. Advances in the research and development of acrylic acid production from biomass. Chin J Chem Eng, 2006; 14: 419–427.

Yamakawa M, Ozaki K, Fujime T, Kakimoto N, Itakura S, Enoki A, Tanaka H. Relationship of phenol oxidase activity and hydroxyl radical generation to wood degradation by white rot basidiomycetes. Biocontr Sci, 2005; 10: 85–90.

Yamamoto H, Horii F. In situ Crystallization of Bacterial Cellulose I. Influences of polymeric additives, stirring and temperature on the formation celluloses Iα and If as revealed by cross polarization/magic angle spinning (CP/MAS) ^{13}C NMR spectroscopy. Cellulose, 1994; 1: 57–66.

Yang J, Wang S, Lorrain MJ, Rho D, Abokitse K, Lau PC. Bioproduction of laurylIactone and 4-vinyl guaiacol as value-added chemicals in two-phase biotransformation systems. Appl Microbiol Biotechnol, 2009; 84: 867–876.

Yaoi K, Kondo H, Hiyoshi A, Noro N, Sugimoto H, Tsuda S, Mitsuishi Y, Miyazaki K. The structural basis for the exo-mode of action in GH74 oligoxyloglucan reducing end-specific cellobiohydrolase. J Mol Biol, 2007; 370: 53–62.

Yaoi K, Kondo H, Hiyoshi A, Noro N, Sugimoto H, Tsuda S, Miyazaki K. The crystal structure of a xyloglucan-specific endo-beta-1,4-glucanase from *Geotrichum* sp. M128 xyloglucanase reveals a key amino acid residue for substrate specificity. FEBS J, 2009; 276: 5094–5100.

Yelle DJ, Wei D, Ralph J, Hammel KE. Multidimensional NMR analysis reveals truncated lignin structures in wood decayed by the brown rot basidiomycete *Postia placenta*. Environ Microbiol, 2011; 13: 1091–1100.

Yelton MM, Hamer JE, Timberlake WE. Transformation of *Aspergillus nidulans* by using a trpC plasmid. Proc Natl Acad Sci USA, 1984; 81: 1470–1474.

Yoder MD, Keen NT, Jurnak F. New domain motif: the structure of pectate lyase C, a secreted plant virulence factor. Science, 1993; 260: 1503–1507.

Yomano LP, York SW, Shanmugam KT, Ingram LO. Deletion of methylglyoxal synthase gene (mgsA) increased sugar co-metabolism in ethanol-producing *Escherichia coli*. Biotechnol Lett, 2009; 31: 1389–1398.

Yoon J, Aishan T, Maruyama J, Kitamoto K. Enhanced production and secretion of heterologous proteins by the filamentous fungus *Aspergillus oryzae* via disruption of vacuolar protein sorting receptor gene Aovps10. Appl Environ Microbiol, 2010; 76: 5718–5727.

Yoshida T, Kato Y, Asada Y, Nakajima T. Filamentous fungus *Aspergillus oryzae* has two types of alpha-1,2-mannosidases, one of which is a microsomal enzyme that removes a single mannose residue from Man9GlcNAc2. Glycoconj J, 2000; 17: 745–748.

Yoshida T, Tsuge H, Konno H, Hisabori T, Sugano Y. The catalytic mechanism of dye-decolorizing peroxidase DyP may require the swinging movement of an aspartic acid residue. FEBS J, 2011; 278: 2387–2394.

Yu C, Cao Y, Zou H, Xian M. Metabolic engineering of *Escherichia coli* for biotechnological production of high-value organic acids and alcohols. Appl Microbiol Biotechnol, 2011; 89: 573–583.

Yu H, Guo G, Zhang X, Yan K, Xu C. The effect of biological pretreatment with the selective white-rot fungus *Echinodontium taxodii* on enzymatic hydrolysis of softwoods and hardwoods. Bioresour Technol, 2009; 100: 5170–5175.

Yu Q, Zhuang X, Yuan Z, Wang Q, Qi W, Wang W, Zhang Y, Xu J, Xu H. Two-step liquid hot water pretreatment of *Eucalyptus grandis* to enhance sugar recovery and enzymatic digestibility of cellulose. Bioresour Technol, 2010; 101: 4895–4899.

Zacchi G, Axelsson A. Economic evaluation of preconcentration in production of ethanol from dilute sugar solutions. Biotechnol Bioeng, 1989; 34: 223–233.

Zamocky M, Furtmuller PG, Obinger C. Two distinct groups of fungal catalase/peroxidases. Biochem Soc Trans, 2009; 37: 772–777.

Zamocky M, Ludwig R, Peterbauer C, Hallberg BM, Divne C, Nicholls P, Haltrich D. Cellobiose dehydrogenase—a flavocytochrome from wood-degrading, phytopathogenic and saprotropic fungi. Curr Protein Pept Sci, 2006; 7: 255–280.

Zapanta LS, Tien M. The roles of veratryl alcohol and oxalate in fungal lignin degradation. J Biotechnol, 1997; 53: 93–102.

Zavrel M, Bross D, Funke M, Buchs J, Spiess AC. High-throughput screening for ionic liquids dissolving (ligno-)cellulose. Bioresour Technol, 2009; 100: 2580–2587.

Zechel DL, Withers SG. Glycosidase mechanisms: anatomy of a finely tuned catalyst. Acc Chem Res, 2000; 33: 11–18.

Zechel DL, Withers SG. Dissection of nucleophilic and acid-base catalysis in glycosidases. Curr Opin Chem Biol, 2001; 5: 643–649.

Zeikus JG, Jain MK, Elankovan P. Biotechnology of succinic acid production and markets for derived industrial products. Appl Environ Microbiol, 1999; 51: 545–552.

Zeilinger S, Mach RL, Kubicek CP. Two adjacent protein binding motifs in the cbh2 (cellobiohydrolase II-encoding) promoter of the fungus *Hypocrea jecorina (Trichoderma reesei)* cooperate in the induction by cellulose. J Biol Chem, 1998; 273: 34463–24471.

Zeng W, Jiang N, Nadella R, Killen TL, Nadella V, Faik A. A glucurono(arabino)xylan synthase complex from wheat contains members of the GT43,GT47, and GT75 families and functions cooperatively. Plant Physiol, 2010; 154: 78–97.

Zhang F, Rodriguez S, Keasling JD. Metabolic engineering of microbial pathways for advanced biofuels production. Curr Opin Biotechnol, 2011a; May 25 [Epub ahead of print].

Zhang L, Pei Y, Xing Z, Ding S, Buswell JA. Comparison of endoglucanase-1 (EG1) induction in the edible straw mushroom *Volvariella volvacea* by lactose and/or cellobiose with or without added sorbose. Appl Microbiol Biotechnol, 2011b; 89: 1939–1946.

Zhang M, Eddy C, Deanda K, Finkelstein M, Picataggio S. Metabolic engineering of a pentose metabolism pathway in ethanologenic *Zymomonas mobilis*. Science, 1995; 267: 240–243.

Zhang X, Qin W, Paice MG, Saddler JN. High consistency enzymatic hydrolysis of hardwood substrates. Bioresour Technol, 2009; 100: 5890–5897.

Zhao Q, Dixon RA. Transcriptional networks for lignin biosynthesis: more complex than we thought? Trends Plant Sci, 2011; 16: 227–233.

Zhao Y, Wang Y, Zhu JY, Ragauskas A, Deng Y. Enhanced enzymatic hydrolysis of spruce by alkaline pretreatment at low temperature. Biotechnol Bioeng, 2008; 99: 1320–1328.

Zhong L, Matthews JF, Hansen PI, Crowley MF, Cleary JM, Walker RC, Nimlos MR, Brooks CL 3rd, Adney WS, Himmel ME, Brady JW. Computational simulations of the *Trichoderma reesei* cellobiohydrolase I acting on microcrystalline cellulose Ibeta: the enzyme-substrate complex. Carbohydr Res, 2009; 344: 1984–1992.

Zhou J, Lee C, Zhong R, Ye ZH. MYB58 and MYB63 are transcriptional activators of the lignin biosynthetic pathway during secondary cell wall formation in Arabidopsis. Plant Cell, 2009; 21: 248–266.

Zhou S, Shanmugam KT, Ingram LO. Functional replacement of the *Escherichia coli* D-(-)-lactate dehydrogenase gene (ldhA) with the L-(+)-lactate dehydrogenase gene (*ldhL*) from *Pediococcus acidilactici*. Appl Environ Microbiol, 2003; 69: 2237–2244.

Zhu JY, Pan X, Zalesny RS Jr. Pretreatment of woody biomass for biofuel production: energy efficiency, technologies, and recalcitrance. Appl Microbiol Biotechnol, 2010; 87: 847–857.

Zimmermann FK, Scheel I. Mutants of *Saccharomyces cerevisiae* resistant to carbon catabolite repression. Mol Gen Genet, 1977; 154: 75–82.

Zimmermann R, Eyrisch S, Ahmad M, Helms V. Protein translocation across the ER membrane. Biochim Biophys Acta, 2011; 1808: 912–924.

Index

accessory enzymes
  acetyl-, methylesterases, 93–94
  CE xylan esterases, 94*t*
  ester linkage, 96*f*
  esterases, feruloyl, p-coumaroyl, 92–93
  glucuronoys esterases, 95–97
  pectin esterases, 95
  phylogenetic relationship, 96*f*
  T.reesei CIP2, structure, 97*f*
acid-catalyzed, pretreatment technologies, 188*f*
acidic, pretreatment technologies, 187
active center, lignin peroxidase, 103*f*
α-fucosidases, 90
α-fucosidases, glycoside hydrolases, 90
AGP, arabinogalactgan proteins, 3
agricultural wastes, 13–14
alkaline, pretreatment technologies, 187–188
anaerobic fungi, 41–44
  carbohydrate degradation, 43*f*
arabidopsis
  arabidopsis CSEA, 18, 23
  cell wall biosysthesis, 25*f*
  thaliana, 16, 22*t*
arabinogalactgan proteins (AGP), 3
arabinose containing substituents, 85–86

bacterial pathway, 125*f*
batch fermentation, ethanol, 206–207*t*
β-elimination, pectin, degradation, 82–84
biofuel alcohols
  isobutanol, 213–214
  isobutanol, metabolic pathway, 214*f*
  isoprenoid fuels, fatty acid pathways, 215
  syngas, 214
  syngas fermentation, 215*f*
biosynthesis, lignin, 24–26, 24*f*
brown rot
  decay mechanism, 37*f*
  free-radical mechanisms, 38*f*
  hydroxyl radicals, 38
  lignocellulose degradation, 35–39
  peptide catalyzed hydrogen peroxide
    formation, 39*f*

carbohydrate binding modules, 53*t*
carbohydrate degradation, 43*f*
carbohydrate polymers, sources
  agricultural wastes, 13
  energy crops, 15
  forest product residues, 14
  weedy lignocellulosic substrates, 15
carbon catabolite repression, 141*f*
carbon skeletons, CEL6A, E2, 60*f*
catabolism pathways, 123*f*
catalytic cycle
  laccases, 112*f*
  lignin peroxidase, 104*f*
  MnPs, manganese peroxidase, 108*f*
CAZy
  clustered organization, 144*f*
  covered modules, 48*t*
  transcripts, 131–132*t*
CAZymes, carbohydrate active enzymes, 48
CE xylan esterases, 94*t*
cell wall
  biosysthesis, 25*f*
  cellulose, 3
  composition, manipulation, 26–27
  hemicelluloses, 8
  layers, 2*f*
  lignin, 11
  pectin, 6

Fungi and Lignocellulosic Biomass, First Edition. Christian P. Kubicek.

cell wall (*Continued*)
 structure, 1
 structure diagram, 2*f*
cell wall biosynthesis, polymers, 16
cellobiohydrolase
 fungal cellulolytic enzymes, 55
 GH6, fungal cellulolytic enzymes, 57
 I proteins, characterized, 55*t*
 II CEL6A, structure, 58*f*
 T.reesei CEL7A, 54*f*
 T.reesei CEL7A, structure, 54*f*
cellobiose dehydrogenase, 116–117
 lignocellulose degradation, reaction, 34*f*
cellulase GH's, 50
cellulase inducer enigma, 129–133
cellulase production
 downstream processing, 182–183
 fermentation, 180–182
cellulase-enhancing protein, CEL61B, 67*f*
cellulose, 3
 chains, 5*f*
 micro, macrofibrils, generation, 4*f*
 polymers, 16–19, 17*f*
 polymorphs, interconversion, 4*f*
 structure, 3*f*
cellulose-binding domains, 52
cellulose-enhancing proteins, 177*f*
chemical hydrolysis, 200
chemical pretreatment, 186–187
chemical processes, refinement
 ferulic acid, production, 226*f*
 furfural, 225
 levulinic acid, 225
 lignin, uses, 225
 lignocelluloses biomass, chemical extraction, 226–227
classical mutageneis, 170–171
clustered genes in fungi, 145*f*
cotranslational targeting of proteins, 150*f*
CWDE, cell wall-degrading enzymes, 40

d-galactose, catabolism, 122–124
d-glucose, d-mannose, metabolism, 121–122
 catabolism pathways, 123*f*
dolichol-phosphate-mannose synthase complex, 160*f*
dye-oxidizing peroxidase, 109–111
 structure, 110*f*

ecological perspectives, 29
endo-β-1,4-glucanases, 59–60
endoplasmic reticulum (ER), 149–152
endoplasmic reticulum quality control, 152–154
endo-polygalacturonases, 81
energy crops, 15
enzymatic, nonenzymatic reactions, 33*f*
ER, endoplasmic reticulum, 149–152
ester linkage, 96*f*
ethanol, yeast strain improvement
 alternatives, 210–212
 batch fermentation performance, 206–207*t*
 global transcriptional machinery, 209*f*
 inhibitor tolerance, 209–210
 metabolic pathway, overview, 205*f*
 process improvement, 212
 transcriptional factors, engineering, 209
 xylose utilization, engineering, 203–204
 xylose utilization, fine tuning, 204–208

FAD-dependent oxidases, 116
fermentation, cellulase production, 180–182
ferulic acid, 226*f*
filamentous fungi, 173*t*
forest product residues, 14
formation pathway, 1,3-propanediol, 223*f*
free-radical mechanisms, 38*f*
fungal cellulolytic enzymes, 51, 56, 61
 β-1,4-glucosidases, 64
 carbohydrate binding modules, 53*t*
 α-carbon skeletons, CEL6A, E2, 60*f*
 cellobiohydrolase (GH6), 57
 cellobiohydrolase I proteins, characterized, 55*t*
 cellobiohydrolase (II CEL6A), structure, 58*f*
 cellobiohydrolase (T.reesei CEL7A), structure, 54*f*
 cellobiohydrolases, 55
 cellulase GH's, distribution, 50
 cellulose-binding domains, 52
 GH7 endo-β-1,4-glucosidaese, 59–60
 GH45 endo-β-1,4-glucosidaese, 62–65
 lignocellulosic degradation, 50*t*
 phylogenetic tree, GH12 endoglucanases, 61f
fungal hexose transporters, 121*t*
fungal lignin, 100*f*
fungal pathway, 126*f*
 hexuronic acids, 127*f*
fungal producer strains
 hydrolytic performances, cellulases, 168*t*
 other strains, 167–169
 thermophilic fungi, 169
 trichoderma reesei, 165–167
fungal secretory pathway
 cotranslational targeting of proteins, 150*f*
 endoplasmic reticulum (ER), 149–152

endoplasmic reticulum quality control, 152–154
golgi apparatus, 154–156
plasma membrane, 156–158
prepropeptide of fungal secretory protein, 151*f*
protein folding, 153*f*
protein secretion, filamentous fungi, 150*f*
protein transport, endoplasmic reticulum, golgi apparatus, 155*f*
unfolded protein response, fungi, 154*f*
vacuoles, 158
vesicles via See4-GTP, 157*f*
fungal secretory pathway, improvement, 161–162
factors influencing choice, 162*t*
folding and UPR, 163
heterologous protein production, 164*t*
fungi, lignocellulose degradation, 30
furfural, chemical processes, 225

galactomannan backbone, degradation, 79–80
enzymes necessary, 79*f*
galactomannans, hemicelluloses, 11
galactose-containing substituents, 87–89
galacturonic acid (GalA), 21*f*
gene deletion, strain improvement, 174*f*
genetic engineering, possible targets, 27*t*
genome accessibility level, 144–145
GH10, T.reesei xylanases, 73
GH11, xylanase, 75–76
GH30, glucuronoxylan xylanohydrolases, 76
GH54, β-xylosidases, xylan backbone, 78
GH61, phylogeny, nonenzymatic proteins, 66*f*
GH3 β-xylosidases, 76–77
GH43 β-xylosidases, xylan backbone, 77–78, 78*f*
GH28 pectinases, 81*t*
GH61 proteins, 65–67
GH61 proteins, nonenzymatic proteins, 65–67
global transcriptional machinery, 209*f*
glucuronoxylan xylanohydrolases (GH30), 76
glycopeptides, 118
glycoside hydrolases (GHs)
arabinose-containing substituents, 85–86
α-D-galactosidases, characterized, 88*t*
α-fucosidases, 90
galactose-containing substituents, 86–89
mechanisms, hydrolyze polysaccharides, enzymes, 47*f*
pectin degradation, 91–92
α-xylosidases, 89–90
glycosyl hdrolase clans, 49*t*
glyoxal oxidase, 116
golgi apparatus, 154–156
GRP, glycine-rich proteins, 3

hemicellulases, inducers, 133–134
hemicellulose biosynthesis, 19–20
lignin, 23–26
pectin, 20–21
phylogenetic identification, 20*f*
polymers, 19–20
hemicellulose degradation, enzymology, 69
hemicelluloses
galactomannans, 11
galactomannans, chemical structure, 11*f*
xylan, chemical structure, 10*f*
xyloglucan, 8
xyloglucan, chemical structure, 9*f*
xyloglucan, one letter code, 9*t*
heterologous enzymes, complementation, 176–178
heterologous protein, production, 164*t*
hexuronic acids, catabolism, 127–128
fungal pathway, 127*f*
hypothetical pathway, 128*f*
HRGP, hydroxyproline-rich proteins, 3
hydrogen peroxide, enzymes generating
FAD-dependent oxidases, 116
glyoxal oxidase, 116
hydrolasis, pectin degradation, 92–93
hydrolysis
chemical, 200
enzymatic, 195–200
enzymes required, 70*t*
lignocelluloses, fungal cellulase mixtures, 199*f*
hydrolytic, pectin, degradation, 80–82
hydrolytic performances, cellulases, 168*t*
hydrolyze polysaccharides, enzymes, 45
CAZy, covered modules, 48*t*
glycoside hydrolases, mechanisms, 47*f*
glycosyl hdrolase clans, 49*t*
oligo, polysaccharide cleavage, 46*f*
hydroxyl radicals, 38
hypothetical pathway, hexuronic acids, catabolism, 128*f*

I proteins, 55*t*
II CEL6A, 58*f*
inhibitor tolerance, ethanol, 209–210
ionic liquid pretreatment parameters, 194*t*
isobutanol
biofuel alcohols, 213–214
metabolic pathway, 214*f*
isoprenoid fuels, fatty acid pathways, 215

laccase-mediator reaction, laccases, 115*f*
laccases, 111
  catalytic cycle, 112*f*
  laccase-mediator reaction, 115*f*
  oxalate formation, basidiomycetes, 115*f*
  oxidase amino acid sequences, joining tree, 113*f*
  structure, 114*f*
lactic acid, 215
  metabolic pathway, 217*f*
  microbial pathway mutants, fuel production, 216*t*
levulinic acid, chemical processes, refinement, 225
light regulation mechanism, 143*f*
lignin
  arabidopsis, cell wall biosysthesis, 25*f*
  biosynthesis, 24–26, 24*f*
  chemical structure, 11*f*
  content manipulation, 27
  degradation, enymology, 99–101
  hemicellulose biosynthesis, 23–26
  uses, 225
lignin peroxidase
  active center, 103*f*
  catalytic cycle, 104*f*
  characterized, 102*f*
  redox potential, 104*f*
  structure, 103f
ligninase gene expression, regulation, 146–147
lignocelllulose biorefinery
  defined, 201–202
  overview, 202*f*
lignocellulose degradation, fungi
  brown rot, 35–39
  cellobiose dehydrogenase, reaction, 34*f*
  enzymatic, nonenzymatic reactions, 33*f*
  pathoginic lifestyles, 40*t*
  soft rot, 39
lignocellulose hydrolysates, other products
  pathway for conversion, 224*f*
lignocellulose-biomass-related glycosyl hydrolase families, 178*t*
lignocelluloses, fermentable sugars, 185
  pretreatment methods, 186*f*
lignocelluloses, RTIL, 193*t*
lignocelluloses biomass, chemical extraction, 226–227
lignocellulosic degradation, 50*t*

manganese peroxidase (MnPs), 105–108
marker gene removal, strain improvement, 175*f*
medicago sativa, 28
metabolic map
  wild-type c.glutamicum, 221*f*
  wild-type e.coli, 219–220*f*
metabolic pathway
  lactic acid, 217*f*
  overview, 205*f*
micro, macrofibrils, generation, cellulose, 4*f*
microbial pathway mutants, 216*t*
MnPs, manganese peroxidase, 105, 108
  catalytic cycle, 108*f*
  characterized, 106–107*t*
  structure, 107*f*
mono, oligosaccharides, uptake, 119–121
  fungal hexose transporters, 121*t*
  yeast sugar transporters, topology, 120*f*
mutations, T.reesei RUTC30 *vs.* QM6a, 172*t*

n-butanol, 212–213
neurospora crassa, 35
n-glycosylation, protein glyosylation, 159–161
nonenzymatic proteins, cellulose hydrolysis
  GH61, phylogeny, 66*f*
  GH61 proteins, 65–67
  swollenin, 67–68

o-glycan synthesis, protein glyosylation, 161
oligo, polysaccharide cleavage, 46*f*
ortho configuration, 28
oxalate formation, basidiomycetes, 115*f*
oxalic acid formation, essential enzymes, 117–118
oxidase amino acid sequences, joining tree, 113*f*
oxidative, pretreatment technologies, 188–189

PAL, phenylalanine ammonia-lyase, 24
pathoginic fungi, pathoginic lifestyles, 40*t*
pathoginic lifestyles, lignocellulose degradation, 40*t*
pathway for conversion, lignocellulose hydrolysates, 224*f*
pectate lyase B, 86*f*
pectate lyases, pectin, degradation, 83*t*
pectin
  arabidopsis thaliana, 22*t*
  biosynthesis, 20–23
  chemical structure, 7–8*f*
  galacturonic acid (GalA), model, 21*f*
  hemicellulose biosynthesis, 19–20
pectin, degradation
  β-elimination, 82–84
  glycoside hydrolases, 91–92

hydrolytic, 80–82
pectate lyases, 83*t*
pectinases (GH28), 81*t*
pectin/pectate cleavage, reaction, 80*f*
polygalacturonate hydrolase, structure, 84*f*
pectin biosynthesis, polymers, 20
pectin esterases, 95
pectinases (GH28), pectin, degradation, 81*t*
pectin/pectate cleavage, reaction, 80*f*
pentoses, catabolism, 127
bacterial pathway, 125*f*
fungal pathway, 126*f*
peptide catalyzed hydrogen peroxide formation, 39*f*
pH, transcriptional regulation, 141*f*
phanerochaete chrysosporium, 35
phylogenetic identification, 20*f*
phylogenetic tree, GH12 endoglucanases, 61f
physical pretreatment, pretreatment technologies, 186
physicochemical properties, cell wall, 3
physicothermal, pretreatment technologies, 189–191
plasma membrane, fungal secretory pathway, 156–158
polygalacturonate hydrolase, 84*f*
polyhydroxyalkanoate, 223
polymers
cell wall biosynthesis, 16
cellulose, 16–19, 17*f*
composition, manipulation, 26–27
hemicellulose biosynthesis, 19–20
pectin biosynthesis, 20
polymorphs, interconversion, cellulose, 4*f*
populus temuloides, 28
prepropeptide of fungal secretory protein, 151*f*
pretreatment methods, lignocelluloses, 186*f*
pretreatment technologies
acid-catalyzed, 188*f*
acidic, 187
advantages, disadvantages, 196*t*
alkaline, 187–188
biological, 193–195
chemical, 186–187, 187*t*
considerations, 195
ionic liquid pretreatment parameters, effects, 194*t*
lignocelluloses, RTIL, 193*t*
oxidative, 188–189
physical pretreatment, 186
physicothermal, 189–191
room temperature iconic liquids (RTILs), 191
solvent, 191–193
processes, refinement, lignocelluloses biomass, 226–227
1,3-propanediol, 222–223
formation pathway, 223*f*
protein engineering, strain improvement, 179–180
protein folding, quality control, 153*f*
protein glyosylation
dolichol-phosphate-mannose synthase complex, 160*f*
n-glycosylation, 159–161
o-glycan synthesis, 161
protein secretion, filamentous fungi, 150*f*
protein transport, endoplasmic reticulum, 155*f*
PRP, proline-rich proteins, 3

redox potential, lignin peroxidase, 104*f*
room temperature iconic liquids (RTILs), 191
rot comparison, 32*t*
RTILs, room temperature iconic liquids, 191

secretory heme peroxidases, properties, 101*t*
signal transduction, transcriptional regulation, 145–146
soft rot, 39
lignocellulose degradation, fungi, 39
solvent, pretreatment technologies, 191–193
specific repressors, transcriptional regulation, 138
strain improvement
cellulose-enhancing proteins, 177*f*
classical mutageneis, strain breeding, 170–171
complementation with heterologous enzymes, 176–178
filamentous fungi, selection makers, 173*t*
gene deletion, 174*f*
lignocellulose-biomass-related glycosyl hydrolase families, 178*t*
marker gene removal, 175*f*
mutations, T.reesei RUTC30 *vs.* QM6a, 172*t*
protein engineering, 179–180
targeted gene manipulation, 171–176
succinic acid, 217–218, 222
metabolic map, wild-type c.glutamicum, 221*f*
metabolic map, wild-type e.coli, 219–220*f*
swollenin, 67–68
syngas, biofuel alcohols, 214
syngas fermentation, biofuel alcohols, 215*f*

targeted gene manipulation, 171–176
thermophilic fungi, 169

transcriptional factors, yeast strain improvement, 209
transcriptional regulation, gene expression, 134
 activators, 135–137
 carbon catabolite repression, 141*f*
 CAZy, clustered organization, 144*f*
 clustered genes in fungi, 145*f*
 genome accessibility level, 144–145
 growth of T.reesei, 135*t*
 light regulation mechanism, 143*f*
 pH regulation, 141*f*
 signal transduction, 145–146
 specific repressors, 138
 wide domain regulators, 138–144
 XlnR/XYR1 protein structure, 136*f*
trans-gauche position, 3
T.reesei, growth, 135*t*
T.reesei α-D-galactosidases, characterized, glycoside hydrolyses, 88*t*
T.reesei CBM1, cellulose-binding module, 79
T.reesei CEL74A, structure, 72*f*
T.reesei CIP2, structure, 97*f*
T.reesei endoglucanase CEL12A, structure, 62*f*
T.reesei XYN2, structure, 75*f*
trichoderma reesei, 165–167

unfolded protein response, 154*f*

vacuoles, fungal secretory pathway, 158
vallonia cellulose, 53
Van der Waals forces, 4
versatile peroxidase, 108
 structure, 109*f*
vesicles via See4-GTP, fungal secretory pathway, 157*f*

weedy lignocellulosic substrates, 15
white rot, 31–35
wide domain regulators, 138–144

XlnR/XYR1 protein structure, 136*f*
xylan, 10
xylan, hemicelluloses, 10
xylan backbone, degradation
 GH11, xylanase, 75–76
 GH54, β-xylosidases, 78
 GH3 β-xylosidases, 76–77
 GH43 β-xylosidases, 77–78, 78*f*
 glucuronoxylan xylanohydrolases (GH30), 76
 xylanases (GH10), 73
 xylan-degrading enzymes, 73*f*
 XYN1, T.aurantiacus, 74*f*
xylanase, (GH11), xylan backbone, degradation, 75–76
xylanases (GH10), xylan backbone, degradation, 73
xylan-degrading enzymes, xylan backbone, degradation, 73
xylitol, 222
xyloglucan, hemicelluloses, 8
xyloglucan hydrolysis, 69–72
 hydrolysis, enzymes required, 70*t*
 subsites, schematic representation, 71*f*
 T.reesei CEL74A, structure, 72*f*
xylose utilization
 fine tuning, 204–208
 yeast engineering, 203–204
XYN1, T.aurantiacus, xylan backbone, 74*f*

yeast sugar transporters, 120*f*